Springer Tracts in Advanced Robotics

Volume 63

Editors: Bruno Siciliano · Oussama Khatib · Frans Groen

Bram Vanderborght

Dynamic Stabilisation of the Biped Lucy Powered by Actuators with Controllable Stiffness

Author

Bram Vanderborght
Vrije Universiteit Brussel
Pleinlaan 2
1050 Brussel
Belgium
E-mail: bram.vanderborght@vub.ac.be

ISBN 978-3-642-13416-6 e-ISBN 978-3-642-13417-3

DOI 10.1007/978-3-642-13417-3

Springer Tracts in Advanced Robotics ISSN 1610-7438

Library of Congress Control Number: 2010928125

Typeset & Cover Design: Scientific Publishing Services Pvt. Ltd., Chennai, India.

Printed on acid-free paper

5 4 3 2 1 0

springer.com

STAR (Springer Tracts in Advanced Robotics) has been promoted under the auspices of EURON (European Robotics Research Network)

European ROBOTICS Research Network EURON

Foreword

Robotics is undergoing a major transformation in scope and dimension. From a largely dominant industrial focus, robotics is rapidly expanding into human environments and vigorously engaged in its new challenges. Interacting with, assisting, serving, and exploring with humans, the emerging robots will increasingly touch people and their lives.

Beyond its impact on physical robots, the body of knowledge robotics has produced is revealing a much wider range of applications reaching across diverse research areas and scientific disciplines, such as: biomechanics, haptics, neurosciences, virtual simulation, animation, surgery, and sensor networks among others. In return, the challenges of the new emerging areas are proving an abundant source of stimulation and insights for the field of robotics. It is indeed at the intersection of disciplines that the most striking advances happen.

The *Springer Tracts in Advanced Robotics (STAR)* is devoted to bringing to the research community the latest advances in the robotics field on the basis of their significance and quality. Through a wide and timely dissemination of critical research developments in robotics, our objective with this series is to promote more exchanges and collaborations among the researchers in the community and contribute to further advancements in this rapidly growing field.

The monograph written by Bram Vanderborght is the third in the series devoted to biped robots. The work speculates on the study of human walking to ensure a suitable compliant behaviour for a pneumatically actuated biped. A complete hardware and software architecture is developed and different planning and control techniques are tested for energy efficient walking.

The monograph is expanded from the doctoral dissertation of the author, which was a finalist for the Eight Edition of the EURON Georges Giralt PhD Award. A very fine addition to STAR!

Naples, Italy
January 2010

Bruno Siciliano
STAR Editor

Preface

This book reports on the developments of the bipedal walking robot Lucy. Special about it is that the biped is not actuated with the classical electrical drives but with pleated pneumatic artificial muscles. In an antagonistic setup of such muscles both the torque and the compliance are controllable. From human walking there is evidence that joint compliance plays an important role in energy efficient walking and running. Moreover pneumatic artificial muscles have a high power to weight ratio and can be coupled directly without complex gearing mechanism, which can be beneficial towards legged mechanisms. Additionally, they have the capability of absorbing impact shocks and store and release motion energy. This book gives a complete description of Lucy: the hardware, the electronics and the software. A hybrid simulation program, combining the robot dynamics and muscle/valve thermodynamics, has been written to evaluate control strategies before implementing them in the real biped.

The current control architecture consists of a trajectory generator and a joint trajectory tracking controller. Two different trajectory generators have been explored. The first is based on an inverted pendulum model where the objective locomotion parameters can be changed from step to step. The second is an implementation of the preview control of the zero moment point developed by Kajita. The joint trajectory tracking unit controls the pressure inside the muscles so the desired motion is followed. It is based on a computed torque model and takes the torque-angle relation of the antagonistic muscle setup into account. With this strategy the robot is able to walk up to a speed of $0.15m/s$. Higher walking speeds are difficult because the robot has to walk flat-feet and the valve system is not fast enough to follow the predescribed pressure courses.

On a single pendulum structure a strategy is developed to combine active trajectory control with the exploitation of the natural dynamics to reduce energy consumption. A mathematical formulation was found to find an optimal compliance setting depending on the trajectory and physical properties of the system. This strategy was not implemented on the real robot because the walking speed of the robot is currently too slow.

Acknowledgements

First of all, I would like to thank my promotor Dirk Lefeber to let me walk freely where I found the road ahead myself, but also for assisting me when needed.

Robotics is a very multidisciplinary research domain, which demands for the most varying competencies. It is thus immensely important to work within a team of enthusiastic people. This was for me one of the key factors for deciding to do my masters thesis and afterwards my doctoral thesis in the Robotics & Multibody Mechanics Research Group. Although a dissertation has only one author, the work was only possible by a close collaboration with the group members. I'm especially grateful to Björn Verrelst and Ronald Van Ham who started the research towards Lucy. Also Michaël Van Damme for his help, particularly on programming issues, Joris Naudet and the other members. Also the technicians Jean-Paul Schepens, André Plasschaert and Gabriël Van den Nest for the crucial help in the construction of Lucy and Thierry Lenoir for keeping the computers alive.

I also want to thank the people of the Vrije Universiteit Brussel, the Faculty of Engineering, the Polytechnische Kring, Liberaal Vlaams Studenten-Verbond, HVI,... I am very grateful to the people that have been with me to share this whole experience. They have stretched me academically and personally. My best friends come out of this splendid period. Thanks Ken, Rianne, Sophie VH, Niels, Sophie H, Jan, Daft, Xtel, Jean-Marc, Nathalie, Nancy, David, Olivier, Rina, Laurent, Jean-Jacques, Eva, Peter, Stéphanie, Emma... You have always been a great support for me.

Thanks to Kazuhito Yokoi I had the opportunity to work for 6 weeks on the humanoid robot HRP-2 at the ISRI/AIST-CNRS Joint Japanese-French Robotics Laboratory. It was a wonderful experience thanks to Olivier Stasse and the other members of the lab.

I would also like to thank the jury members of my PhD: Jean-Paul Laumond, Martijn Wisse, Yvan Baudoin, Rik Pintelon, Jacques Tiberghien, Philippe Lataire and Patrick Kool for their comments and suggestions to improve the dissertation.

Sidney Appelboom learned me as swimming coach to push my boundaries and helped me to overstep my own limits. His training schedules were sometimes extremely hard, some said they were crazy. Although I never reached the top, they were important lessons for me. So many thanks Sid!

Finally, my most profound thanks are to my parents and the other members of the family, especially my brother, oma, opa and bompa for their love, support and understanding throughout the years. My mother has spent many hours to tinker with me; my father helped me a lot building model boats and airplanes. They both were convinced one cannot start explaining science and technology at an early enough stage. Probably these were the first steps towards robotics. They showed that science and technology can be fun and I hope the work done for Lucy and my efforts explaining it to children has inspired many to start a career in exact or applied sciences.

Brussels
January 2010

Bram Vanderborght

Fig. 0.1 Visit of His Majesty King Albert II of Belgium.

Contents

Chapter 1
Introduction

1.1 Motivation

Perhaps humans always have been envious of the power, speed and beauty of certain animals. By taking their names and wearing their skin people thought they would acquire the same performances. It is only by our intelligence that we were able to survive in the animal world. By creating tools we were able to compensate our weak power and became the most dominant species of the world. The machines that have been built surpass in many ways the possibilities of animals. Airplanes are faster than most of the birds; submarines can dive deeper than whales. Our cars are faster than cheetahs. In many cases the animal world has been the model for the designers and engineers. The design of a swimsuit for competition swimmers is inspired by the skin of sharks. Aircraft collision detection and avoidance is inspired by the sophisticated echolocation of bats. Fireflies utilize compounds to emit cold light that is so efficient it emits no heat as LEDs do. In view of this it is strange that walking, so common and normal for most of us, is still so difficult for robots. Even the most advanced walking robots can only attain a few kilometers per hour with the highlight Asimo who can attain an astonishing $6km/h$ running (175) and the Toyota running humanoid who can reach $7km/h$ (390). This is still much slower than a human and surely than most of the animals. Also walking on rough terrain, one of the advantages of using legs over wheels, is still impossible for bipeds. Another deficiency of walking robots in comparison with their biological counterparts is the high energy consumption. The continuous operating time of for example Asimo is $1hour$ (175). To be ever useful in a real application the autonomy must definitely increase. This can be done by developing better power sources as batteries and increasing the efficiency of walking. The most energy efficient bipeds are the so called passive walkers, they don't need any actuation at all to walk down a slope. The slope is used as a source of energy to compensate the friction and impact losses and can be replaced by actuation. The energetic cost (amount of energy used per meter traveled per unit of weight) of these robots is between one and two orders of magnitude smaller than the energetic cost of actively controlled humanoids (100). The passive walkers are designed to exploit the natural dynamics of the system while walking.

B. Vanderborght: Bipedal Walking by the Pneumatic Robot Lucy, STAR 63, pp. 1–40.
springerlink.com

Unfortunately they are of little practical use: they cannot start and stop and they cannot change their gait due to the fixed dynamics. This is in contrast with the actively controlled bipeds as for example Asimo and HRP-2. They do precise joint-angle control and are consequently very versatile. For example these robots are able to walk among obstacles (282), step over obstacles (379), climb stairs (212) and manipulate objects while walking (461). These capabilities are still impossible for the actuated passive walkers. The optimal is probably somewhere in between those two approaches as shown in figure 1.1: a combination of active control to be able to perform different tasks while still exploiting the passive dynamics to reduce energy consumption. Most of the research trying to incorporate energy efficient locomotion is performed on the side of the passive walkers, so on the right hand side of figure 1.1. The robots developed on the left hand side of the figure are usually built to evaluate task driven applications with as final goal to have enough capabilities for close cooperation with humans in a home or office. The goal of this work is to investigate how fully actively controlled robots can improve their energy efficiency while maintaining their versatility. Consequently this work is situated on the left hand side of figure 1.1. The control strategy is a combination of calculating dynamic stable trajectories for the different joint links which are tracked by actively controlling the actuators in the different joints and an extra controller to reduce the energy consumption. Essential for this research is the use of adaptable compliant actuators so the natural dynamics of the system can be controlled. Adaptable compliance is also important for human walking and running. During walking electromyograhical data shows there is little muscle activity during for instance the swing phase of the leg. So this means this motion is mainly passive. When running, motion energy is stored in the Achilles tendon and released in the next hop. The compliant muscles allow also to absorb impact shocks. An interesting actuator, introducing such compliance for robotic mechanisms, is the pleated pneumatic artificial muscle, because in an antagonistic setup both the torque and the compliance of the joint can be controlled.

This book is a summary of the PhD work of the author and is dedicated to the elaborate control aspects demanded by compliant actuation mechanisms. The work, which emphasizes mostly on real experiments, discusses the complete concept of the biped Lucy which has been built for this study. The robot is a planar walking robot actuated by pleated pneumatic artificial muscles (105). In this chapter the different aspects of the motivation will be discussed in more detail.

This chapter starts with an overview of some trends (section 1.2) that can be observed in modern robotics where humanoid robots are one of the emerging research topics. Probably they will play an important role in our daily lives in the next two or three decades. Nowadays, humanoid robots are mostly found in research centres. To have an idea about where research is performed on humanoid robots an overview is given in section 1.2.1. This book focus on compliant actuation for legged robots which is the reason why the biped Lucy has only legs and an upper body, without e.g. arms and a head. A motivation why legs can be more interesting than wheels, is provided in section 1.3. Nature has always been a source of ideas and it is worth looking on how nature has solved the locomotion issue. In section 1.4.1 the biological aspects of walking and running are discussed and a major role in this story is the

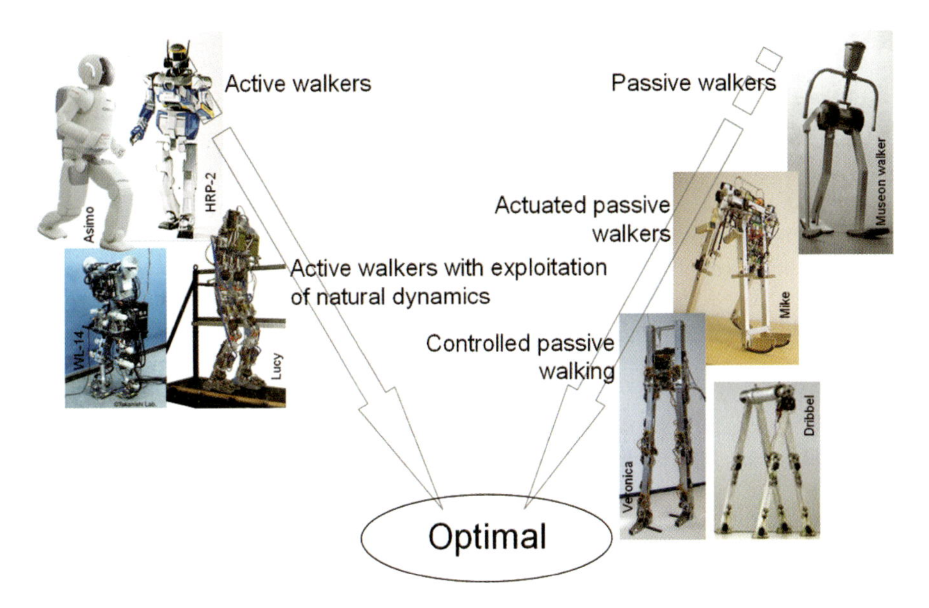

Fig. 1.1 Location of the biped Lucy among the active and passive bipeds.

compliance of a human joint. For biological aspects the adaptable compliance is an important factor minimizing the energy consumption, however introducing adaptable compliance in a mechanical actuator is fairly new in robotics. Section 1.4.2 gives an overview of the different designs of passive compliant actuators. Such actuators are currently implemented in a number of bipeds, in section 1.4.3 the author's research will be positioned among the other research concerning energy-efficient bipedal locomotion. Because Lucy is a pneumatic biped, also an overview of these robots is given in section 1.4.3.3. At the end the goal and approach of the Lucy-project are described in more detail.

1.2 Robotics

The roots of robotics can be tracked back to the Egyptians. Egyptians invented the idea of thinking machines: citizens turned for advice to oracles, which are statues with priests hidden inside (256). Also the Greek mythology had some ideas of robots: Daedalus, whose son Icarus flew too close to the sun, created animated statues that guarded the entrance to the labyrinth in Crete according to Aristotle (38). This philosopher also wrote "If every tool, when ordered, or even of its own accord, could do the work that benefits it... then there would be no need either of apprentices for the master workers or of slaves for the lords." (39), already dreaming what robots can do for humans. The first recorded design of a humanoid automaton is credited to the famous engineer and painter Leonardo da Vinci around 1495.

It was a mould of a human scale armored knight, using metal linkages and a lot of mechanical gears. The design of Leonardo's robot was not rediscovered until the 1950s (353). The word robot was introduced in the play R.U.R. (Rossum's Universal Robots) which opened in Prague in January 1921 and was written by the Czech playwright Karel Capek. "Robot" is the Czech word for forced labour or slave. The word "robotics" was first used in Runaround, a short story published in 1942, by Isaac Asimov. "I, Robot", a collection of several of these stories, was published in 1950. Asimov also proposed his three "Laws of Robotics" (40), and he later added a "zeroth law" (41). These laws had a big influence on how to look at robot behaviour and their interaction with humans (96; 97).

In 1961 the first industrial robot, Unimate, joined the assembly line at a General Motors plant to work with heated die-casting machines. Unimate was a robotic arm which took hot metal die castings from machines and performed welding on car bodies; tasks that were hated by the factory workers. Unimate was built by a company "Unimation" which stands for universal automation and this was the first commercial company that made robots. Robots are currently widespread in factories assembling cars and other consumer goods, but robots plays also a major role in for example the Human Genome Project to unravel human DNA, to explore Mars and many others. An important trend that can be observed nowadays is that robotics is about to enter and will change our daily lives, in the next coming years. The robotic vacuum cleaners such as Roomba and entertainment robots as Aibo, Robosapien, Furby, Lego (see figure 1.2) Mindstorms NXT are already very popular. However these robots demand for new requirements compared to industrial robots: first of all safety, a comfortable human-robot interaction, intelligence, but also legislation, ethics and social issues must be addressed alongside the research and technology development (432). There are several reasons why robots are now at our doorstep. First of all the demographical situation will create large markets for robots. The Japanese population is aging very rapidly with 28.1% of the population expected to be over age 65 by the year 2020 (20). This trend can also be witnessed in other highly industrialized regions as Europe and America. This is why Japan is investing heavily in robotics R&D, both by the government as by industry. Also the technology becomes affordable: there exists a wide range of cheap sensors, communication technology is ubiquitous and computers are still following Moore's law (406). The Japan Robot Association predicts the robotics will be the next digital revolution (202). Predictions of the Japan Robot Association state that the market of robots by the year 2010 will reach a turnover of about $ 35 billion each year, a sales number which exceeds the current Japanese PC market (27). As correctly stated by Bill Gates, despite all the excitement and promises, the robot industry for service robots lacks critical mass and practical applications are relatively rare (134). He compared it with the computer industry during the mid-1970s. Probably the first market that will succeed in becoming profitable is that of specialized niche products. This will pave the way for more sophisticated costly general-purpose systems such as a complete autonomous and intelligent humanoid.

Fig. 1.2 Lego Mindstorms NXT is often used to educate science to children, here during RoboCup Junior.

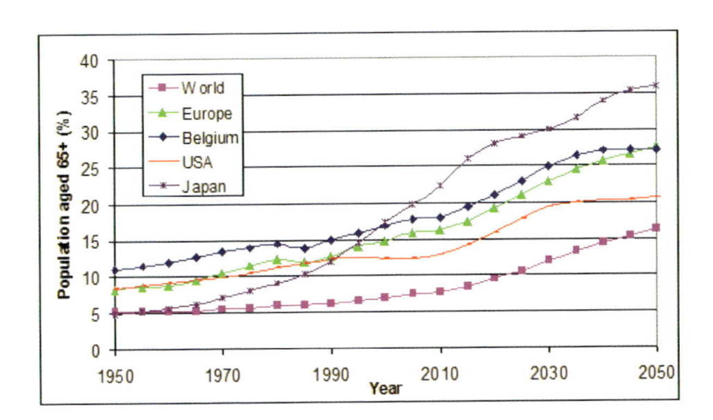

Fig. 1.3 Population aged 65+ (source (20)).

1.2.1 Humanoid Robots

A humanoid robot is not only an anthropomorphic robot with the overall appearance based on that of the human body (so a torso with a head, two arms and legs), but is also able to interact like a person. Even some research groups are focussing on emotions and facial expressions (198), something that was unthinkable 10 years ago. Famous social robots in this field are Kismet (61) and Leonardo (62), the android Repliee series (284), WE-4RII (199), Robot head ROMAN (52) and Probo (357). The six basic emotions of the huggable robot Probo are shown in figure 1.4. But why would we build machines that look and act like us? In many religions people

Fig. 1.4 The 6 basic facial expressions of the huggable robot Probo (happy, surprise, sad, anger, fear and disgust).

think that god created man in his own image and maybe roboticians want to do the same. Of course there are more reasons to develop humanoid robots. People have to adapt their way of communication to interact for example with a computer or mobile phone. To control such a device the user has to touch some keys and scroll through menus. This is a difficult task for many, especially elder persons. An important challenge will be how robots can be adapted to humans instead of the other way. And it seems that people may be better able to relate to robots that look like us and that can also communicate in a similar manner (219) (64) (60). Communication is not only verbal but supported by body and gestures (177). Humanoids can operate in environments designed for humans. Our offices and homes contain stairs and other obstacles which are difficult to negotiate for wheeled robots. Humanoids can for example open and close doors, reach switches, crawl or go through narrow spaces. Humanoids can also use tools made for humans. Well-known is the Robonaut, a robotic astronaut built by NASA. The robot is designed to work within existing corridors and use the same tools as space walking astronauts (58) (59).

Although the most advanced humanoids are already very astonishing, their capabilities are not sufficient to engage them in society yet. They have a disappointing lack of mobility against the severity of the real environment and also their intelligence is low. Mobility is often compared with that of an 80 year old person and intelligence is comparable to the one of a three year old child. Even Honda, one of the pioneer companies investing in humanoid robot technology, is not yet in the position of commercializing their humanoids. Honda is only doing some renting business for sales promotion and similar events. There is little demand for humanoid robots at the moment because they are extremely expensive, not flexible and not intelligent enough to be of practical use. But hopefully this will change soon, as explained by H. Hirukawa in the Financial Times (3.06.2006 p8): "a bipedal robot costs now more than a Ferrari, if one can find a nice application and sell a million of them, the price would fall to that of a cheap car." As Niels Bohr once said: "It is difficult to make predictions, especially about the future". But if one analyzes the evolution of an airplane, some 100 years old, a personal computer 30 years and the mobile phone 15 years it is not so unreasonable to think that in the nearby future humanoid robots will have a major role in our society.

In the next section an overview is given of the most advanced humanoid robots in the world.

1.2.2 Overview of Humanoid Robots

1.2.2.1 Humanoid and Walking Robots in Japan

The country with most of the humanoid robotic research is certainly Japan. Some say that Japanese people have grown up while watching robot animations (for example human-like robot character Astro-Boy, an idol for many Japanese children) or try to find explanations in their religion (257). Probably the release of the humanoid Honda robot P2 in 1996, started the snowball effect triggering a technological race to develop humanoids in Japan (161).

1.2.2.2 Honda

Honda is a private company with the longest history in humanoid robotics. After 10 years of secret research the P2 was unveiled in 1996 (159). The robot walked stable and could climb stairs. This astonished both the robotic researchers as the society. Since then many other projects started in Japan. To come to the P2 a lot of prototypes were built E0 (1986) for examining the principles of two-legged locomotion, E1-E2-E3 (1987-1991) for realizing rapid two-legged walking, E4-E5-E6 (1991-1993) for completing the basic functions and P1 (1993) to perform research on completely independent humanoid robots. Since P2, Honda presented P3 (158) in 1997 and in 2000 Asimo (355). Asimo is the acronym of “Advanced Step In MObility”. Honda has chosen the height of $120cm$ as ideal for a robot to operate in a human living space. Different version of Asimo exist. Asimo’s first evolutionary phase was released in 2002. The robot had a more advanced communication ability thanks to recognition technology (355). In 2004 a second evolutionary phase was made public with a “posture control” technology making it possible to run in a natural human-like way (17). The latest version of Asimo is able to run at $6km/h$ on a straight line while $5km/h$ running in a circular pattern ($2.5m$ radius) (175). This makes Asimo the fastest human sized robot in the world.

Asimo (see figure 1.5) receives a lot of media exposure, every place the robot visits a lot of fans attend the robot and VIP persons shake hands with Asimo. But this is certainly not the main objective of the project. The scale of the project is too big to explain the motivation only by the publicity (161). The goal of the Honda project is to realize a humanoid robot that can work at home and office: a partner for people, a new kind of robot that functions in society.

Fig. 1.5 Asimo for a Belgium TV show to convince children to start a career in science and technology.

1.2.2.3 Sony

A new way of entertainment for Sony was robotics. After the success of AIBO, a commercial robotic dog, Sony proposed the small-scale humanoid SDR-3X in 2000 (197), SDR-4X in 2002, and SDR-4XII in 2003, better known as QRIO with stands for "Quest for Curiosity" (figure 1.6). SDR-4X has a Real-time Integrated Adaptive Motion Control so the robot can walk on uneven surface and make an adaptive motion control against external forces. Also the falling-over control of the robot is realized by this controller.

QRIO can perform a lot of amusing things: walk on a wobbling surface, fall down and get up again, throw a ball, dance, and it was the first humanoid that was capable of running (21).

Its moving parts consists of totally 38 degrees of freedom, and each joint (except the joints in the hands and the neck) is driven by the actuator unit called ISA (Intelligent Servo Actuator) which has a motor driver and communication circuits built-in (196). Plain flat gears were used instead of harmonic drives to have back-drivability of the gears. When the robot is pushed by an external force, the robot adapts quickly.

SDR Motion Creating System is used to realize dynamic and elegant motion performances (254). This software system with GUI is composed of two parts: the Motion Editor to edit upper body motion and whole body motion, and the Foot Trajectory Editor to create stable lower body motion.

In March 2006 Sony ended the robotics project: end of AIBO and no further development of QRIO (1). The robotics community was very disappointed and many asked why. Probably the bad situation of their main business and the current market size of robots lead to this discussion. However the Intelligence Dynamics Laboratory, housing the developers of their robot technology, is still active.

Fig. 1.6 QRIO in a dance act to show the SDR Motion Creating System.

1.2.2.4 Toyota

For the world expo in Aichi 2005 Toyota developed "out of the blue" an orchestra of robots capable of playing trumpets and other instruments. A biped humanoid Partner Robot, a Segway-type robot and a wheeled robot were developed (18). The reliability was shown by the 2000 shows they gave during the 185 expo days without problems. Toyota wants to continue its robotics division. The first practical application of their robots is to set them in Toyota leasing garages. Also the ifoot was showed, a two-legged walker with an egg-shaped cockpit for a seated human rider (18) and the first wire-driven bipedal robot (404) (see figure 1.7). Most multi-jointed bipedal robots are actuated using direct-driven motors and gears placed at the individual joints. For this robot the motors are placed in the torso and the forces are transferred to the joints by wires. This design reduces the weight and moment if inertia of the robot leg and makes the robot safer for human interaction in symbiotic environments. By including an additional spring in the wires a more safe robot is envisaged. Toyota sees this robot as new type of mobility besides their cars. Their last creation is a one-legged hopping robot (391) and a running humanoid robot (390).

1.2.2.5 Humanoid Robotics Project (HRP)

Not only private Japanese companies are investing heavily in humanoid robots, also the Japanese government provides a large budget for the development of humanoid robots. The "Humanoid Robotics Project (HRP)" of METI (=Ministry of Economy, Trade and Industry) was the largest in scale with a budget of 40M USD. It had been launched in 1998 and ran for five years. Main goal of the project was to use the humanoid for tasks in industrial plants and services at home and offices (194). HRP-1 was the first robot developed in the HRP project and is an enhanced version of the

Fig. 1.7 Wire driven robot of Toyota.

Honda P3 robot. The legs and arms of HRP-1 were controlled separately so the robot had to stop walking when it wanted to use it arms. A lot of practical applications were consequently impossible. This was solved by replacing the main control CPU and the software, which resulted in HRP-1S (459). Successively, Kawada Industries developed HRP-2L (where "L" stands for legs alone) (220), HRP-2P (where "P" stands for prototype) and HRP-2 (figure 1.8) (164). Its appearance was designed by a professional designer Yutaka Izubuchi for humanoid heroes in animation. Also a software platform OpenHRP (221), to perform dynamic simulations, was developed in contribution with General Robotix. Different research institutes and universities are currently using the HRP-2. The only HRP-2 outside Japan is used in JRL-France (Joint Japanese-French Robotics Laboratory) in Toulouse. The author had the opportunity to work on HRP-2 in the sister laboratory JRL-Japan in Tsukuba in the ongoing research "Dynamically Stepping Over Large Obstacles by the Humanoid Robot HRP-2". More information about this topic can be found in section 3.7. NEDO (=New Energy and Industrial Technology Development Organization) has sponsored the further development of HRP and the first prototype HRP-3P is water and dust proof, so the robot is capable of working outdoors and the hardware has been improved (223). For the world expo 2006 the knowledge of HRP was used to develop two biped dinosaur robots (figure 1.9) (258). During the development of cybernetic human HRP-4C is focussed on a realistic figure and head of a human being (227).

Because the price to lease a HRP-2 is too high for many research institutes and universities, it costs 400.000 euro to lease it for 4 years, AIST (National Institute

Fig. 1.8 HRP-2 playing the drum.

of Advanced Industrial Science and Technology) and 4 companies revealed in May 2006 the small-sized HRP-2M Choromet (5). It is $33cm$ tall, weighs $1.5kg$, has 20 DOF and costs 3.000 euro. It has an accelerometer, gyro and force sensors in the feet. The movements and especially the walking pattern is much smoother than for example other servo-controlled humanoids as for example a Kondo-robot (12).

1.2.2.6 Waseda University

"The Bio-engineering group", consisting of four laboratories in the School of Science & Engineering of Waseda University, started the WABOT (WAseda roBOT) project in 1970. The WABOT-1 (1970-1973) was the first full-scale anthropomorphic robot developed in the world (231). It was equipped with a visual recognition system, a verbal communication system and a quasi-static walking controller. The WABOT-1 walked with his lower limbs and was able to grip and transport objects with hands that used tactile-sensors and consisted of the WAM-4 (as its artificial hands) and the WL-5 (its artificial legs). The research towards bipedal locomotion started already in 1966 with the development of the lower limb model WL-1 (6). Waseda University has consequently one of the longest histories on the development of human-like robots. An impressive row of prototypes have been built. Nowadays the Takanishi Laboratory is in charge of the research on biped walking mechanisms. Wabian-2 (figure 1.10) (305) has a special designed waist and the legs having 7 DOF. The main purpose is to investigate stretched-leg walking and its consequences towards energy-consumption. Merging the technology of Wabian-2 and the expressive

Fig. 1.9 Out of the technology of the HRP series two dinosaurs were built for the World Expo 2006 in Aichi, Japan.

head WE-4RII (287), Kobian is developed capable to express human-like emotions and gestures (464).

Also famous is their multi-purpose locomotor WL-16RII (Waseda-leg no 16 Refind II) able to walk up and down stairs carrying a human (382). The legs consist of 6 DOF parallel mechanisms instead of the more common articulated legs.

1.2.2.7 Tokyo University

The humanoid robot H5 was created by Jouhou System Kougaku (JSK) Laboratory of the Tokyo University for research on dynamic bipedal locomotion (206). An offline algorithm generated dynamically equilibrated motions. Together with an online ZMP compensation method the robot was able to walk, step down and so on. To achieve full body motion H6 was developed (301; 210). H7 was built by Kawada Industries and University of Tokyo and is an improved version of H6 (209). The internal design with used motors, gear reductions and many other useful information is nicely described in the Japanese book Robot Anatomy (193). This differs a lot from the other Japanese robots for which the design is most of the time secret.

The main research topic of JSK is Dynamic Walking Pattern Generation (299). To improve their locomotion strategies comparisons with humans were made (207). This robot is also used to investigate robot motion planning (208).

Fig. 1.10 Wabian2 with 7DOF legs.

1.2.2.8 Small-Scale Humanoids

A group of robots that is often used for research and hobby purposes are the robots driven by hobby servomotors. Famous in Japan and Korea is the ROBO-ONE robot competition which centers the battle of two walking robots (4). The robots used for this competition are usually Kondo KHR-1 (12), Hitec's Robonova (10) or similar robots. Mostly an interface architecture is sold with the robot kit so users can develop their own programs easily. The HOAP (Humanoid for Open Architecture Platform) series is developed by Fujitsu Laboraties for researchers (368). ZMP inc. (26) is the creator of several small humanoid robots: the open-source PINO (454), the experimental Morph 3 (399), and the consumer robot Nuvo (16).

1.2.2.9 Humanoids and Walking Robots in Korea

1.2.2.10 KAIST

KAIST (Korea Advanced Institute of Science and Technology) developed a series of walking robots. KHR-0 (KAIST Humanoid Robot) was developed in 2001 and has 2 legs without an upper body. The actuator requirements were studied by using the robot. Afterwards a complete humanoid KHR-1 was developed (241). KHR-2 has an updated design in the mechanical and electrical architecture. The joint stiffness and the joint angle ranges have been improved, and the appearance of the robot has become more human-like, and human friendly (318) (242). This robot is Windows operated because it is familiar to many software developers and thus more easy to maintain and improve the system. In 2004 KHR-3 (or also HUBO) was finished and is an upgrade compared to KHR-2 (317) (316). For one of their robots of HUBO KHR-3 the stock head was replaced with an animatronic replica of Albert Einstein's head and is called Albert HUBO (310). Also a human-riding biped Hubo FX-1 was developed comparable with the ifoot of Toyota (259).

The AIM laboratory of KAIST has developed the humanoid robot AMI2 to study biped locomotion and social interaction (455). It is the successor of the wheeled robot AMI (205).

1.2.2.11 KIST

NBH-1 (network based humanoid) has been built by the Korean Institute of Science and Technology (KIST) and Samsung and can walk at a speed of up to $0.9km/h$ (15). MAHRU (male) and AHRA (female), were born in March 2005, and recently in 2006 new models have been developed. The robot is connected to an external server through a network and sends images or voice data to the external server and the external server analyses and processes the data and sends back commands. Recently KIST Babybot was presented (311).

1.2.2.12 Humanoids and Walking Robots in the Rest of the World

1.2.2.13 China

BHR-1 is a major project for the Beijing University of Science and Engineering under China's High and New Technology Research and Development Program. The humanoid robot BHR-1 Huitong consists of a head, upper body, two arms and legs, and has in total 31 DOF. There are two computers built in BHR-1's body, one is for motion control, the other for information processing (261) (326). The newest version is BHR-2 (449).

1.2.2.14 Russia

Russia has two humanoids developed called ARNE and ARNEA. ARNE is the male version and ARNEA the female. The name ARNE is an abbreviation of "Anthropomorphic Robot of New Era". They were built by company at St. Petersburg called New Era and students from the Polytechnic University in St. Petersburg. Main goal for these robots is to play soccer (377).

1.2.2.15 United Arab Emirates

REEM-A is a humanoid robot developed by PAL Technology (19), a company of the United Arab Emirates but the research team is located in Spain. The robot can walk up to 1.1 km/h. REEM-B has the ability to autonomously navigate in indoor environments while avoiding obstacles and the complete control software is integrate within the robot itself (401).

1.2.2.16 Germany

The robot Johnnie has been developed by the Institute for Applied Mechanics at the Technical University of Munich (TUM) (265) (263). The main objective was to realize an anthropomorphic walking machine with a human-like, dynamically stable gait. Because the research was focused on walking, the robot only consists of 17 joints and is able to walk at $2.2km/h$. The robot is also equipped with a visual guidance system developed by the Institute of Automatic Control Engineering of the Technical University Munich (103). With this vision system, the robot is able to detect obstacles and to decide whether to step on, over or walk around these obstacles. The project was supported by the Deutsche Forschungsgemeinschaft (German Research Foundation) within the Priority Program Autonomous Walking. The successor of the Johnnie robot is the humanoid robot Lola with enhanced performances, which is currently under developed by the University of Munich (266) (71). Significant increase in walking speed and more flexible gait patterns are the main focuses of the new design. The researchers intend the robot to reach the average human walking speed of approximately $5km/h$.

The university of Hannover created BARt-TH for which the motion is restricted to the sagittal plane (136). It was built to investigate the technical requirements for bipedal service robots in the human environment. The successor LISA (Legged Intelligent Service Agent) has twice the number of DOF and is an experimental robot to perform research in the field of autonomous bipedal walking (169). The hip joint has a spherical parallel manipulator with three degrees of freedom.

1.2.2.17 France

The robot Rabbit is the result of a joint effort by several French research laboratories and is a testbench for studying dynamic motion control (89). The lateral stabilization is assured by a rotating bar, and thus only 2D motion in the sagittal plane is considered. The robot is a five-link, four-actuator bipedal robot and has consequently no feet, so the robot is under-actuated. This was done to demonstrate that actuated ankles are not absolutely necessary for the existence of asymptotically stable locomotion patterns. A bipedal running controller, based on hybrid zero dynamics (HZD) framework, was developed and the robot Rabbit executed six consecutive running steps (293). The observed gait was remarkably human-like, having long stride lengths (approx. $50cm$ or 36% of body length), flight phases of significant duration (approx. $100ms$ or 25% of step duration), an upright posture, and an average forward rate of $0.6m/s$ (293).

The BIP project is a joint French project, started in fall 1994, which involves four laboratories (43) (42). BIP is an anthropomorphic walking robot with 15 DOF

Fig. 1.11 The author with the humanoid iCub and his science popularizing book about robots.

designed for the study of both human and artificial bipedal locomotion. Sherpa is a biped robot with directdrive capabilities and parallel manner actuation (314).

In 2005 the first French company dealing with humanoid robotics, Aldebaran Robotics, was launched (2). The Nao humanoid robot aims to be a robot with an affordable price. The 55*cm* tall robot will use the URBI (Universal Real-time Behaviour Interface) scripting language (44).

1.2.2.18 Italy

Within the RobotCub project (358), funded by the European Commission, the goal is the development of an embodied robotic child iCub (see figure 1.11) with the physical and ultimately cognitive abilities of a 2,5 year old human baby (403). The "baby" robot is designed for crawling and the selection of motors to power the lower body were done through simulations of crawling motions of different speeds and transitions from sitting to crawling pose and vice versa.

1.2.2.19 Spain

Rh-1 (75) is a humanoid robot with 21 DOF designed and constructed by the Robotics Lab in University Carlos III of Madrid and is a redesign of the humanoid robot Rh-0 (262). It has a height of 1200*mm* (without head) and weighs 50*kg*. Each leg has 6 joints, three in the hip and one in the knee and two in the ankle.

The SILO-2 robot (73) of the Industrial Automation Institute, Madrid (IAI-CSIC) is a 14 DOF biped robot powered by SMART actuators (Special Mechatronic Actuator for Robot joinTs) (74). A SMART drive is a nonlinear actuator with variable reduction ratio and implemented using a four-bar linkage mechanism. It is considered efficient for humanoid robot locomotion (72). This drive is characterized by

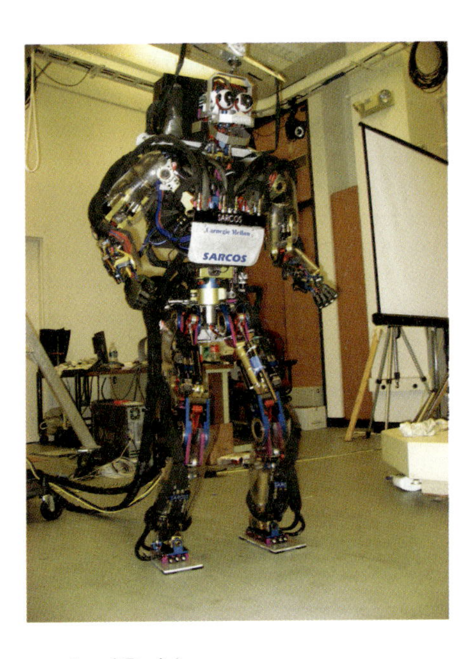

Fig. 1.12 Sarcos CB (Computational Brain).

the change of the transmission ratio from some value in the medium part of a trajectory ad infinitum at its end positions. An enhanced version is the DUAL SMART drive combining continuously changing transmission ratio and dual properties for realization of slow motion of a heavy robot body and quick motion of the robot's leg (124).

1.2.2.20 United States of America

SARCOS is a Utah based company selling different robot applications (23). Their newest research humanoid robot is called CB (Computational Brain, see figure 1.12) and is made so also walking can be studied (86). The predecessor DB (Dynamic Brain) was not able to walk (359). Both robots are located at the Advanced Telecommunications Research Institute International (ATR) in Japan. The robot has 50 DOF and the controller provides full position/velocity/force sensing and control at $1kHz$, allowing the flexibility in deriving various forms of control schemes. Most of the DOF are driven by hydraulic servo actuators. The first experiments revealed the robot is able to keep self-balance under unknown disturbances, future work includes more advanced full-body human-humanoid interaction as well as dynamic locomotion, such as walking and running (191).

After the successful running and hopping robots developed by Marc Raibert (341), the Massachusetts Institute of Technology (MIT) Leg Lab developed the series elastic actuator (333). An elastic element was placed between the output of the gear reduction and the load. This actuator was used in the planar biped robots

Spring Turkey and Spring Flamingo and two autonomous robots named Troody and M2 (331). The company Boston Dynamics founded by Raibert presented the quadruped BigDog (69), a rough-terrain robot that walks, runs, climbs and carries heavy loads and Petman, an anthropomorphic bipedal robot for testing chemical protection clothing used by the US Army. The successor of Spring Flamingo and M2 is currently developed by Pratt et al. (338) and is called M2V2. It is a 12 degree of freedom three dimensional walking robot using Series Elastic Actuators to achieve force control. The walking and balance recovery controllers will use the concepts of Capture Points and the Capture Region in order to decide where to step (347).

1.2.2.21 RoboCup and FIRA

RoboCup is an international research and education initiative. Its goal is to foster artificial intelligence and robotics research by providing a standard problem where a wide range of technologies can be examined and integrated (22). Their dream is "By the year 2050, develop a team of fully autonomous humanoid robots that can win against the human world soccer champion team" (70). The main focus of the RoboCup activities is competitive football amongst robots as shown in figure 1.13 and the robots are split up in different categories. One of the leagues is the humanoid league and was introduced in 2002. The robots are grouped in two size classes: KidSize ($30cm <$ Height $< 60cm$) and TeenSize ($65cm <$ Height $< 130cm$) and the humanoid robots play in "penalty kick" and "1 vs. 1", "2 vs. 2" matches. Another organization for football robot competition is "The Federation of International Robot-soccer Association (FIRA)" with the Humanoid Robot World Cup Soccer Tournament (HuroSot) (7).

For these competitions commercial robots sometimes are used while other groups developed their own robots like Toni (49), Bruno (prototype HR18 of Darmstadt Dribbllers & Hajime Team) (155) and Robotinho (50). Several skills have to be

Fig. 1.13 Soccer player robot in action during RoboCup game.

programmed for successful football. Effective and powerful kicking is for example a challenging task because of balance. During the period of kicking, the kicking leg moves very fast and therefore the dynamics should not be ignored (398).

1.3 Focus on Legs

An important motivation for research and development of legged robots is their potential for higher mobility. Legged robots are often grouped depending their number of legs. Since these machines only need a discrete number of isolated footholds, their mobility in unstructured environments can be much higher than their wheeled counterparts, which require a more or less continuous path of support (340). Preparing a special arrangement such as ramps to allow them moving around is not required for legged robots (457). Legged robots therefore are suitable for rough terrain like minefields (420) (160) (302), volcanoes (249) (45), disaster zones and so one. Payload can be traveled smoothly despite pronounced variations in the terrain using an active suspension that decouples the path of the body from the paths of the feet (341). Controlling bipeds is more difficult than using more legs due to stability problems, but this is a technological problem and may not be the reason to abandon the research towards bipedal robots. Humanoids are also more dextrous and have higher motion flexibility in human environments with obstacles compared to other legged robots. A human environment is optimised for humans so the best machine that will be able to work in such an environment is a human-shaped robot. Yi also argues that building a biped robot is more cost effective than the other legged robots since the cost of actuators is considerable; a biped has less actuators than other legged robot (457).

The second reason to study legged machines is to gain a better understanding of human and animal locomotion. Such insights are required to build protheses and orthoses. Well-known innovative examples are the exoskeleton Lokomat to rehabilitate paraplegic persons (203) or the robot suit HAL(233) (see figure 1.14) and BLEEX (235) to augment human strength and endurance during locomotion. A nice overview o f robotic exoskeletons is given by Guizzo and Goldstein (146). At the Robotics & Multibody Mechanics Research Group the ALTACRO project (figure 1.15) (54) has emerged from the insights gained during the Lucy project. The goal is to design, build and test a novel step rehabilitation robot with adaptable compliant actuation. Another project by the same group aims at the development of an intelligent transtibial prosthesis IPPAM powered by PPAMs (figure 1.16) (431) and the passive AMPfoot (figure 1.17) (430).

Despite their potential for high mobility, most of the bipeds have never been outside a laboratory. The fastest biped built until now is Asimo which can run at speeds up to $6km/h$. A huge technological achievement but still rather slow compared to humans. Also walking on rough terrain is not yet achieved. When a show is given with Asimo the floor surface is required to have irregularities of at most $2mm$ and a horizontal deviation of at most 1°, no slippery or springy floors are allowed. HRP-2 can cope with slightly uneven terrain, the surface may have gaps smaller than $20mm$ and slopes $< 5\%$ (224). The problem is mainly due to the fact that the control of

Fig. 1.14 HAL exoskeleton.

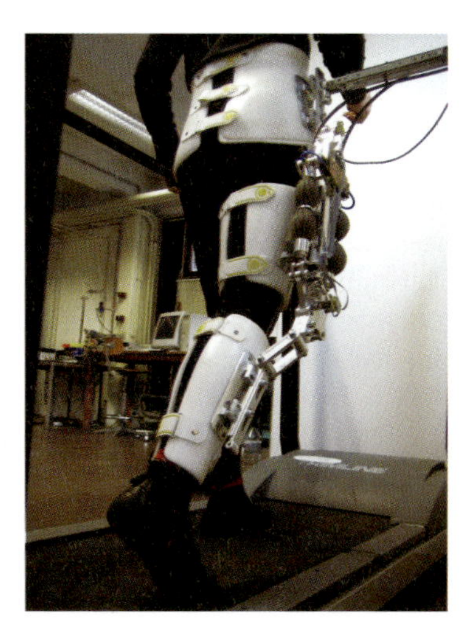

Fig. 1.15 Knee joint of the Altacro exoskeleton powered by PPAMs.

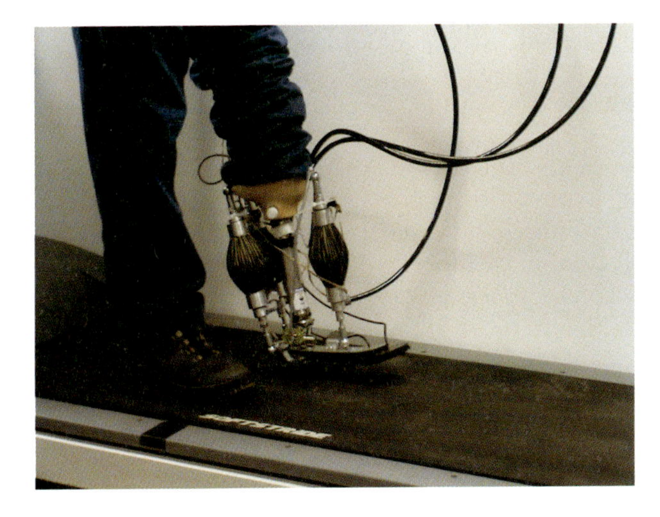

Fig. 1.16 IPPAM ankle-foot prosthesis powered by PPAMS.

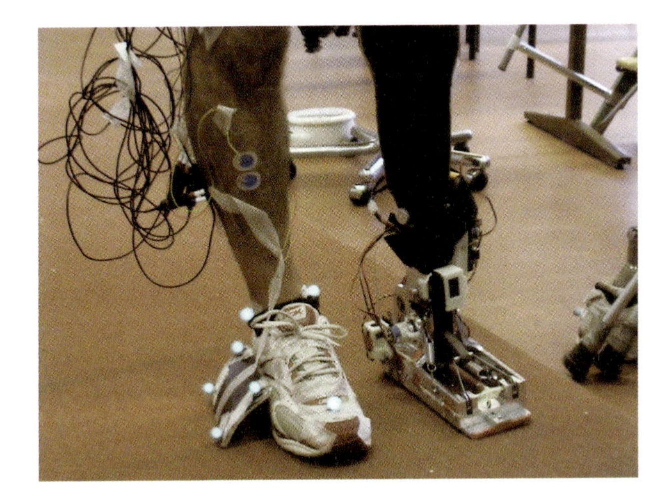

Fig. 1.17 AMPfoot prosthesis powered by modified MACCEPA actuator.

a legged machine is intrinsically a complex issue. Some major difficulties for the control system are the following (348) (335):

- The robot kinematics and dynamics are non-linear and difficult to accurately model. Robot parameters such as centers of mass, moments of inertia, etc are not known exactly.
- The dynamics of the robot depend on which legs make contact with the ground. In other words, the dynamics change whenever the robot makes a transition between a single support phase and a double support phase or a flight phase, and

vice-versa. Moreover, exchange of leg support is accompanied by an impact disturbing the robot's motion.
- A legged robot is submitted to intermittent holonomic and nonholonomic constraints.
- The environment is unknown and dynamic. The surface might be elastic, sticky, soft or stiff.
- Vertical contact forces on the surface are unilateral, meaning that they can not pull the robot against the surface.
- The goal of keeping dynamic balance is difficult to decompose into actuator commands.
- Many degrees of freedom, which have to be controlled real-time.

To summarize one can conclude that controlling a bipedal walking robot that is able to negotiate different terrains and walk/run at high speeds is still an unsolved problem. Mastering this technology would be beneficial for many applications as eg service robots.

1.4 Compliant Actuation

Most of the robots are actuated by electrical drives since these actuators are widely available and their control aspects are well-known. The rotational speed of the shaft of an electrical motor is high and the torque is low, while a robot joint generally requires a fairly low rotation speed but with high torques. Thus a transmission unit is often required. Harmonic drives are very popular transmission units because they combine zero backlash, high precision, high single-stage transmission ratio, compact dimensions and a high torque capacity (372) (242) (301) (220). Because the transmission ratios are high (1/160 for HRP-2L (220), 1/100-1/160 for KHR-3 (HUBO) (316), 1/160 for Johnnie (328)), they are non-backdrivable and this is very inconvenient for shock absorbance and stiff actuators cannot store energy. For manipulator robot implementation, stiff joints have always been preferred over compliant joints since they increase tracking precision. For legged robots however, tracking precision is not that stringent as overall dynamic stability. Elastic joint properties on the contrary might be used for shock absorbance and be exploited to store energy and reduce control effort. As will shown in the next section compliance plays an important role for human walking and running.

1.4.1 Biological Aspects of Walking and Running

Locomotion of humans and other mammals is richly studied, but due to the complexity not yet completely understood. When building bipedal walking robots it is however worth looking to the research performed by biomechanicists. The interconnection of muscles, sensors, spinal cord and brain intelligence seems to have an overwhelming perfection because humans are able to cope with most of the surface

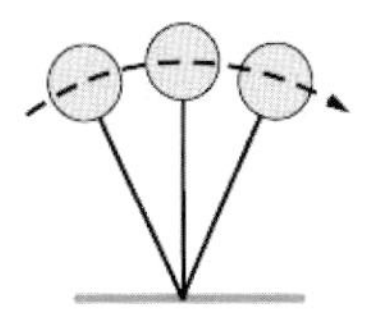

Fig. 1.18 Basic model for walking.

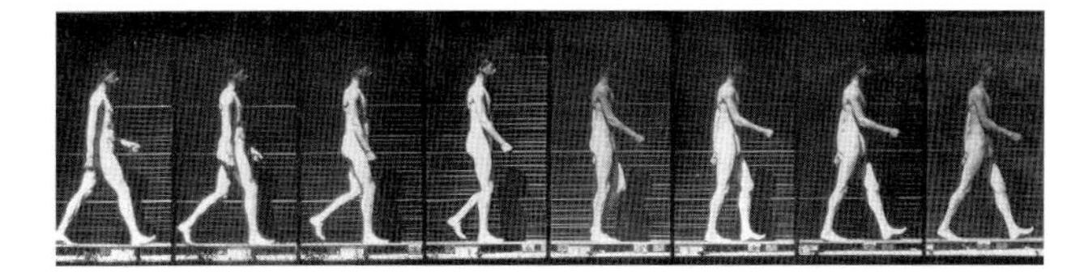

Fig. 1.19 Eadweard Muybridge's sequence of walking (407).

structures. By training our body and mind, humans are even able to achieve outstanding performances which can be witnessed at the Olympic Games, at a show of Cirque Du Soleil and so one.

However the biological solution is not always the best solution because it is a product of evolution and is consequently a combination of historical, functional and structural constraints (the so called Seilacher's triangle (364)). Out of this it is impossible to have an optimal design because then the design should only have functional constraints (365). Evolution comes up with a partially optimal solution so it is fit enough to survive the current environment. Some go even further: "If there were no imperfections, there would be no evidence to favor evolution by natural selection over creation." said by Jeremy Cherfas (87) or "The proof of evolution lies in imperfection." of Stephen Jay Gould (142).

Evolution came up with excellent solutions and it makes sense to study them and transfer the underlying ideas and principles into technology, not one to one but in a reasonable, technology-oriented way.

1.4.1.1 Walking and Running

Walking is classically defined as a gait in which at least one leg is in contact with the ground at all times (156). In contrast, running involves aerial phases when no feet are in contact with the ground. So during walking there are no aerial phases, while in running there are aerial phases. There are more differences still.

During walking the stance leg is almost completely stretched in the single support phase (32). As a consequence the head goes up and down with an amplitude of about $4cm$. This motion can be seen in figure 1.19. By doing this the kinetic energy and gravitational potential energy of the center of mass are approximately 180° out of phase. At mid-stance in walking, the gravitational potential energy is at its maximum and the kinetic energy is at its minimum. During the first half of the stance phase of walking, the center of mass loses kinetic energy but gains gravitational potential energy. During the second half of the stance phase, the center of mass loses gravitational potential energy but gains kinetic energy. As a result, the energy transfer mechanism used in walking is often referred to as the "inverted pendulum mechanism" (figure 1.18). At intermediate speeds up to a maximum of about 65% of the mechanical energy required to lift and accelerate the center of mass is recovered

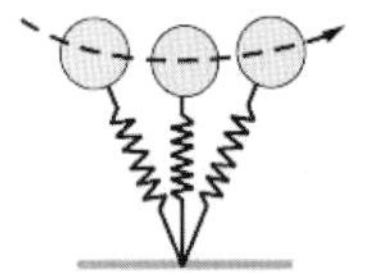

Fig. 1.20 Basic model for running.

Fig. 1.21 Eadweard Muybridge's sequence of running (407).

by this energy transfer mechanism. The mechanical energy-savings fall toward zero at very low and very high walking speeds (82).

At a certain speed (which is about 2-2.5m/s) a person starts running instead of walking. It is reasonable to think that gravity causes the walk-run transition. The center of mass m describes a circular arc with speed v around the ankle with a radius which is the length of the leg L. The required centripetal force is mv^2/L with may not be bigger than gravitational force which is mg. Or $v < \sqrt{gL}$. When making a quick calculation with $g = 10m/s^2$ and $L = 0.9m$ gives $v = 3m/s$. So walking with a higher speed is impossible based on this calculation. However a race walker can reach up to $4m/s$. The reason is the strange movement of the hip: the center of mass is slightly lowered when the stance leg is vertical, this to reduce the vertical movements of the body's centre of mass. The centre of mass rises and falls less than the hip joints so the inverted pendulum model is not valid anymore.

The ratio between the centrifugal force and the gravitational force is the Froude number (v^2/gL) and has consequently to stay under 1 for walking. In normal gravity, humans and other bipeds with different leg lengths all choose to switch from a walk to a run at different absolute speeds but at approximately the same Froude number (0.5) (32). In (248) it was found that, at lower levels of gravity, the walk-run transition occurred at progressively slower absolute speeds but at approximately the same Froude number. This supports the hypothesis that the walk-run transition is triggered by the dynamics of an inverted-pendulum system. However, it remains unclear why the transition occurs at that particular dimensionless speed.

In contrast with walking, the stance limb during running is compliant so that the joints undergo substantial flexion during the first half of stance and extension during the second half of stance (figure 1.21). Due to the compliance of the stance limb the center of mass reaches its minimum height at mid-stance. As a consequence, the kinetic energy and gravitational potential energy are nearly in phase and the mechanism of saving energy based on the "inverted pendulum" is not valid anymore. The pattern of movement of the center of mass has been proposed as the defining difference between a walking gait and a running gait (277).

Running uses another strategy to conserve energy by storing and releasing energy in elastic tissues. Because the movements of the center of mass during running are similar to a bouncing ball, running is often referred to as a "bouncing gait" (figure 1.20). Because it is often compared to a bouncing rubber ball that moves forward and upward with each ground contact. The efficiency in running has been calculated to be about 40-50% (83). The most important spring in the legs to store motion energy is the Achilles tendon (245). The Achilles tendon is a fibrous tissue that connects the heel to the muscles of the lower leg: the calf muscles. When running the impact forces are about 2.7 times the bodyweight. The reaction force takes place about $116mm$ in front of the ankle joint. This is a free rotating joint and because the Achilles tendon is about $47mm$ after the ankle joint, the Achilles tendon has to hold 7 times the bodyweight. For a man of $70kg$ the force is about $5000N$, for a woman of $50kg$ this is $3500N$. This force is enough to stretch the tendon for about 6%. The Achilles tendon has a length of about $250mm$, so its stretches for about $15mm$. A second important spring is the ball of the foot (33).

The metabolic power increases with speed for both walking and running (32). The curve of walking and running crosses each other at $2m/s$. Slower than $2m/s$ the preferred gait is walking, at higher speeds humans choose to run. The energy consumption is measured by measuring the O_2 consumption. At high running speeds the anaerobe muscles starts working. Humans also try to avoid the neighborhood of the crossing of the two curves. The optimal walking speed is about $1.34m/s$, this is the minimum when the metabolic energy consumption is normalized by the distance traveled (36).

As a summary one can state for walking:

- No aerial phase
- Straight supporting leg
- Potential energy and kinetic energy out of phase
- Energy storage by interchange of gravitational potential energy and kinetic energy
- Behaves like an inverted pendulum

and for running:

- Aerial phase
- Bent legs
- Potential energy and kinetic energy in phase
- Energy storage by elastic properties of the joints
- Behaves like spring-mass

Remarkably, these basic mechanisms of energy conservation have been demonstrated in a wide variety of animals that differ in leg number, posture, body shape, body mass, or skeleton type, including humans, kangaroos, dogs, lizards, crabs, and cockroaches (114). Geyer et al. (137) showed that not stiff but compliant legs are essential to explain walking mechanics. With a bipedal spring-mass model that includes the double support as an essential part of its motion, they could reproduce the characteristic dynamics of walking that result in the observed small vertical oscillation of the body and the observed out-of-phase changes in forward kinetic and gravitational potential energies. This model combines the basic dynamics of walking and running in one mechanical system, but also show these gaits to be just two out of many solutions to legged locomotion offered by compliant leg behavior and accessed by energy or speed.

1.4.1.2 Role of Compliance in Walking and Running

One of the most remarkable characteristics of a muscle is the large range over which the stiffness of a muscle and hence joint can be controlled (184). Kearney and Hunter (236) have measured up to 50-fold changes in human ankle stiffness resulting from triceps surae activation.

For walking, the compliance is needed to let the limbs swing as a pendulum mechanism. Electromyographic measurements show that nearly no muscle activity is present in the swing limb at some walking speeds (46). It is thought that the limb swings forward passively after the muscles start the limb into motion during the period of double support. Observation of animals also partially validates passive-dynamic approaches. For example, electromyographic muscle signals (EMG) recorded by Basmajian and Tuttle (47) show a low level of muscular activity in human and gorilla legs during walking, as compared to other movements. Especially during single support phase low muscle activity can be observed suggesting a natural adaptation of the structure of the body to enable stable gait. There is more activity during double support probably to achieve sufficient propulsion to continue the motion. It appears that the mechanical work for step-to-step transitions, rather than pendular motion itself, is the major determinant of the metabolic cost of walking (116) (252). Experiments by Whittington et al. (441) provide quantitative support for the suggestion that passive elastic mechanisms about the hip are utilized during human walking. This mechanism reduces the amount of the pre-swing hip power burst that must be generated actively to initiate leg swing.

These observations have led to the development of the so called "passive walkers". In 1990, Tad McGeer (275) showed for the first time that a mechanical structure, without sensors, motors or control, could walk on its own down a slope. This was a totally different approach for the biped robot community who had for years built elaborate robots with many sensors and motors and complex control. The idea was to put the intelligence not in the control of the robot but in the mechanics. Since then research groups have built several simple passive dynamic walkers. To be able to walk over level ground minimal actuation is provided just enough to overcome friction when walking over level ground like the Cornell biped and the Delft biped Denise (100). To be able to incorporate passive dynamics, compliant actuation is required.

Few research has been done on how the compliance changes. Hansen et al. discovered that the slopes of the moment versus ankle angle curves (called quasi-stiffness) during loading appeared to change as speed was increased (149).

For running the compliance is used to store energy during early stance and then recovering it near the end of stance. This makes it possible for the whole body to operate at an efficiency of 40-50% during running while the maximum efficiency of a contracting muscle is 25% (83).

When humans run, the overall musculoskeletal system behaves like a single linear spring, the so called "leg spring" (126). Experiments have shown that leg stiffness is independent of forward speed (122). This seems illogical on first sight. The only difference is that the angle swept by the leg spring is greater while running at high speeds. Because of the greater angle swept, there is an increased compression of the leg spring and an increased force in the leg spring. Consequently, the vertical displacement of the mass during the ground contact phase is smaller and the ground contact time decreases. When measuring joint stiffness it was found that the stiffness was constant ($7Nm/deg$) in the ankle joint and increased from 17 to $24Nm/deg$ in the knee joint for increasing running speed (251). The leg stiffness is also independent of simulated gravity level (153). However leg stiffness is adjusted to achieve different stride frequencies at the same speed (122) both for hopping in place as during running. Between the lowest and highest possible stride frequencies, the stiffness of the leg spring changes more than twofold.

Studies showed that leg stiffness is adjusted to accommodate surfaces with different properties. Experiments were both performed during running (123) as hopping in place (126). When animals run on a compliant surface, the surface acts as a second spring in series with the leg spring. In this case, the mechanics of a bouncing gait depend on the combined stiffness of the leg spring and the surface spring. The leg spring stiffness increases to accommodate compliant surfaces, thus offsetting the effects of the compliant surface on the mechanics of locomotion. Both the experimental and theoretical results support the observation that runner's change their leg spring stiffness in an attempt to not disrupt their overall running dynamics. If runners do not adjust their leg stiffness when running on different surface stiffnesses, then their ground contact time and center of mass displacement will increase as surface stiffness decreases. It is believed that the adjustment in leg stiffness is an attempt to keep the overall stiffness of the system (runner and ground) constant.

The importance of the compliance is also well-understood in the development of high-tech prosthetics. The double-leg amputee sprinter Oscar Pistorius set a new world record during the men's 200-meter race at the Athens 2004 Paralympic Games, probably fast enough to qualify for the able-bodied Olympic Games. The core of the L-shaped prosthese is a carbon-fiber composite. This forms an extremely efficient spring, returning nearly all of the energy stored (176).

One can conclude that compliance is important for human running. What about this topic in running robots? Raibert has studied different running robots (344) (343) (see figure 1.22). These famous hopping robots were driven by pneumatic and hydraulic actuators and performed various actions including somersaults (329). Important was the spring mechanism to store kinetic energy during running cycles. The biologically-inspired hopping robot Kenken has an articulated leg composed of three links, and uses two hydraulic actuators as muscles and linear springs as a tendon (190). The robot has succeeded in running of several steps in a plane. KenkenII has two legs to realize not only hopping, but also biped walking and running (189). To improve stability a tail was added in KenkenIIR. In (188) Hurst and Rizzi showed by the construction of the Electric Cable Differential Leg (ECD Leg) that a running robot is a unified dynamic system comprising electronics, software, and mechanical components, and for certain tasks such as running, a significant portion of the behavior is best exhibited through natural dynamics of the mechanism.

Fig. 1.22 Different robots of Raibert are on display at the MIT Museum.

Also the most common known robots as QRIO, Asimo and HRP-2 try to jump and run, but without spring elements in the actuators. The compliance consists of the result of the compliance of the ground, the sole and the non negligible compliance of the robot servos working at high speeds. However, the main problem is the huge impact forces when the foot hits the ground. These forces can damage the hardware of the robot and need to be reduced. This has as consequence that the flight phases and jumping heights of these robots are consequently still rather short.

QRIO demonstrated running with $0.23m/s$ whose flight phase is approx. $20ms$ (21), HRP-2LR realized a steady hopping motion of $60ms$ flight phase, $0.5s$ support phase and $3mm$ footlift (215). The maximum force of about $1000N$ was generated at touchdown and this is within acceptable limits for the mechanical strength of HRP-2LR which is more than three times the robot's weight. Asimo is up to now the fastest running biped. His top speed is $6km/h$ and he has a flight phase of $80ms$.

To reduce the impact forces, an impact absorbing control is needed. Different approaches exist to tackle this problem. When excessive vertical force is detected for HRP-2, the controller shortens the legs: when the total vertical force exceeds the threshold of $410N$, the foot is lifted with a speed to shorten the legs (216). When the total vertical force becomes smaller than the threshold of $410N$, the foot returns to its original height. A proper switching function is responsible for smooth changes to avoid chattering.

In (304) the impact at landing is minimized by inducing almost zero absolute vertical velocity of the foot at landing. However the flight time on OpenHRP simulation is often bigger than expected and therefore the landing happens in an unexpected instant, producing large impacts at landing. So all these methods are only patches to minimize the problem. And this is a key objective, because the large impacts at landing are the major obstacles to perform aerial phases in human size humanoid robots. The use of actuators with inherent compliance can solve this problem.

Putting the necessary compliance in the controller rather than in the hardware is also insufficient. Rigid actuators, as for example an electric gearmotor, are not feasible for three reasons: bandwidth limitations with respect to impacts, power output limitations, and energetic efficiency (185). The bandwidth limitation of an electric motor is due, in large part, to the high reflected inertia of the motor linked rigidly to the robot leg via the gearbox.

Another problem for robots without spring elements is that most of the kinetic energy is lost at each hop in an inelastic collision with the ground (187). After impact the foot should stick to the ground on impact without chattering, implying an inelastic collision, because during this stance phase the robot needs sufficient time to apply the necessary control forces.

So during the next hop all the necessary energy has to come from the joint torques. One study on the HRP1 robot found that the motors would have to be 28-56 times more powerful without increasing the weight of the robot, to make the robot run at $10.4km/h$ (217) because the joint torques are between 7.3 and 9.2 times higher compared to those in walking. A comparison between the robot and a human running with $9.8km/h$ shows that the robot consumes power which is about ten times higher than the human runner. When the actuator has elastic elements the energy can be stored in the springs and the efficiency of the robot increases dramatically.

1.4.2 Passive Compliant Actuators

Introducing compliance in the actuators is a fairly new trend in robotics. Traditional robotics focuses on the interface between motor and loads which is "as stiff as possible". This rule of thumb arose because stiffness improves the precision, stability and bandwidth of position control. A welding robot for example needs stiff actuators. But a precise tracking of a trajectory is not that stringent anymore for walking and running robots. New requirements arise to let a robot walk or run and those can be found in compliant actuation: good shock tolerance, lower reflected inertia, more accurate and stable force control, less damage during inadvertent contact, and the potential for energy storage (332).

An important contribution in the research towards compliant actuators has been given by Pratt with the development of the "series elastic actuator" (333). It consists of a motor drive in series with a spring and has been successfully implemented in the two legged robot "Spring Flamingo" (335). In this setup the stiffness is fixed. Nowadays research is focused on actuators with adaptable compliance. Takanishi developed the two-legged walker WL-14 (1998) (450), where a complex non-linear spring mechanism makes predefined changes in stiffness possible. Hurst et al. of the Robotics Institute at Carnegie Mellon University developed the "Actuator with Mechanically Adjustable Series Compliance" (AMASC) (186). It has fiberglass springs with a high energy storage capacity. The mechanism has two motors, one for changing the position and one for controlling the stiffness. The electro-mechanical Variable Stiffness Actuation (VSA) (55) motor developed by Bicchi and Tonietti of the university of Pisa is designed for safe and fast physical human/robot interaction. A timing transmission belt connects nonlinearly the main shaft to an antagonistic pair of actuator pulleys connected to position-controlled backdrivable DC motors. The belt is tensioned by springs. Concordant angular variations control displacements of the main shaft, while the opposite variations of the two DC motors generate stiffness variations. The Biologically Inspired Joint Stiffness Control (by Migliore et al. of the Georgia Institute of Technology, USA) (283) can be described as two antagonistically coupled Series Elastic Actuators, where the springs are made non-linear. Instead of using an antagonistic setup of non-linear springs, different designs exist based on the manipulation of the effective structure of a spring element. Hollander and Sugar developed different prototypes. A first design consists of an elastic beam with a rectangular cross section surrounded by a spring to avoid buckling (172). The compliance can be changed by rotating the beam. For the Jack Spring Actuator the active coil region is changed by inserting an axis in the spring (173). Other designs based on the principle of changing the effective length of a compliant element is the "Mechanical Impedance Adjuster" (MIA) developed by Morita et al. (292; 200). By pressing together many different sheets a passive element with variable mechanical impedance is obtained (234), both vacuum techniques (234) and electrostatic forces (386) can be used. At Northwestern University the "Moment arm Adjustment for Remote Induction Of Net Effective Torque" (MARIONET) (384) is being developed by Sulzer et al. This rotational joint uses cables and a transmission to vary the moment arm such that the compliance and equilibrium position is controllable. Special is that this device does not use an elastic element, the system moves against a conservative force field created by a tensioner.

Most mechanisms with variable compliance mentioned above are relatively heavy and large to be used in mobile robots. An elegant way to implement variable compliance is to use pneumatic artificial muscles. Well-known pneumatic muscles are the McKibben muscles (78). These muscles can only pull, thus in order to have a bidirectionally working revolute joint, one has to couple two muscles antagonistically. Using artificial muscles, the applied pressures determine position and stiffness. Research at the Vrije Universiteit Brussel focuses on the Pleated Pneumatic Artificial Muscle (PPAM) (105). A recent development by the same lab is the "Mechanically Adjustable Compliance and Controllable Equilibrium Position Actuator" (MACCEPA) (414) and MACCEPA 2.0 with a stiffening characteristic (417). The actuator is a straightforward and easy to construct rotational actuator, of which the compliance can be controlled separately from the equilibrium position. To prove MACCEPA 2.0 is beneficial for energy storage and return the hopping robot Chobino has been constructed as can be seen in figure 1.23. An overview paper of

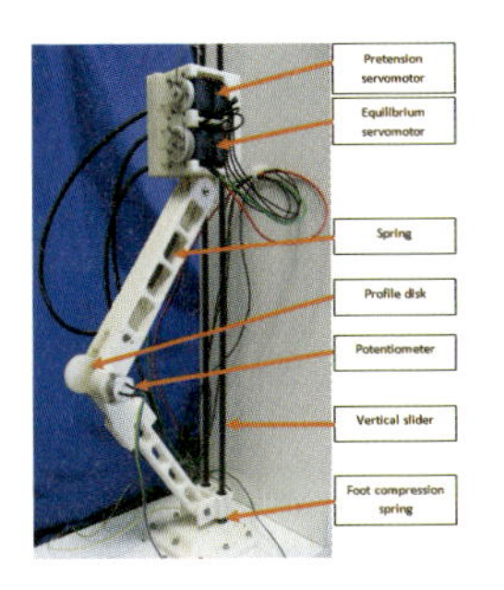

Fig. 1.23 Picture of Chobino1D, a 1DOF hopping robot actuated with the MACCEPA 2.0.

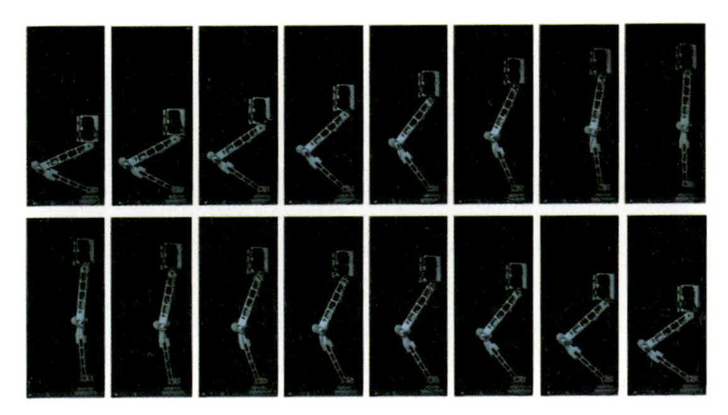

Fig. 1.24 A sequence of photos of a jumping motion of Chobino1D. The images were taken every 40ms with a high speed camera.

passive compliant actuators is presented is presented in (413) where a categorization for them is presented.

1.4.3 *Use of Compliant Actuators in Robotics*

1.4.3.1 Compliant Actuators for Safe Human-Robot Interaction

One reason to use the actuators compliance is for safe human-robot interaction (HRI). The biggest danger present when working in close proximity with robotic manipulators is the potential for large impact loads resulting from the large effective inertia (or more generally effective impedance) of many robotic manipulators (466). These authors reported about an empirical formula "Head Injury criteria (HIC)" for the thread of serious damages or injuries after collision.

For the PUMA 560 industrial robot an impact velocity of $1m/s$ produces a HIC greater than 500, more than enough to cause injury. Ways to reduce impact loadings is to reduce the arm effective inertia or decrease the interface stiffness by for example adding an amount of compliant material (467). For this example the required thickness of a compliant cover is more than $5inch$, which is very impractical. Therefore it is important to produce manipulators possessing naturally low impedance in order to achieve natural safety in the mutual interaction man-robot (84).

When the robot moves slowly the joint can be made stiff for improved positional precision. For fast movements the stiffness can be decreased so the inertia of the motor is disconnected from the arm. An impact in both cases will result in a relatively low impact force (56). On the other hand, Van Damme et al. (409) studied the effect of joint stiffness on safety. The maximum impact force turned out to be almost independent of joint stiffness over a wide range of stiffnesses. This indicates that the impact is mainly determined by the inertia of the impacting link.

In the same context the 2 DOF planar pneumatic "softarm" (see figure 1.25) is developed by Van Damme et al. (410). The assistive manipulator will interact

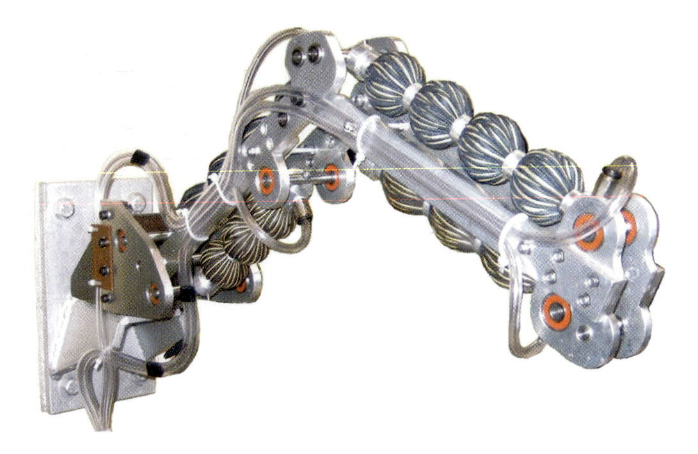

Fig. 1.25 Softarm actuated by PPAMs to study safe human-robot interaction.

directly with an operator in order to assist him handling heavy loads. The project provided two important conclusions (409). The first is that safety oriented control is crucial. Due to its low weight (around $2.5kg$) and the inherent compliance of the pneumatic muscles, the investigated system had excellent hardware safety characteristics. In spite of this, when its safety was numerically quantified, it was found that it could be quite dangerous to humans in some circumstances when using standard control techniques (PID with gravity compensation in this case). Much better results, both with respect to safety and tracking performance, were obtained with a modified form of Proxy-Based Sliding Mode control (PSMC) (240). The second conclusion is that passive compliance is a double-edged sword with respect to robot safety. Passive compliance can greatly improve robot safety in cases of impact between a human and a robot. It also has the ability to store energy (very similar to loading a spring). However, when this energy is suddenly released, it can result in high speed motions of the robot, and correspondingly in a high risk for humans in case of collision. PSMC limits the energy build-up by introducing a type of sensorless active compliance, which illustrates again that safety on the hardware level has to be complemented by safety on the software level.

The role of passive compliance in active force control is investigated by De Schutter (111). He showed that all active force control methods require a comparable degree of passive compliance to yield a comparable execution speed and disturbance rejection capability. It is intuitive that a very low stiffness joint is more capable of applying a constant force in the face of position disturbances than a rigid joint. For different force control tasks, different joint stiffnesses may be desirable (185).

1.4.3.2 Compliant Actuators for Bipedal Locomotion

In section 1.4.1 the biological aspects of walking were discussed and it was shown that compliance plays an important role to walk energy-efficient. Most of the

Fig. 1.26 Building and tuning the simple 2D passive dynamic ramp-walker Elvis during Dynamic Walking conference.

research concerning energy-efficient locomotion is performed on the right hand side of figure 1.1, the side of the passive walkers. Most of these robots have compliant actuation because compliance is needed to exploit the natural dynamics of the system. To have the same versatility as actively controlled robots the goal of Lucy is to be situated on the left hand side of figure 1.1. On this side usually stiff actuators are used because they are excellent in precise tracking of a desired trajectory. To be able to combine trajectory tracking with the exploitation of the natural dynamics compliant actuation is required.

Passive walkers are robots basing their locomotion on natural dynamics solely. No energy is supplied and to overcome friction and impact loses they have to walk down a slope. The inertial properties and compliance characteristics are designed in such a way that they can walk within a certain rhythmic motion, but its dynamically feasible walking patterns are situated within a small range of possible motions. Depending on the stability of the system, a certain error is allowed. Since biped locomotion has periodic gaits, Poincaré maps can be used to analyze the problem. A Poincaré map samples the flow of a periodic system once every period (239) (443).

The simplest passive walker consists of two rigid legs interconnected by a passive hinge. There are some possible ways to allow the leg to swing from the back to the front. The first uses special designed feet to make the body toddle and causes the feet to be lifted out of the sagittal plane. This principle is used in passive walking toys, dating back to the 1800s (121). Another possibility is to equip the slope with blocks placed at the footholds of the robot as shown in figure 1.26. This way, the leg can swing freely between the blocks and end up on top of the next block.

Articulated legs can avoid the blocks. The knee joint creates a double pendulum leg. The first robot of this type was built by McGeer in 1989 (274). A mechanism is placed in the knee to prevent the lower leg from overstretching, otherwise it is not possible to support the weight of the robot during stance phase.

The passive walker developed at the Nagoya Institute of Technology contains a fixed point, kind of a stopper to limit the forward motion of the leg. Experiments showed the results with and without stopper. It is difficult for the passive walker without stopper to walk for more than 4 steps. While the passive walker with stopper can walk for many steps with a best record number of 4010 steps (192).

To make passive walkers walk on level ground, a minimal actuation is required. Adding actuation also increases stability. To maintain the capability of the joints to swing passively, compliant actuation is required. The hip actuation by McKibben muscles increases the robustness of 2D motions as shown by experiments with the robot Mike (448). If one puts the swing leg fast enough in front of the stance leg the robot will not fall forward (445).

Current research tries to improve the performances of passive walkers. Max, built by Wisse, contains a passive upper body connected to a bisecting mechanism at the hip (447). A skateboard-like ankle joint is a solution for 3D stability as shown in the robot Denise (446). Meta is the first robot with electric actuation of the University of Delft. A Maxon DC motor, in series with a spring, is used to actuate the hip and is PD position controlled (361). Their latest walking robots, Flame (166), and the RoboCup soccer robot TUlip (165), are the first steps towards a more versatile walking robot.

Besides hip actuation to restore energy losses during walking, also a push-off can inject energy in the system. The Cornell biped (99) has only actuated ankle joints. A small motor stores energy in a spring during leg swing which is released to perform a push-off. This provides a powerful impulse while minimizing motor power requirements. This is remarkably effective with a mechanical cost approximately equal to the human value and far more better than the estimated mechanical cost of the robot Asimo (34).

Osaka University constructed several bipeds actuated with McKibben muscles. Pneu-man is a 3D biped robot that has 10 joints driven by antagonistic pairs of McKibben actuators: 1 DOF arms and 4 DOF legs (178). The only sensors on the robot are several touch sensors on the soles. Que-kaku (396) has three joints: 1 hip and both knees and is a planar walker. The opening time of the expel air valve of the hip muscle is controlled in order to operate the hip joint passivity. The results showed that the passive walker changes its behaviour by the hip passivity (395). The successor has also actuated joint ankles and is able to jump and run up to 7 steps (179). It could not remain stable during running because the robot does not have feedback from external sensors. In (180) a pneumatic biped robot is presented that can change its joint compliance for different locomotion modes as walking, jumping, and running.

The problem of the passive walkers mentioned above is the fix walking speed due to the fix dynamics. The work of Van Ham et al. (414) focus on overcoming

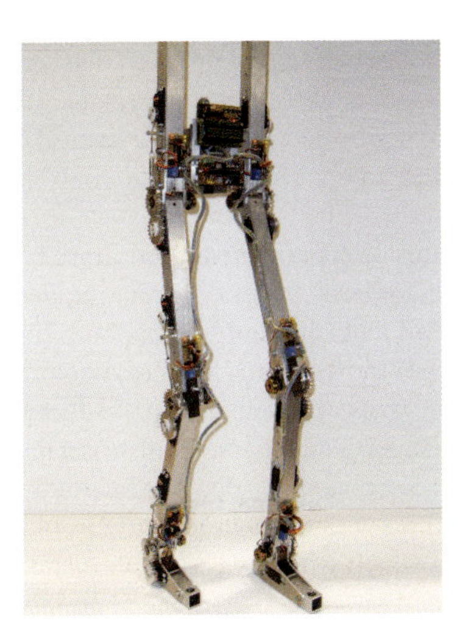

Fig. 1.27 Veronica powered by MACCEPAs.

this problem by using a range of compliances instead of a discrete number of compliances, in order to be able to vary the walking speed. The biped Veronica (figure 1.27) has been actuated with 6 MACCEPA actuators (414). The control concept of the robot is based on "Controlled Passive Walking". Instead of continuously controlling the biped, the sets of control parameter for each joint, meaning equilibrium position and compliance, are only changed a few times during walking. A transition to a next phase is triggered by an event as footswitches, joint angle values or time delays. Between the discrete control actions the motion is passive due to the compliant actuators. So this robot behaves like an actuated passive walker. Because the compliance is adaptable, different natural frequencies can be selected. This means that the walking motion is not restricted to a fixed walking speed (412). A similar biped Achilles has been built by the Tsinghua University and is controlled to walk with a phase transforming based reinforcement learning method (270). For the robot Dribbel the behaviour of a joint with springs was emulated by a geared motor with torque sensor (113). By changing the setpoint and controller gain, the walking gait of the robot was influenced so the robot was able to walk with different gaits at different speeds.

Unfortunately these passive walkers are of little practical use because they lack the versatility of the actively controlled robots. Another problem of passive walkers is that the robot is launched to walk by a human operator and that consequently the variance in these experiments of the initial condition is quite large (178). Besides starting to walk, also stopping is difficult or impossible for a passive walker. 3D passive walkers are currently under development, but also these robots can only walk on a straight line and are not able to make a controlled curve.

To cope these problems the strategy elaborated in this thesis is to start from dynamically stable trajectories which are tracked by the different joint actuators.

The advantage is that the robot can stand still, start and stop by its own and walk at different walking speeds and with different step lengths. In a second stage the compliance of the joints will be controlled. Hereby the natural dynamics will be adapted as a function of the imposed trajectories, such that control activity and energy consumption will be reduced.

To the author's knowledge only two other bipeds have been built for which a strategy has been developed combining trajectory tracking and compliance adaptation. Waseda University built two anthropomorphic walking bipeds having antagonistic driven joints, WL-13 and WL-14. The typical characteristics of the antagonistic driven joints using nonlinear spring mechanism are to vary the stiffness of joints over a broad range (452). By varying the joint stiffness a reduction of 25% of energy consumption during the swing phase was observed in the walking WL-14 compared to the case when the stiffness was not varied actively (450) (451). It is a pity the authors did not mentioned which stiffness was chosen and why.

For this thesis the pleated pneumatic artificial muscle was chosen as compliant actuator. The muscles are implemented in a biped, called Lucy. 12 pleated pneumatic artificial muscles actuate 6 DOF: the hip, knee and ankle of both legs. Using compressed air as power source is not so common, only a few pneumatic bipeds exist of which an overview is given in the next section.

1.4.3.3 Overview Pneumatic Bipeds

Pneumatic bipeds can further be grouped into robots using pneumatic cylinders and pneumatic muscles. Pneumatic muscles are distinguished from the pneumatic cylinder by their changing force to length relationship, this is due to their shape changing property. Pneumatic muscles can also generate higher forces compared to cylinders. Pressure control, instead of mass flow control, is the key to control for this kind of actuator (104). Another major difference is weight: whereas cylinders are typically made of aluminum or plastics, a muscles core element is its membrane which is, obviously, extremely lightweight. Another disadvantage of pneumatic cylinders is friction and stickslip effects.

The number of pneumatic legged robots built worldwide up till now is rather limited compared to the amount of robots with electrical actuation. Probably because the control is much more difficult. One of the first to incorporate pneumatics is the Japanese pioneer for legged locomotion Kato. During the sixties and seventies he has built several statically balanced walking bipeds such as WAP I, II and III (232). These machines where actuated by different types of pneumatic artificial muscles and were able to move very slowly.

Another pioneer in legged robotics is Raibert, who has built several hopping and running machines during the eighties at the Massachusetts Institute of Technology. His mono-, bi- and quadruped robots used a combination of hydraulic actuators with pneumatic cylinders to actuate the telescopic legs (342). Raibert implemented control algorithms focussing on controlling hopping height, forward running speed, and body posture. Energy is stored in a pneumatic spring in the legs and is modulated to manipulate hopping height, forward speed is controlled by positioning the legs during the flight phase and body attitude is regulated during the stance phase. Based on this principle a one-legged robot hopping in 2D was constructed (343), a one-legged robot hopping in 3D (344), a running robot on four legs (345), a bipedal robot running and being able to execute a forward flip in 2D (167) (168).

Einstein and Pawlik (117) constructed a statically balanced pneumatic walking robot machine at the polytechnic institute of Czestochowa in Poland. The COG was positioned above one foot, the other foot was lifted and then the entire body was rotated around the vertical axis, passing through the area covered by the standing foot. 3 different EP-WAR (Electro-Pneumatic-Walking-Robot) robots have been built at LARM (Laboratory of Robotics and Mechatronics) in Cassino using pneumatic cylinders. The first prototype EP-WAR was able to walk straight, to turn right and left (128). The second prototype EP-WAR2 was also able to climb stairs (129). Then, EP-WAR3 has been designed, built and tested at LARM in order to descend stairs (130). Because the pneumatic cylinders are used in a binary way, only a discrete number of postural positions are possible. The walking stability of the biped robot is obtained by using suction-cups, which are installed on the sole of each foot. Guihard et al. designed BIPMAN (140). Each leg has three rotational joints actuated by four-way servo valves. The control architecture is composed of two main levels. The upper one, called the "Coordinator" level, maintains the robot stability by correcting on-line its center of mass acceleration and distributing correctly the forces on each limb. The lower level, called the "Limb" level, is devoted to the control of each limb according to the desired position and force trajectories given by the Coordinator level. Spampinato and Muscato constructed a 10 DOF biped called DIEES actuated by pneumatic pistons (145; 143). The swing leg is controlled through a set of parabolic trajectories generated during the gait. The stance leg is actuated through a simple but efficient force control approach based on a different interpretation of the Virtual Model Control strategy (375). The robot is able to perform walking motions when assisted by a wheeled device to prevent the robot from falling. Step is a 5-link electro-pneumatic biped robot developed by the Laboratoire de Robotique de Versailles consisting of 2 legs and a free trunk, so the robot is underactuated. A sliding mode control scheme was implemented but experiments showed there were still some problems leading the robot to fall (294). Festo developed a full scale humanoid TronX which is actuated with pneumatic cylinders. Although this robot has two actuated legs, it is not able to walk. Anybots, a technology company of California, has developed Dexter driven by air cylinders (3). This is the only robot from this paragraph which is able to walk without support.

The Shadow Walker is a wooden leg-skeleton powered by Shadow air muscles and built by the Shadow Company (439). Twenty-eight air muscles (14 on each leg) act across the eight joints, enabling a total of twelve degrees of freedom. This project is no longer active and the robot was never able to walk. A 4 DOF planar robotic leg actuated with McKibben artificial muscles was designed, constructed, and controlled by the Case Western Reserve University (98). There main research interests are building robots using insights gained through the study of biological mechanisms. Their most famous robots are the cockroaches (243) and whegs series (339). The group of Caldwell, at the university of Salford, developed the biped Salford Lady (77) actuated with McKibben artificial muscles. The local joint control directly calculates desired pressure levels with a PID position feedback loop. The pressure itself is regulated with fast switching pulse-width modulated on/off valves. For the PANTER biped, studies are performed to implement the Festo muscles for "elastic locomotion" (237). The walking performances of these robots are very limited and they are not able to show the advantages of muscles over more traditional actuation as electrical drives, the reason why these robots were initially built.

The most successful use of pneumatic muscles is to actuate the passive walkers which are described in section 1.4.3.2. However the control of the muscles is much simpler compared to a tracking-controller. For example when a foot touches the ground certain valves are opened during some time. These parameters are usually tuned by hand. For a robot with many DOF or when one wants the robot to walk with different gaits, learning algorithms are an interesting alternative because the search space is big.

1.5 Goal of the Lucy Project

The goal of this work is to give an answer to the following questions:

Can pneumatic artificial muscles be used for dynamic balanced bipedal locomotion in a trajectory controlled manner?

- The control of pneumatic muscles and also more general compliant actuators to power bipeds is not well-known. Currently the most successful use of them is in passive walkers were the control scheme is rather simple. To tackle the problem in a trajectory controlled manner is not shown yet.

How to control the adaptable compliance of a joint powered by passive compliant actuators?

- By using compliant actuators an extra parameter can be controlled: the compliance. Mostly an arbitrary value is chosen. The concept to control the compliance is also special. Moreover in robotics the motto was for a long time "the stiffer the better".

1.6 Approach

A lot of attention is gone to actually prove the proposed control architecture is working on a real biped. The bipedal walking Lucy is built containing the necessary sensors, actuators and processing power. A trajectory generator and joint trajectory tracking controller is developed to prove pneumatic muscles are able to power the biped. The trajectory generator uses the zero moment point (ZMP) concept as stability criterion (434). The joint trajectory tracking controller uses the dynamics of the robot and the characteristics of the muscles and joint to control the pressure inside the muscles so the desired trajectory is tracked. Besides the discussion of all the graphs, probably the most convincing to prove the strategy actually works is to see the robot in motion. So please see the video **http://lucy.vub.ac.be/phdlucy.wmv**.

To study the compliance of pneumatic muscles and how to control this extra parameter, a simple pendulum setup has been built. First sine trajectories were studied and a mathematical formulation has been developed to select an optimal compliance for reduced energy consumption. A strategy for more complex trajectories is proposed too. Different designs of compliant actuators were compared.

1.7 Outline

In chapter 2 the robot Lucy is described. Because the muscles play a major role, the chapter starts with the design and force characteristic of the muscle. Two muscles in an antagonistic setup powers a joint with a certain torque and compliance. Muscles also require adjusted mechanics and control hardware. The second half of the chapter is devoted to an extensive description of the mechanical and electronic design. Also a virtual Lucy has been built in simulation, written in Visual C++. Mathematical models of the different units are described. Special is that mechanics and thermodynamics are put in one dynamical simulation.

The current control architecture for Lucy can be split into two components: a trajectory generator and a joint trajectory tracking controller. The trajectory generator, given in chapter 3, uses objective locomotion parameters (which are step length, intermediate foot lift and mean velocity) to calculate dynamically stable trajectories which can be changed from step to step. Two methods are developed. The first method is based on the principles of inverted pendulum walking, modeling the robot dynamics as a single point mass. Disadvantage is that not the complete multibody mechanics with distributed masses and inertias is taken into account, causing a difference in desired and real ZMP especially at higher walking speeds. A second approach describes the implementation of a preview controller to control the Zero Moment Point (ZMP). Special for this controller is that also future information of the motion is exploited. The dynamics of the robot are represented by a cart-table model. Because both methods use the ZMP as stability criterion, this concept is repeated at the beginning. The author performed research on the humanoid robot HRP-2 at the Joint Japanese/French Robotics Laboratory (JRL) in AIST, Tsukuba (Japan) in the ongoing research "Gait Planning for Humanoids Robots: Negotiating Obstacles". A quick overview of the strategy and results are provided because the base is also the preview controller.

Chapter 4 handles about the joint trajectory tracking controller. The goal of this unit is to control the pressures inside the muscles so that a prescribed trajectory is tracked. The joint trajectory tracking controller is divided into an inverse dynamics unit, a delta-p unit and a pressure bang-bang controller. With the combination of the trajectory generator and the joint trajectory tracking controller the robot Lucy is able to walk. The experimental results for walking are discussed.

In chapter 5 the role of compliance is discussed. The concept of controlling compliance is fairly new in robotics. The first part deals about the ability of compliant actuators to adapt the compliance to exploit the natural dynamics and how this can be used for reduced energy consumption. This study was not performed on the biped Lucy, but on a single pendulum structure powered by pleated pneumatic artificial muscles. A strategy is proposed to find an optimal compliance were the energy consumption is minimal dependent on the trajectory and physical properties of the pendulum. Also the energy consumption of other designs of compliant designs with a spring element are compared. In the last part of chapter 5 some preliminary experiments of the robot Lucy performing jumping motions are shown, to present compliant actuators can absorb impact shocks.

Finally, in chapter 6 the overall conclusions and future work are given.

Chapter 2
Description of Lucy

2.1 Introduction

The main goal of the construction of the biped Lucy is to investigate the use of Pleated Pneumatic Artificial Muscles (PPAM) as an interesting alternative to the electrical drives generally used in walking robots. Hereby will be focussed on the exploitation of compliance characteristics in combination with trajectory tracking.

Lucy is a sagittal walking robot with a mechanical structure that is representative for human walking. One of the biggest challenges of building and controlling humanoid robots is the complexity due to the large number of degrees of freedom. This means large and complex equations, a lot of computational power needed, many electronics, sensors and actuators to control the various joints. Usually a fully 3D biped has 6 DOF for each leg (3 for the hip, 1 in the knee and 2 in the ankle), while a planar biped only has 3 DOF per leg. This is the main reason why Lucy is a planar biped (with 7 links). Moreover, it has been shown that for biped walking the dynamical effects in the lateral plane have a marginal influence on the dynamics in the sagittal plane (48). To study the essence of bipedal locomotion a planar approach is a first important step. In the frontal plane there is mainly the lateral stabilization and the exchange of support motion. The complexity can further be reduced by eliminating the active ankles, resulting in 5 links and 4 actuated joints as in the biped Rabbit (89). The contact with the ground is just a point and not an area as is the case with a foot. Only a reaction force and not a torque can be applied to the ground so the robot is underactuated during the single support phase. This option has not been taken because feet improve the stability (115).

The upper part is a single part without arms and head. Elftman (118) calculated the angular momentums arising as a consequence of the arms motion during the gait. Based on the energy efficiency was shown that the arms by their motion annuls the vertical (yaw) component of the angular momentum that appears at the body gravity center. For a planner robot this is not a problem so arms are not required for this study.

This chapter starts with a description of the pleated pneumatic artificial muscle and the antagonistic muscle set-up. They are incorporated in a modular unit. The

B. Vanderborght: Bipedal Walking by the Pneumatic Robot Lucy, STAR 63, pp. 41–91.
springerlink.com

different modular units are linked to each other and form together with the feet a complete robot. To prevent the sagittal robot from falling sidewards a guiding mechanism is chosen consisting of a horizontal and a vertical rail. These rails are mounted on a frame which also incorporates a treadmill so that the robot is enabled to walk longer distances.

The next section describes the electronics to control the robot. An important element is the communication between the robot and a central PC. An overview of the interface program is provided.

The last section handles about a "virtual Lucy", a hybrid simulator combining the mechanics and dynamics. For all the essential parts of the robot a mathematical model is given.

2.2 Pleated Pneumatic Artificial Muscle (PPAM)

A pneumatic artificial muscle (PAM), also called a fluidic muscle, an air muscle or pneumatic muscle actuator, is essentially a volume, enclosed by a reinforced membrane, that expands radially and contracts axially when inflated with pressurized air. Hereby the muscle generates a uni-directional pulling force along the longitudinal axis. Different designs exist. Daerden et al. (105) classified the pneumatic muscles under Braided muscles (contains the McKibben muscle and Sleeved Bladder Muscle), Netted Muscles (Yarlott Muscle, RObotic Muscle Actuator, Kukolj Muscle) and Embedded Muscles (Morin Muscle, Baldwin Muscle, UnderPressure Artificial Muscle, Paynter Knitted Muscle, Paynter Hyperboloid Muscle, Kleinwachter torsion device). The McKibben muscle is the most popular and is made commercially available by different companies such as the Shadow Robot Company (24), Merlin Systems Corporation (13), Hitachi Medical Corporation (9) and Festo (8). The

Fig. 2.1 Photograph of 3 contraction levels of the 1st generation of PPAM.

McKibben muscle contains a rubber inner tube which will expand when inflated, while a braided sleeving transfers tension (78). Inherent to this design are dry friction between the netting and the inner tube and deformation of the rubber tube.

The McKibben muscle however has some drawbacks: moderate capacity of contraction, hysteresis as a result of friction between an outer sleeve and its membrane and a threshold behavior (104). Consequently this muscle is difficult to control. Besides this, friction reduces the life span of this actuator. Daerden developed a new pleated PAM, the Pleated PAM , to cope with those disadvantages (104). The muscle has a high stiffness membrane that is initially folded together and unfold upon inflation. This leads to a strong reduction in energy losses with regard to the classical types and, hence, develops stronger forces and higher values of maximum contraction. Verrelst et al. developed a second generation of the PPAM to extend the muscle lifespan and to simplify the construction process of the muscles (425). In the first generation, the tension is transferred by the stiff longitudinal fibres, spread all over the surface of the membrane. This results in the pileup and crumple of the fibres near the end fittings, since the deformation is different for a fibre at the top and at the bottom of a fold. When only a strand of stiff fibres is placed at the bottom of each fold, while the rest of the folded membrane is made out of a more flexible airtight material, each strand has the same deformation. Doing so ensures a more equal unfolding of the membrane, which is clear when comparing figure 2.1 and figure 2.2. As a result, the lifetime of the muscle increases drastically: during a durability test more than 400.000 cycles were achieved moving a payload of $130kg$ up and down. The first generation only attained 3000 cycles. Currently the third generation PPAM are developed, for which specifically the Kevlar fibres of the pleats are rearranged. Using toothed ABS parts and a continuous high tensile fibre, it is possible to fold the membrane at the same time that it is fixed to the end fittings which simplifies

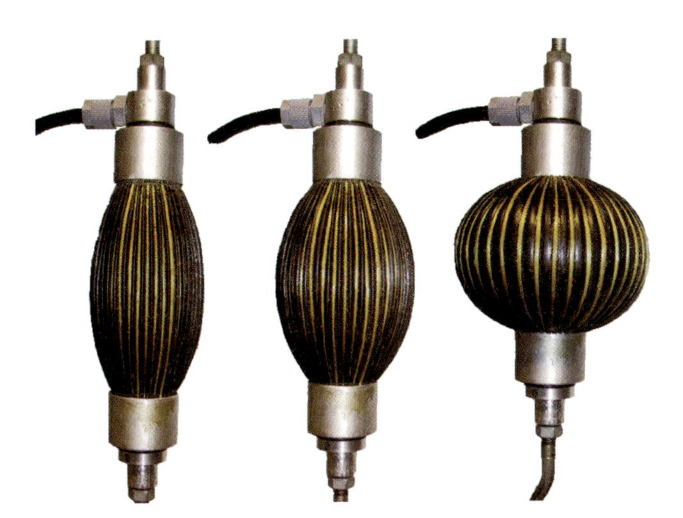

Fig. 2.2 Photograph of 3 contraction levels of the 2nd generation of PPAM (used for Lucy).

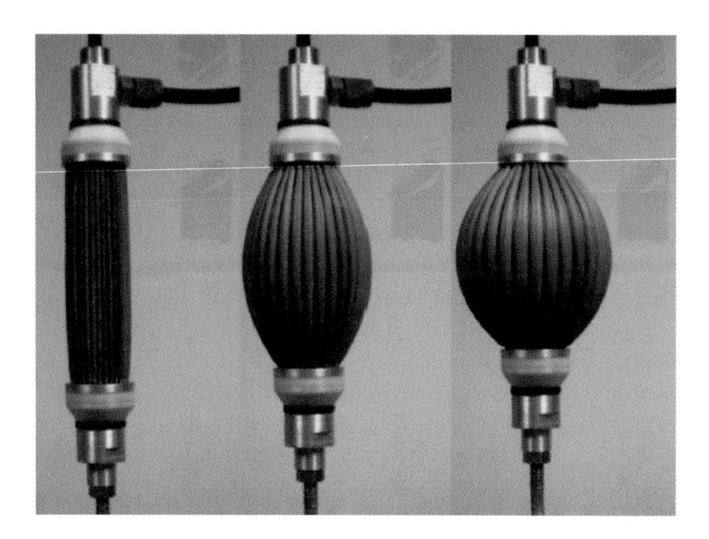

Fig. 2.3 Photograph of 3 contraction levels of the 3rd generation of PPAM.

drastically the production process and reduces the muscles weight even further. Besides, the new production technique makes it relatively easy to fill the dead volume with a cylindrical tube during the production process. A disadvantage is that the PPAM expands more radial than the McKibben muscle. It is also more difficult to use a bundle of PPAMs as the McKibben muscles used in the Shadow Hand (351) or to twist the PPAM around something. Two McKibben muscle were for example twisted around the radius and ulnar bone of the forearm to produce pronation/supination (79). This is the motion to move the palm facing down and up.

Besides the pneumatic artificial muscles other forms of muscle technology exist, which use active materials like shape memory alloys and polymeric actuators. They are not yet sufficiently developed to be used in walking machines; their speeds of operation are very low, with time constants in the order of tens of seconds (174; 80) and they generate weak forces (94). Primary application of this technology is focused in micro-actuators and micro-manipulation. Exception is for example a small-sized biped actuated by an antagonistic pair of IMPMCs (ionic polymer-metal composite) (453).

2.3 Concept

The working principle of a PPAM is that, when inflated, the pleats of the membrane unfold and the muscle contracts while generating high pulling forces. The flexible fabric is a simple woven polyester cloth, which is made airtight by a polymer liner. This structure is folded and in each crease a yarn of high-tensile Kevlar fibres is responsible for transferring the large axial tension. Figure 2.4 depicts the complete straightforward construction of the new muscle. The end fittings have a treated hole

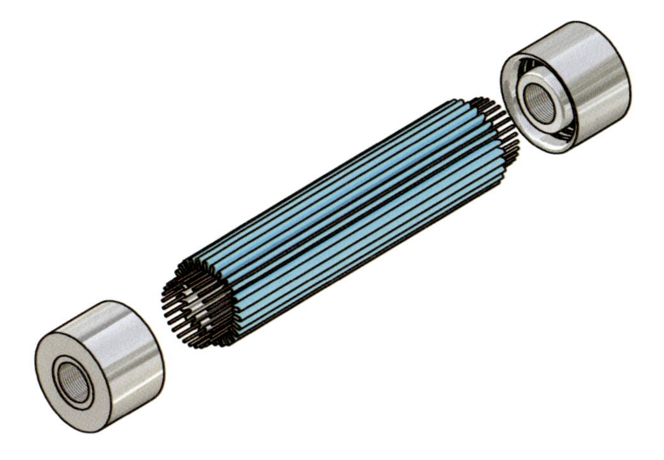

Fig. 2.4 Composition of the new muscle prototype.

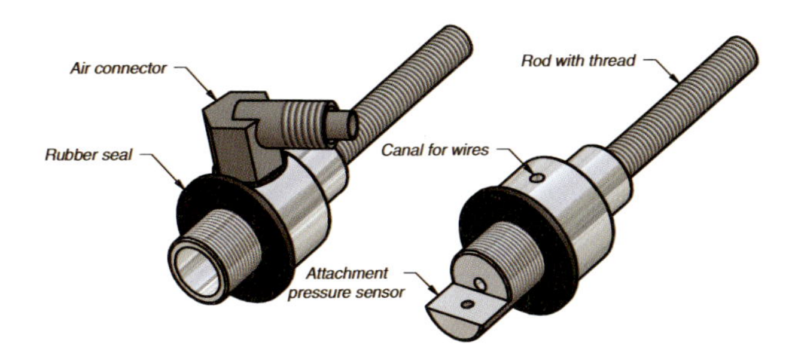

Fig. 2.5 Drawing of the two muscle end connectors.

in which additional muscle connectors can be screwed. These connectors incorporate three functions: guiding the pressurized air in and out the enclosed volume, creating the interface for the connection to the specific application frame, and providing an attachment for a pressure sensor positioned inside the muscle. Figure 2.5 shows the two different connectors to be fixed at each side of the muscle. The left side drawing of figure 2.5 shows the connector which allows the air to flow in and out of the muscle, while the right side drawing depicts the connector with the attachment for a pressure sensor.

2.4 Force Characteristic

When neglecting the membrane's material deformation and the inertial muscle properties, the generated force is expressed as:

$$F = -p\frac{dV}{dl} \tag{2.1}$$

Table 2.1 Coefficients of the force fitting function

f_4	f_3	f_2	f_1	f_0
−2.04130	171.623	−7178.93	128611	146099

with p the gauge pressure inside the muscle, dV enclosed muscle volume changes and dl actuator length changes. Comparing the force-length expression to that of pneumatic cylinders, dV/dl is defined as the actuator's "effective area" (325). The volume of the actuator increases with decreasing length until a maximum volume is reached. At maximum contraction, forces become zero, and at low contraction the forces can be very high. The changing force as a function of contraction at constant pressure is essentially different compared to standard pneumatic cylinders, for which the generated force does not change at constant pressure. For these devices the generated force is proportional to the piston area on which the internal pressure works, consequently the force does not change with piston position at constant pressure.

Verrelst et al. (425) provide a mathematical model for the muscle and this model describes the shape of the muscle bulging at each contraction level, and gives essential characteristics such as muscle traction and enclosed volume. Static load tests validate this model such that the developed functions can be used for dimensioning purposes. Additionally, a fit on the measured force data was carried out, because it is easier to work with than the numerical solution derived from the mathematical model. Here, only the main equations are shown.

The influence of elasticity can be omitted for the high tensile strength material used for the fibres. The generated force is given by:

$$F = pl_0^2 f\left(\varepsilon, \frac{l}{R}\right) \tag{2.2}$$

where p is the applied gauge pressure, l_0 the muscle's full length, R its unloaded radius and ε the contraction. The dimensionless force function f depends only on contraction and geometry. This force function can be approximated by the following fitting function:

$$F = pl_0^2\left(f_4\varepsilon^3 + f_3\varepsilon^2 + f_2\varepsilon + f_1 + f_0\varepsilon^{-1}\right) \tag{2.3}$$

The coefficients of the fitting process for the force function, f_0 to f_4, following the structure of equation (2.3), are given in table (2.1). This table is made for a muscle of $l_0 = 110mm$ and $R = 16mm$, the size of the muscles used for the biped Lucy. The values are valid when the generated force F is expressed in N, the initial muscle length l_0 in m, the pressure expressed in bar and the contraction ε expressed in %.

The graph in figure 2.6 gives the generated force for different pressures of this muscle. At low contraction, forces are extremely high causing excessive material loading, and for large contraction the generated forces become very low. So the contraction range is set for this application between 5 and 35%. The generated forces are

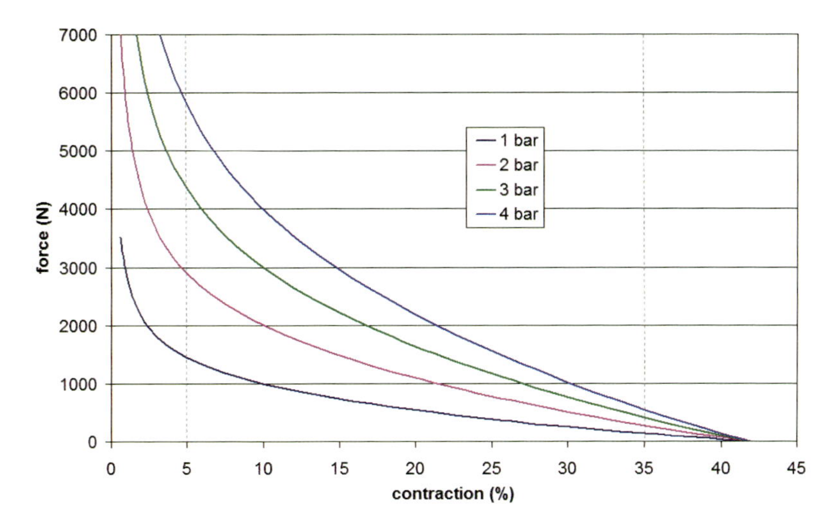

Fig. 2.6 Generated force as a function of contraction at different pressure levels ($l_0 = 110mm$ and $R = 16mm$).

much higher at lower pressure levels compared to the McKibben and Festo muscles. For example a McKibben muscle with a diameter of $22mm$ can generate maximum $300N$ at $4bar$ (278). A Festo muscle with an internal diameter of $20mm$ and a nominal length of $200mm$ can generate up to $1500N$ at $3.5bar$ (127). The force of a PPAM used for Lucy is maximum $6.000N$ at $4bar$. The PPAM's maximum muscle contraction is 40%, much higher compared to McKibben and Festo muscles which can contract typically up to 25%. For a hysteretic model of the PPAM is referred to (411), where a Preisach model for hysteresis was introduced. Good agreement between predictions and measurements were achieved in an important range of contractions.

2.5 Volume Characteristic

The fitting for the enclosed muscle volume is performed on the theoretical data (figure 2.7) (425):

$$V(\varepsilon) = l_0^3 v(\varepsilon) = l_0^3 \left(v_5 \varepsilon^5 + v_4 \varepsilon^4 + v_3 \varepsilon^3 + v_2 \varepsilon^2 + v_1 \varepsilon + v_0 \right) \tag{2.4}$$

In table 2.2 the coefficients of the volume fitting v_0 to v_5, following equation (2.4), are given. The values are valid for the volume given in ml, the initial length expressed in m and the contraction ε expressed in %. The data in table 2.1 and 2.2, together with equations (2.3) and (2.4), can also be used to generate an approximation of the force and volume characteristics for muscles with lengths different from $l_0 = 110mm$. But the values in these tables are only valid for muscles with a specific slenderness ($l_0/R = 110/16 = 6.9$). So, whenever the fitting is used for a muscle

Table 2.2 Coefficients of the volume fitting function

v_5	v_4	v_3	v_2	v_1	v_0
0.02254	−2.6296	113.82	−2386.3	30080	71728

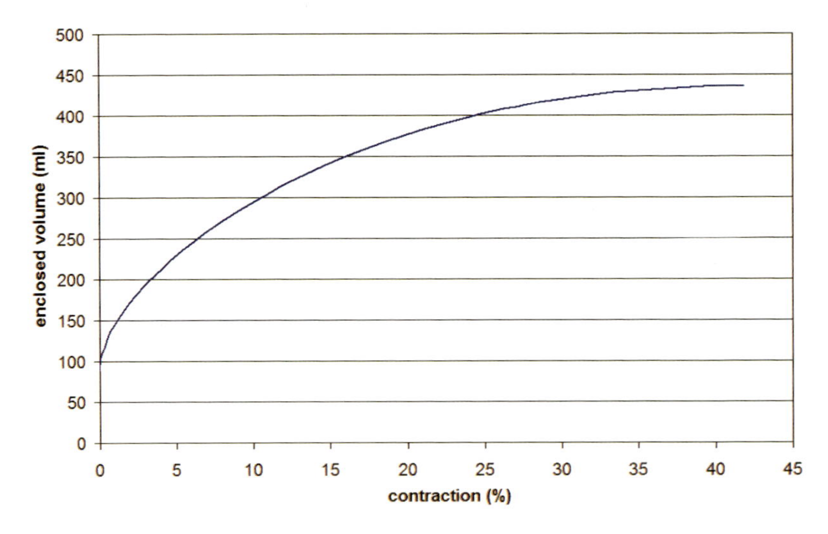

Fig. 2.7 Theoretical enclosed muscle volume as a function of contraction ($l_0 = 110mm$ and $R = 16mm$).

with different initial length, the unloaded radius of that muscle has to be adapted, otherwise the force and volume approximations are not valid.

2.5.1 Antagonistic Muscle Setup

2.5.1.1 Kinematics

Pneumatic artificial muscles can only pull. In order to have a bidirectionally working revolute joint one has to couple two muscles antagonistically. In fact only one muscle e.g. in combination with a mechanical return spring could be used, but in order to be able to control joint compliance, this option is not chosen (see section 2.5.2). The antagonistic coupling of two muscles could be achieved with either a pulley mechanism or a pull rod and leverage mechanism. The latter is chosen since the lever arm can be varied such that the highly nonlinear force-length characteristic of the PPAM is transformed to a more flattened torque-angle relation. A scheme of the basic configuration of the pull rod and leverage mechanism is depicted in figure 2.8 and figure 2.9 shows the implementation in a modular unit.

Two muscles, muscle 1 and 2, are connected at one side of the system to a fixed base in the points B_1 and B_2 respectively. The other ends of the muscles are attached

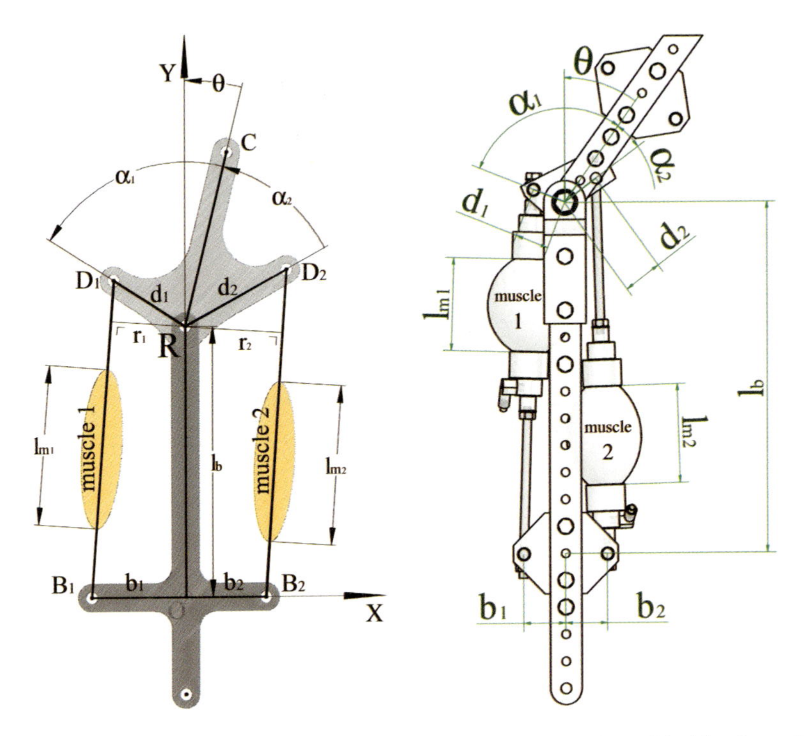

Fig. 2.8 Schematic overview of the antagonistic muscle pull rod system.

Fig. 2.9 CAD drawing with side view of a modular unit, showing the kinematical joint design parameters.

to a pivoting part at the points D_1 and D_2, of which the rotation axis passes through a point R. The rods are assumed to be rigid.

To determine the kinematic expressions of the joint system, an orthogonal X,Y-coordinate system is defined. The X-axis is aligned with the base points B_1 and B_2, while the vertical Y-axis intersects the physical pivoting point R and lies along the base suspension bar of the pull rod mechanism. The essential parameters to be determined during the design process of the joint are the following:

- b_i is the distance between the origin O and the point B_i.
- d_i is the distance between the pivoting point R and the point D_i.
- α_i is the angle between the vector $\overline{RD_i}$ and $\overline{RC}$, with C a point on the rotating part. (α_i is not oriented and always positive)
- l_{m_i} is the actual length of muscle i
- l_b is the length of the base suspension bar, measured between the origin O and the pivot point R.
- θ represents the rotation angle, measured between $\overline{RC}$ and the Y-axis. (θ is oriented, counter-clockwise is positive)

The vectors $\overline{B_iD_i}$ and $\overline{RD_i}$ are expressed in the proposed coordinate system as follows:

$$\overline{B_1D_1} = [b_1 - d_1 \sin(\alpha_1 - \theta), l_b + d_1 \cos(\alpha_1 - \theta)] \quad (2.5)$$

$$\overline{B_2D_2} = [d_2 \sin(\alpha_2 + \theta) - b_2, l_b + d_2 \cos(\alpha_2 + \theta)] \quad (2.6)$$

$$\overline{RD_1} = [-d_1 \sin(\alpha_1 - \theta), d_1 \cos(\alpha_1 - \theta)] \quad (2.7)$$

$$\overline{RD_2} = [d_2 \sin(\alpha_2 + \theta), d_2 \cos(\alpha_2 + \theta)] \quad (2.8)$$

The expression for $r_i(\theta)$ can then be found as:

$$r_i(\theta) = \frac{\left|\overline{B_iD_i} \times \overline{RD_i}\right|}{\left|\overline{B_iD_i}\right|} \quad (2.9)$$

The muscle contraction ε_i relates to the rotation angle θ as:

$$\varepsilon_i(\theta) = 1 - \frac{l_{m_i}}{l_{0_i}} = \varepsilon_i^c + \frac{l_{m_i}^c - l_{m_i}}{l_{0_i}} = \varepsilon_i^c + \frac{|\overline{B_iD_i}^c| - |\overline{B_iD_i}|}{l_{0_i}} \quad (2.10)$$

The contraction $\varepsilon_i(\theta)$ is defined with respect to ε_i^c, which is the contraction of muscle i at a chosen central reference position θ^c. The parameters ε_i^c and θ^c are defined during the joint design process.

By changing the position of the connections of the muscles, the torque and compliance characteristics of the joint and the joint angle range can be influenced.

2.5.1.2 Torque Characteristics

Taking into account equation (2.2) then with r_1 and r_2 the lever arm of the agonist and antagonist muscle respectively, the joint torque is given by the following expression:

$$\begin{aligned} T = T_1 - T_2 &= p_1 l_1^2 r_1 f_1 - p_2 l_2^2 r_2 f_2 \\ &= p_1 t_1(\theta) - p_2 t_2(\theta) \end{aligned} \quad (2.11)$$

with p_1 and p_2 the applied gauge pressures in the agonist and antagonist muscles respectively with lengths l_1 and l_2. The dimensionless force functions of both muscles are given by f_1 and f_2. The functions t_1 and t_2 are determined by the choices made during the design and depend on the angle θ. Thus the joint torque is influenced by weighted differences in gauge pressures of both muscles.

2.5.2 Compliance Characteristics of an Antagonistic Muscle Setup

The compressibility of air makes the PPAM compliant. Joint stiffness, the inverse of compliance, for the considered revolute joint can be obtained by the angular derivative of the torque characteristic in equation (2.11):

$$K = \frac{dT}{d\theta} = \frac{dT_1}{d\theta} - \frac{dT_2}{d\theta}$$
$$= \frac{dp_1}{d\theta}t_1 + p_1\frac{dt_1}{d\theta} - \frac{dp_2}{d\theta}t_2 - p_2\frac{dt_2}{d\theta} \tag{2.12}$$

The terms $dp_i/d\theta$ represent the share in stiffness of the pressure changing with contraction, which is determined by the action of the valves controlling the joint and by the thermodynamical processes. If polytropic compression/expansion with closed valves is assumed, then the pressure changes inside the muscle will be a function of volume changes:

$$P_iV_i^n = P_{i_o}V_{i_o}^n \tag{2.13}$$

with:

$$P_i = P_{atm} + p_i \tag{2.14}$$

leading to:

$$\frac{dp_i}{d\theta} = -n\left(P_{atm} + p_{i_o}\right)\frac{V_{i_o}^n}{V_i^{n+1}}\frac{dV_i}{d\theta} \tag{2.15}$$

with P_i, V_i the absolute pressure and volume of muscle i, P_{i_o} the absolute initial pressure, V_{i_o} the initial volume when the valves of muscle i were closed and p_i, p_{i_o} the gauge pressure and initial gauge pressure. n is the polytropic index and P_{atm} the atmospheric pressure.

Taking the torque characteristics as an example, the following reasoning can be made for muscles with closed valves. An increase of the angle θ will result in an increase of the torque generated by the agonistic muscle while its volume will decrease. Thus $dt_1/d\theta > 0$ and $dV_1/d\theta < 0$. For the antagonistic muscle the actions will be opposite. Combining equation (2.12), (2.13) and (2.15) with this information gives:

$$K = (k_1 p_{1_o} + k_2 p_{2_o} + k_{atm} P_{atm}) \tag{2.16}$$

with:

$$k_1 = t_1 n\frac{V_{1_o}^n}{V_1^{n+1}}\left|\frac{dV_1}{d\theta}\right| + \frac{V_{1_o}^n}{V_1^n}\left|\frac{dt_1}{d\theta}\right| \quad > 0$$
$$k_2 = t_2 n\frac{V_{2_o}^n}{V_2^{n+1}}\left|\frac{dV_2}{d\theta}\right| + \frac{V_{2_o}^n}{V_2^n}\left|\frac{dt_2}{d\theta}\right| \quad > 0$$
$$k_{atm} = k_1 + k_2 - \left|\frac{dt_1}{d\theta}\right| - \left|\frac{dt_2}{d\theta}\right|$$

The coefficients k_1, k_2, k_{atm} are a function of the joint angle and are determined by the joint and muscles geometry. From equation (2.16) the conclusion is drawn that a passive spring element is created with an adaptable stiffness controlled by the weighted sum of both initial gauge pressures when closing the muscle.

Since stiffness depends on a weighted sum of gauge pressures while torque is determined by a weighted difference in gauge pressure, the torque and stiffness can be controlled simultaneously.

2.6 Modular Unit

A convenient elements in the design phase was modularity and flexibility regarding the ability to make changes to the robot configuration during the experimental process. The complete robot consists of 6 identical modular units and two feet. A modular unit is a link of the robot driving one joint. The upper body, consisting of two units, drives the hip joints, the upper leg drives the knee joint and the lower leg drives the ankle joint. The mechanical setup of a modular unit incorporates a basic frame, two muscles attached to the frame via a pull rod mechanism, a leverage mechanism creating the interface to the neighboring unit and two pneumatic valve systems which regulates the pressure inside both muscles.

2.6.0.1 Basic Frame

The basic frame is a pull rod and leverage mechanism to position two muscles in an antagonistic setup and is depicted in figure 2.10. The CAD drawing shows both assembled and exploded view of the basic frame. The modular unit is made of two slats at the side, which are connected parallel to each other by two linking bars. A joint rotary part, provided with roller bearings, is foreseen for the connection with an other modular unit. The fixed base for the pull rods mechanism includes two rotary axes at which the muscles are attached. The small rotations of these axes are guided by plain bearings positioned in the frame. As can be seen in the exploded view, the basic frame is created by assembling several elementary parts. All these parts are made of a high grade aluminium alloy, AlSiMg1, apart from the bolts and nuts, required to assemble the frame. The cross sectional dimensions of the frame are determined to withstand buckling due to the load set by the muscles in the antagonistic setup. Forces generated by the muscles can easily go up to $5000N$.

Figure 2.11 shows a CAD drawing with the muscles attached to the frame by the pull rods and lever mechanism. The muscles are positioned crosswise to allow complete bulging. At one side they are attached to the frame via the fixed rotary base and at the other side the interface to the next modular unit is provided via the leverage mechanism. The parameters b_1, b_2 and l_b are the same for all the joints. Two connection plates, joined together with two rotary axes, are fixed to the next modular unit and incorporate the leverage mechanism. Again plain bearings are used to guide the rotations of both rotary axes. The position of the rotation points determine the dimensions of the leverage mechanism and consequently joint torque characteristics. The connection plates incorporate the parameters α_1, α_2, d_1 and d_2 of the leverage mechanism for both muscles. Since these parameters have a large influence, the connection plate system is the one which can be changed easily to alter the joint torque characteristics. The muscle contraction parameter ε^c which is defined at a chosen mid angle θ^c, is also adaptable. This parameter is associated with the length of the threaded rods which form the interface between muscle and leverage mechanism. This length can be altered with the nuts that cling the rods to the rotary muscle axes (see figure 2.11).

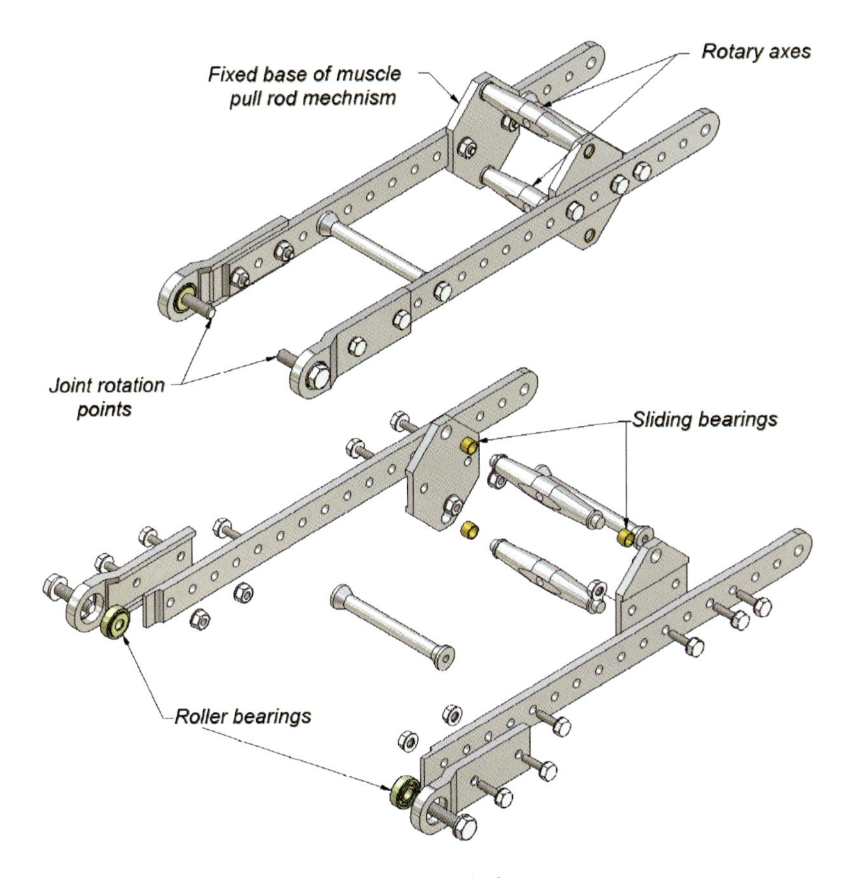

Fig. 2.10 CAD drawing of the modular unit's basic frame.

When determining the joint characteristics, a lot of requirements should be taken into account. Such as static torques required for standing still and more important, the dynamic torque values for walking. Of course, the latter are strongly related to the walking speed and the control strategies. The ranges of angular motion in combination with the torque values should be determined as well. These also depends on the various movements the robot should perform. For the biped Lucy, another design factor, associated with natural dynamics, has to be taken into account. The kinematic joint parameters in combination with muscle dimensions determine the range in which the compliance of the joint can be altered. Of course, if this compliance variation is intended for energy minimization, the range in which it should vary depends on the walking speed and on the specific control strategy. This all clearly indicates that a good joint design is hard to make in an initial design. As indicated before, during the evolution of the experimental and theoretical knowledge concerning the different aspects of controlling Lucy, the design parameters can be altered based on the gained insights in this complex matter, by changing the leverage mechanism.

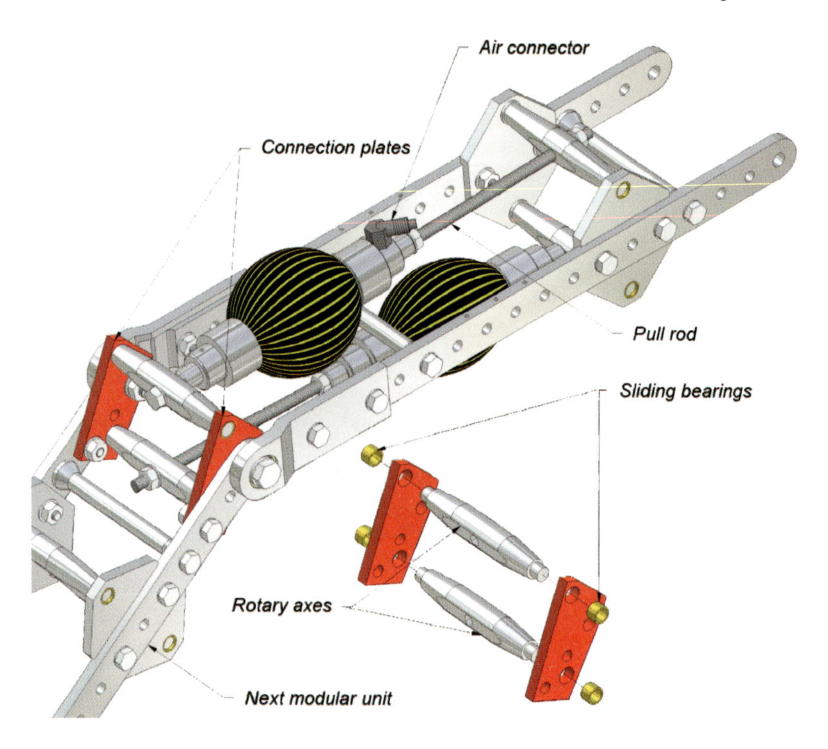

Fig. 2.11 CAD drawing of the modular unit's basic frame with muscles and connection plates.

Currently, the first design of the parameters has been made rather empirically and based on simulations performed by Vermeulen (421) and some analogy with human walking. In figure 2.12 the specific oriented relative ankle, knee and hip angles are defined (counterclockwise positive). In the shown posture angles β_1 and β_3 are consequently negative. The ankle angle β_1 varies with respect to the lower leg between -30° and 25° (-15° and 10°). The knee is not able to stretch completely and the specific joint angle ranges from 15° to 65° (8° and 68°). The upper body should be able to rotate more to the front than to the rear as is the case for humans. The range of angular motion for the hip joint is therefore set between -35° and 15° (-30° and 18°). The values between brackets are typical for a walking human (352). The joint range for the human ankle is remarkable smaller. Reason is that the plantigrade human foot rolls over the ground during each walking step, roughly analogous to a wheel (29) whereas the flat sole of Lucy cannot roll.

The generated torque at $3bar$ is designed to be able to generate 70 up to $80Nm$ at the extreme positions, which generally require the largest joint torques for static postures. In the first design attempt, the torque generation is taken symmetrical for both flexor and extensor muscle of a joint. This is not always necessary. The flexor of the knee joint for example generally does not require the same torque as the extensor muscle. The knee extensor muscle has to carry the weight of the robot,

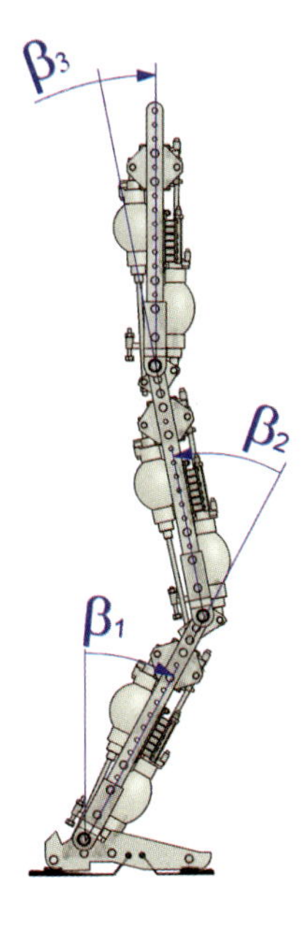

Fig. 2.12 Definition of the oriented relative joint angles (counterclockwise positive, for shown posture angles β_1 and β_3 are negative, β_2 is positive).

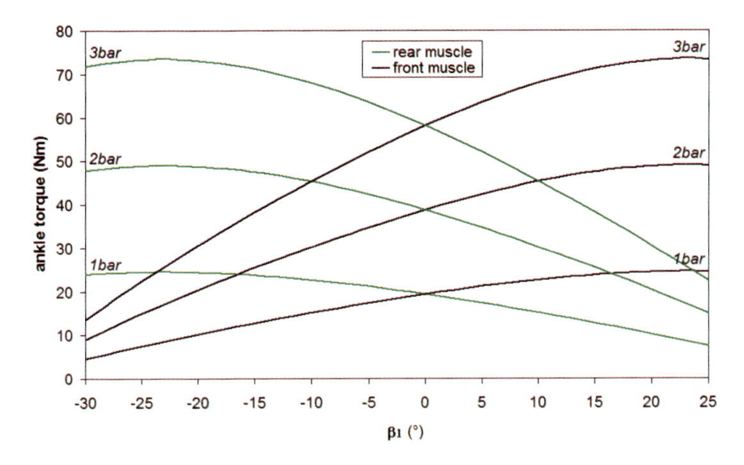

Fig. 2.13 Generated flexor and extensor torque in the ankle joint for 1, 2 and 3 bar.

while the flexor is required to lift the lower leg when the specific leg is in a swing phase. So the torque characteristics where designed with $3bar$ gauge pressure, but it is taken into account that higher pressures up to $4.2bar$ can be set in the muscles (see alarm pressure sensor in 2.7.0.1).

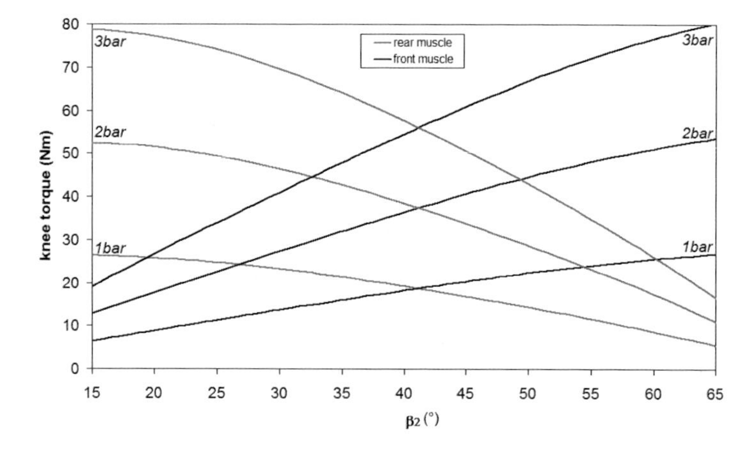

Fig. 2.14 Generated flexor and extensor torque in the knee joint for 1, 2 and 3 bar.

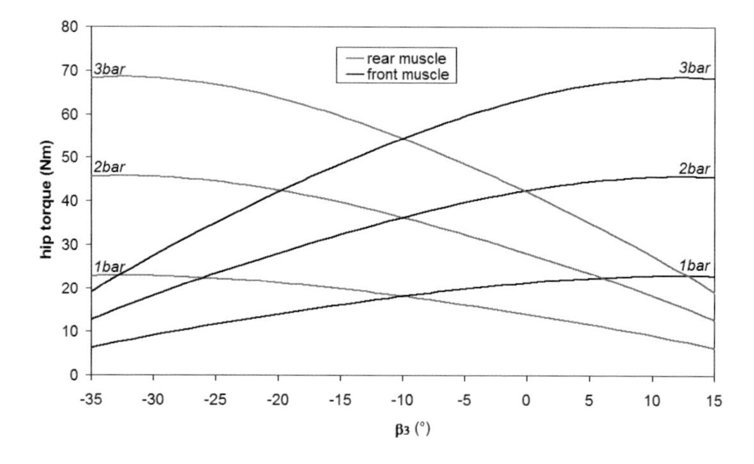

Fig. 2.15 Generated flexor and extensor torque in the hip joint for 1, 2 and 3 bar.

So whenever the tracking controller demands higher torques, it can apply higher pressures in the specific muscle. The actual torque characteristics currently determined for Lucy are depicted in figure 2.13, 2.14 and 2.15. The graphs on these figures give extensor and flexor torques respectively for ankle, knee and hip at 1, 2 and 3*bar* muscle gauge pressure. The muscle contraction range associated with the angle range for each joint are approximately between 7 and 30%. This means that the angular ranges still can be extended when required, the nuts of the angular position limiters in the joints can be fine-tuned to set exact ranges.

2.6.0.2 Valve System

Pneumatic artificial muscles have a high power to weight ratio which makes them suitable for legged robots (106). For a pneumatic system the weight of the pressure control device should be taken into account to evaluate this ratio if the valve system is on board of the robot. Placing the valves close to the muscle is preferable so the tubes can be as short as possible. All the air in the tubes does not contribute to the generated force, but consume compressed air. So the weight of the valves controlling the muscles should be taken as low as possible without compromising too much on performance. Since most pneumatic systems are designed for fixed automation purposes where weight is not an issue at all, most off-the-shelf proportional valves are far too heavy for this application.

In order to realize a lightweight rapid and accurate pressure control, fast switching on/off valves are used. The pneumatic solenoid valve 821 2/2 NC made by Matrix weighs only $25g$. It has a reported switching time of about $1ms$ and flow rate of $180Std.l/min$. Figure 2.16 shows a picture of the selected valve. The valves come with two different types, one with and one without return spring which acts on the air flow interrupting flapper inside the valve.

To pressurize and depressurize the muscle which has a varying volume up to $400ml$, it is best to place a number of these small on/off valves in parallel. Obviously the more valves used, the better the pressure tracking, but also the higher the electric power consumption, price and weight will be. Simulations of the pressure control on a constant volume led to the compromise of 2 inlet and 4 outlet valves. This asymmetrical situation is introduced since asymmetrical pneumatic conditions exist between exhaust and inlet. The orifice airflow though a valve is characterized by the pressure difference over the valve. The gauge pressure inside the muscle generally varies between 0 and $3bar$, while the pressure of the inlet is set at 6 to $7bar$. This means that the maximum pressure difference over the exhaust valves is $3bar$ and over the inlet valves 6 to $7bar$. Consequently, the orifice airflow through valves with the same opening section is much lower for exhaust compared to the inlet.

Fig. 2.16 Picture of the pneumatic solenoid valve 821 2/2 NC made by Matrix.

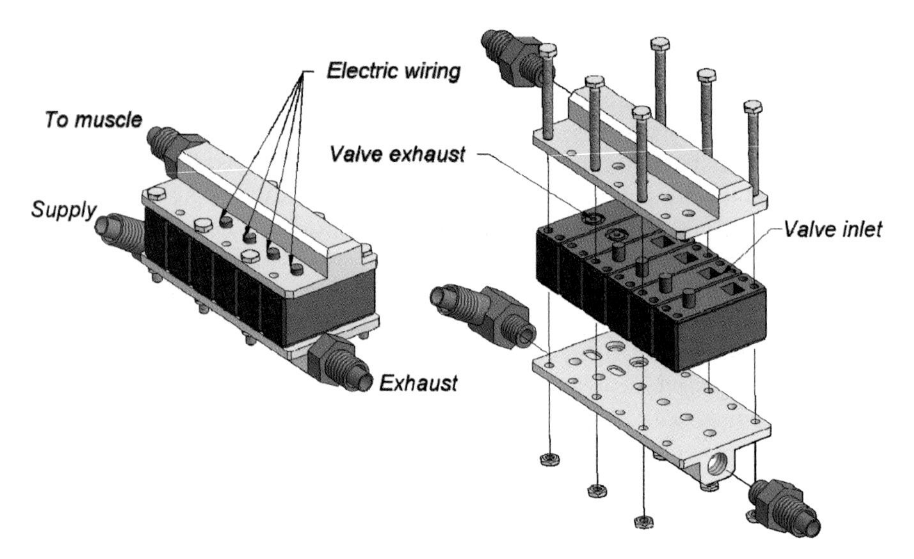

Fig. 2.17 CAD drawing of the valve island.

This means that the time required to set the pressure inside a muscle differs significantly between inflation and deflation of a muscle. In order to level this difference, the number of exhaust valves has been doubled. Of course, increasing the number of valves and reaction levels ameliorates and fastens the pressure tracking, but on the other hand increases the weight of the pneumatic valve system and the electronic power consumption, required to switch the valves. Simulations and tests on a robot arm with one pair of comparable artificial muscles, which are not discussed in this work, have led to the current compromise of 2 inlet and 4 outlet valves. The 6 valves are brought together in a valve island with special designed inlet and outlet collectors after removing parts of the original housing material. A CAD drawing of the valve island is given in figure 2.17. The total weight of this device is less than $150g$. The two valves at inlet are without spring, while the four valves responsible to deflate the muscle have an internal return spring. The pressure difference over the valves are minimum $4bar$ for the inlet valves and $0bar$ for the exhaust valves. Removing a spring significantly decreases opening times of the valve, while on the other hand the presence of the spring decreases closing times of the valves. On the contrary, a large pressure difference over the valves increases opening times, while a small pressure difference increases closing times of the valves. So, due to the opposite pressure difference conditions over the inlet and exhaust valves, both situations concerning the return spring are exploited positively. The valves are controlled by a multilevel bang-bang controller with dead zone as described in section 4.1.4. In this section also experiments showing the tracking performances are shown. For more detailed experimental information on this topic one is referred to (415).

2.6.0.3 Complete Mechanical Setup of a Modular Unit

In figure 2.18 a final CAD drawing is given of the complete modular unit. The two valve islands are mounted at each side of the frame. The muscles are connected with the valves and the latter with a compressed air buffer. This buffer is required to avoid the pressure fluctuations in the compressed air supply tubes while controlling the complete biped. The volume of this buffer is comparable to the volume of one muscle. In normal operation, only one muscle of the antagonistic setup is inflated. The other muscle is deflated, except when the controller decides to increase the stiffness of the joint by increasing the mean pressure of both muscles. Additionally, a silencer is added at the exhaust of each valve island of the modular unit. Without a silencer, the immediate expansion to atmospheric conditions of the compressed air at the exhaust creates a lot of noise. A silencers consists of a closed permeable tube which makes the pressurized air leave the volume slowly, resulting in a

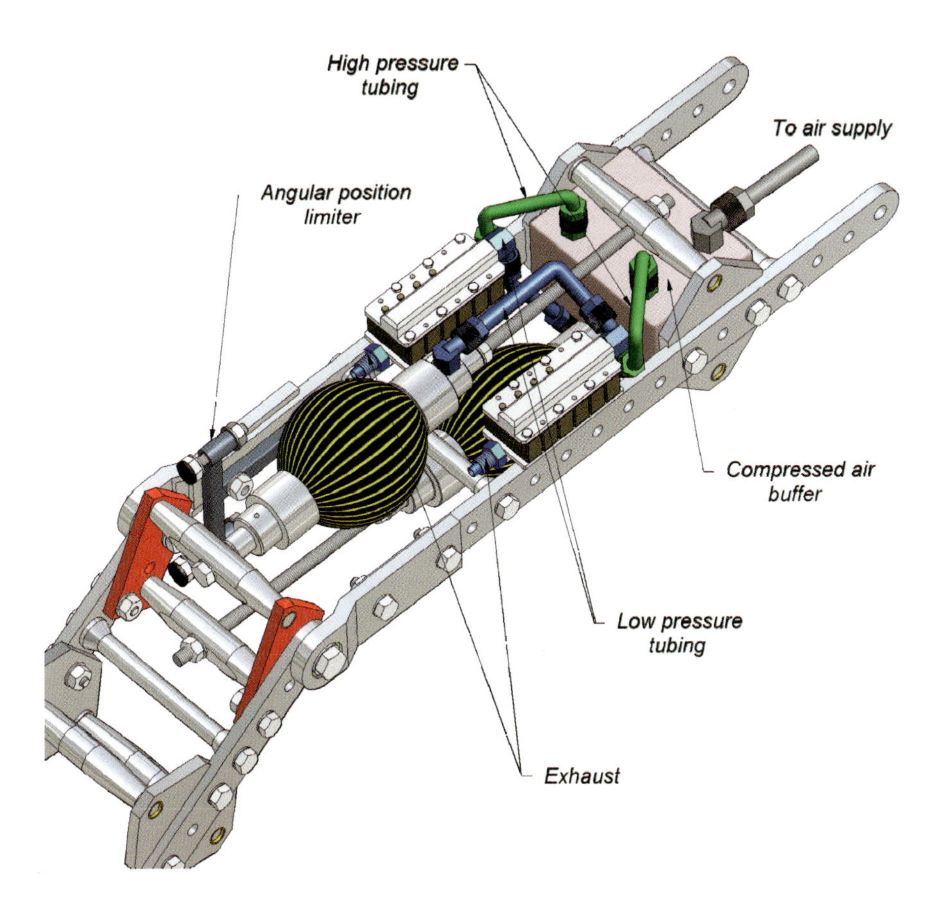

Fig. 2.18 CAD drawing of the complete modular unit.

strongly reduced noise generation. But generally, a silencer also obstructs the dynamic performance of muscle deflation, since a pressure rise in the silencer lowers the exhaust airflow. It is therefore important to use sufficiently large silencers with good permeable material adapted to the volume of the muscle.

In order to set the joint rotation range, an angular position limiter is provided. This device is equipped with two screws to regulate separately the maximum and minimum joint angle. The limits of the angular position are provided to avoid singular joint configurations in the pull rod and leverage mechanism. Such configuration occurs when the axis of the muscle is in line with the joint axis and the muscle attachment point in the leverage mechanism. In this situation the muscle can seriously damage the leverage mechanism when increasing pressures would by applied by the controller. This angular position limiter is also used to bound the muscle contraction range. As was argued in section 2.4, this range lies between 5 and 35%. Finally, this limiter can also be used to create a joint locking state by means of one muscle driving the joint to its extreme position. This can be exploited for example in the knee during stance, in order to induce a simple inverted pendulum motion over the stance foot (443; 331). Figure 2.19 shows a photograph of the modular unit.

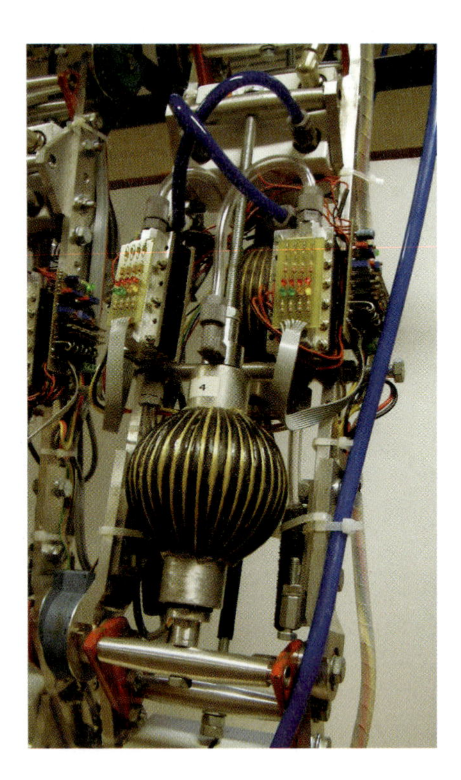

Fig. 2.19 Photograph of a modular unit.

2.6.1 Complete Robot

Six modular units, as discussed in the previous sections, are combined to create the complete biped. A CAD drawing of the mechanical configuration of the complete robot is given in figure 2.24. Figure 2.25 gives a photograph of the real robot, including the electronic components. The upper body of the robot consists of two modular units which are rigidly connected to each other. The left and right antagonistic muscle pairs of the upper body drive the left and right hip joint respectively. Each leg has two modular units, which form the upper leg and the lower leg. The muscle pair of the modular unit in the upper leg actuates the knee joint and the muscles in the lower leg drive the ankle joint. The latter forms the connection to the foot, which is the only link with a configuration different from the modular unit setup.

The feet do not have any form of toes and do not explicitly have a heel shape rounding at the rear. Thus currently, Lucy can only walk with the feet kept parallel to the ground at touch-down and foot lift-off. The sole of the foot consists of two rotating plates as can be seen in figure 2.20. Each foot has two loadcells to measure ground forces and two switches to detect if the foot is on the ground or not. A ball is placed between the force sensor, which is attached to the sole plate, and the rest of the foot so only unidirectional forces are applied to the sensor. A picture of the feet is given in 2.21.

Figure 2.22 gives an overview of the pneumatic circuit, which is used to control the different muscles of the robot. The pneumatic scheme shows the 6 identical pneumatic circuits of which each of them drives one antagonistic flexor/extensor muscle pair. This scheme contains the local reservoir from which the two valve islands are supplied with compressed air. The valve island separately shows inlet and exhaust, each of them represented by two "2/2 electrically actuated" valve symbols. These two symbols represent the 2 reaction levels of the valve system. The number of actual valves which are included in each configuration are depicted as well.

All reservoirs of the modular units are connected to the pressure regulating unit at the central pneumatic distributor by separate tubes. The pressure regulating unit consists of two supply circuits with different pressure levels. One for the normal operating high pressure supply and an other one for a lower reference pressure supply. The latter circuit is used for the calibration of the pressure sensors (2.7.0.1) each time the robot is initialized. Two mechanical pressure regulating units determine the pressures in the high and low pressure circuits respectively, and each circuit is interrupted with an electrically actuated valve. The reference circuit uses a 2/2 valve, while the high pressure circuit is interrupted by a 3/2 valve. The exhaust of this high pressure valve is connected to an electrically actuated 2/2 depressurizing valve in order to deflate the complete robot. The air supply is buffered and an airflow sensor is positioned in the supply line of this reservoir.

Since the robot can only walk in the sagittal direction, a kind of supporting structure has to be provided to avoid turning over in the frontal plane. Several configurations can be used for this purpose. One such configuration is a rotating boom mechanism attached to the hip and a central rotating point as was used for e.g. the biped Rabbit in France (89) and Spring Flamingo at MIT (335). This solution

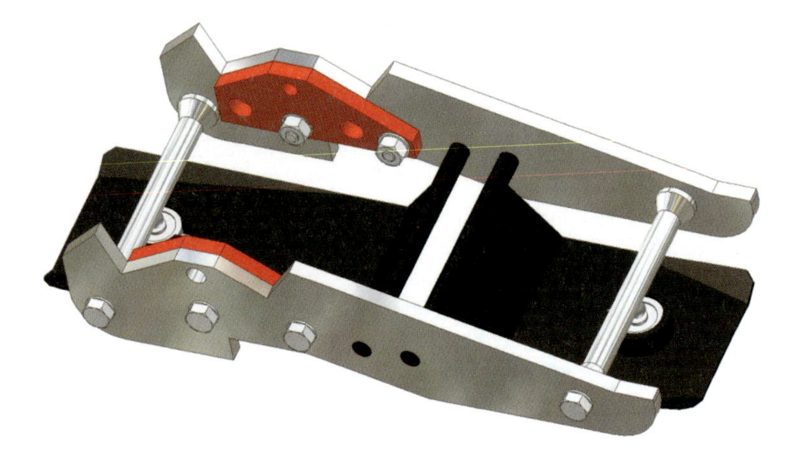

Fig. 2.20 CAD drawing of the foot.

Fig. 2.21 Photograph of the feet with electronic boards for ground force sensors.

requires a lot of space since the boom mechanism has to be large in order to mimic planar walking. Laterally extended feet are another possible configuration, such that the projection of the COG on the ground in the lateral plane lies within the supporting feet area. This for example has been used for the robot BARt-UH in Germany (268). The extended feet however require a large distance between the legs such that they can never hit each other and some positions of the feet are not possible

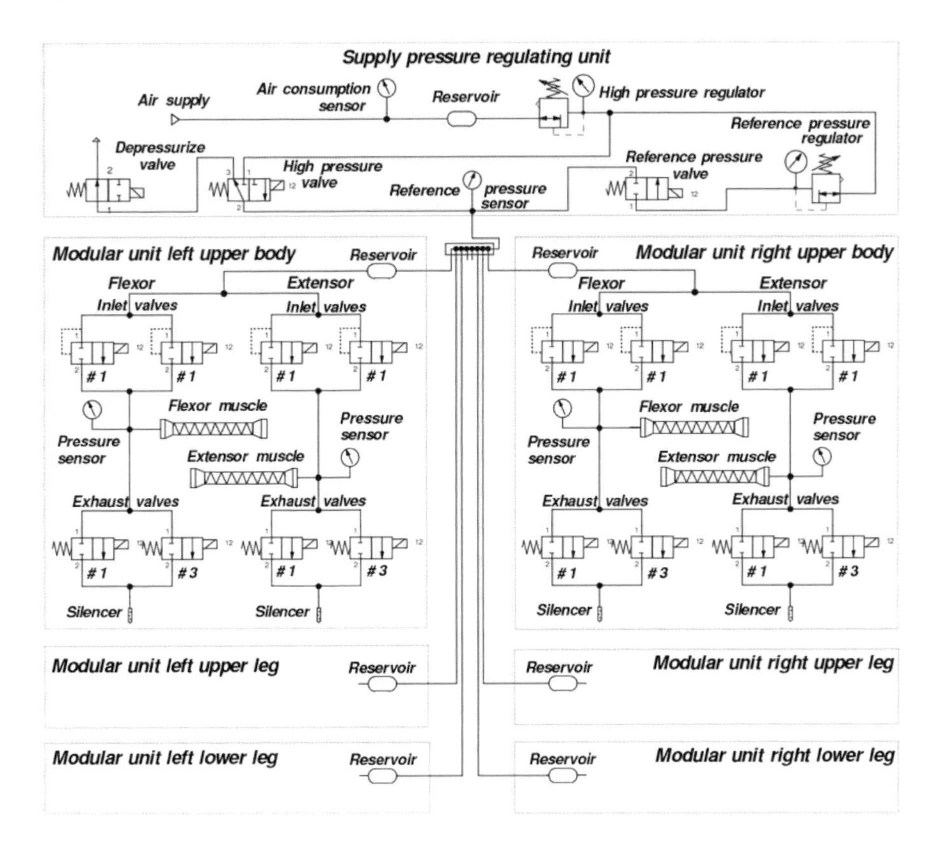

Fig. 2.22 Schematic overview of the complete pneumatic circuit.

anymore. For Lucy, it has been decided to use a vertically positioned XY-frame (figure 2.23), to which the hip points of the robot are attached with two ball bearings. The XY-guiding mechanism is of high quality for smooth sliding of the frame, in order not to disturb the robot motion in the sagittal plane too much.

The guiding mechanism is mounted on the same frame which incorporates the treadmill for enabling the robot to walk for longer distances. The treadmill consists of a wooden deck and a belt, driven by a 3 phase synchronous AC Motor. The speed of the motor is reduced by a transmission gearbox in series with a belt guide. Space for the central computer is provided so the whole experimental setup robot-guiding mechanism-treadmill-computer is integrated in one frame.

2.7 Electronics

Each modular unit has its own low-level control hardware in order to control the pressure inside the muscles in order to generate a desired motion. An overview of this hardware and its function is given in figure 2.26. Pressures are measured with

Fig. 2.23 Guiding mechanism consisting of two rails connected to the hip.

absolute pressure sensors and the angular position and velocity is captured with an incremental encoder. The valves of the two valve islands are controlled by digital micro-controller signals after being transformed by the speed-up board in order to enhance the switching speed of the valves. In the next sections detailed information is given about the different elements of the low-level control hardware.

2.7.0.1 Pressure Sensor

To have an accurate dynamic pressure measurement, the sensor is positioned inside the muscle (see figure 2.27). Since this sensor is inside the muscle volume, an absolute pressure sensor is used. In order to pass through the orifice of a muscle, the size of the sensor and its electronics has to be small ($12mm$). An absolute pressure sensor, CPC100AFC, from Honeywell has been selected for this purpose. The sensor measures absolute pressure values up to $100psi$ ($6.9bar$) and has an accuracy of about $20mbar$.

The principe of the electronics, which conditions the millivolt output of the pressure sensor, is depicted in figure 2.28. The complete electronic scheme can be found in appendix C.3.

The output of the pressure sensor is amplified by a differential amplifier, and in order to avoid noise disturbance as much as possible, the amplified pressure signal

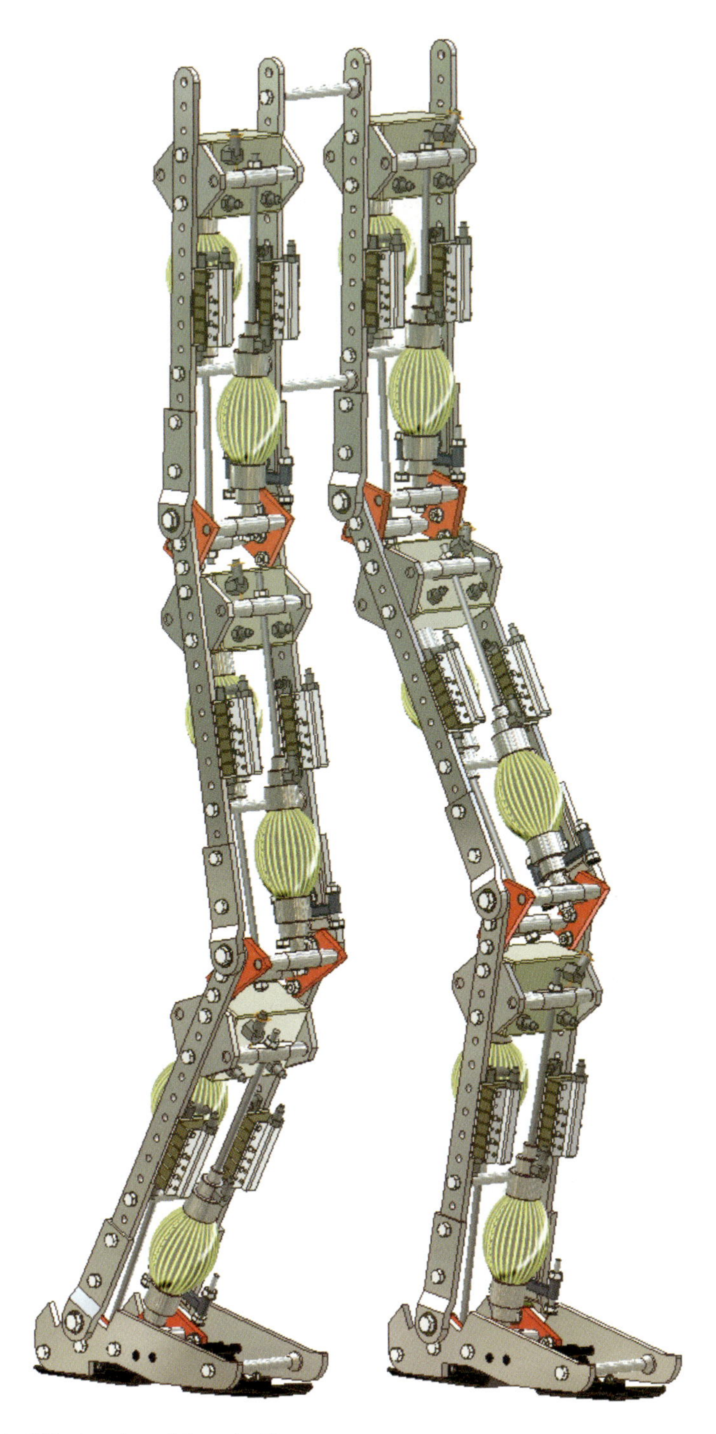

Fig. 2.24 CAD drawing of the robot Lucy.

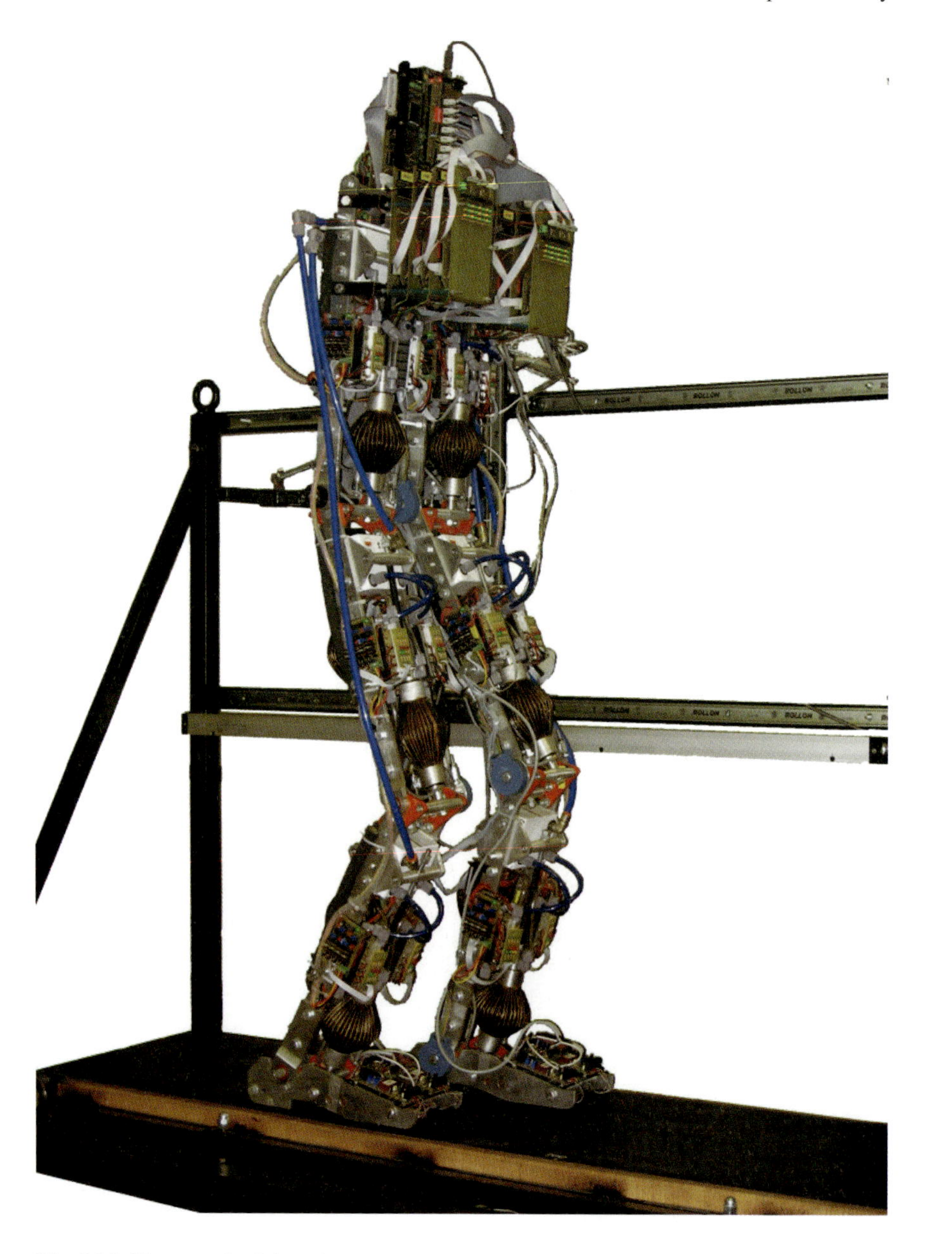

Fig. 2.25 Photograph of the robot Lucy.

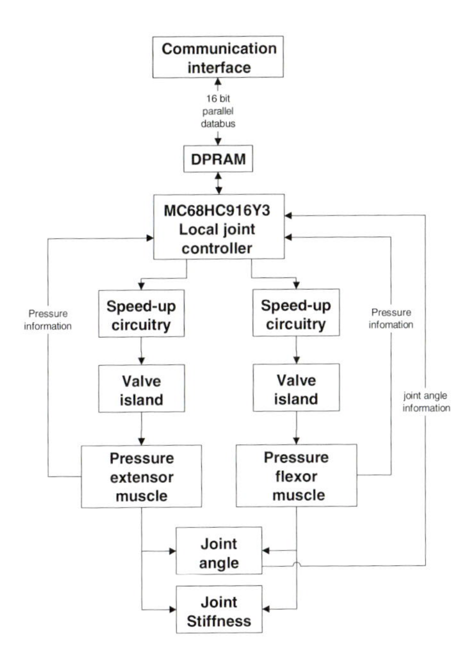

Fig. 2.26 Overview of the low-level control hardware.

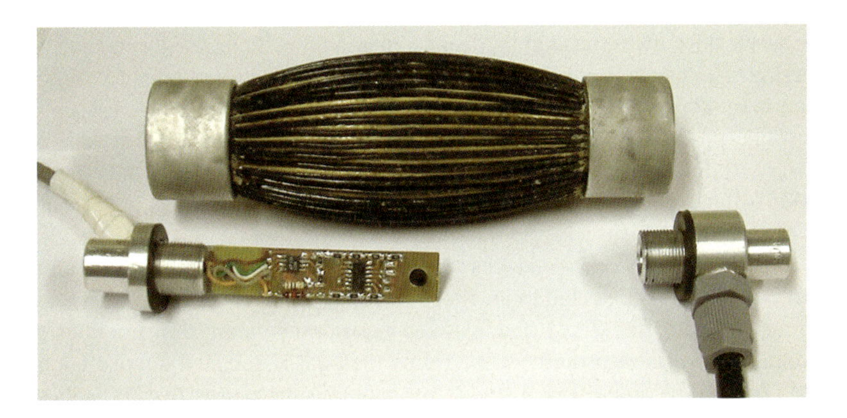

Fig. 2.27 Pressure sensor to be positioned inside the muscle.

is immediately digitized by a 12*bit* analog to digital converter. This chip communicates with the micro-controller unit by a serial peripheral interface (SPI), which is typically used for communication between chips and micro-controllers. A comparator is provided to generate an alarm signal in order to protect the muscle against pressure overload and consequently extend its lifespan. This alarm signal is not treated by a micro-controller, but immediately acts on the central pressure supply

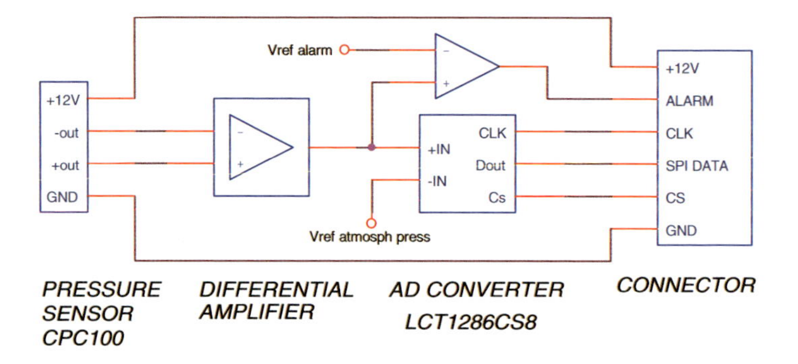

Fig. 2.28 Block diagram of the pressure sensor electronics.

valve (see 2.7.1.2). Whenever the muscle gauge pressure exceeds approximately $4.2bar$, the pressure supply is cut-off.

2.7.0.2 Encoder

The HEDM6540 incremental encoder is used for reading the joint position and velocity. An incremental encoder is a disk with two sets of regularly-spaced slots set along concentric circles. The passage of a slot in front of a light beam produces a pulse; the net number of pulses in a given direction, multiplied by the (constant) angle between slots, gives the angular displacement in that direction during the counting period. The two sets are in quadrature, so that it is possible to deduce direction of motion by knowing which pulse train leads the other. Because the encoder loses its absolute position when the electricity is cut off, an initialization at start-up is necessary to find the reference point of the encoder, provided by a third line. The HEDM6540 has 2000 pulses per revolution, the micro-controller is able to detect the 4 flanks, so this gives a resolution of $0.045°$.

A rotary optical encoder produces two square waves in quadrature. Each transition of both waves is detected as an encoder line. The measurement of the velocity can either be done by a fixed-time or fixed-position method (260). In either case, only one variable is measured. A fixed-time method, also called a pulse-counting or “M” method, estimates the velocity by counting the pulses over a fixed sampling period. A fixed-position method, also called a pulse-timing or “T” method, estimates the velocity by measuring the time for one encoder pulse using a high-frequency auxiliary clock signal. It is well known, however, that the fixed-time and fixed-position methods are inaccurate respectively at low and at high velocities.

In essence, the velocity is estimated by performing an approximate derivative operation on the discrete data. Many designs of discrete-time derivative filters exist today; unfortunately, most of these are unsatisfactory for control applications as the delay inherent to these derivative filters adversely affects stability (65). Furthermore, it is well known that derivative operators tend to magnify errors.

In (65) several velocity estimator algorithms using discrete position versus time measurements were discussed for microprocessor-based systems with a discrete position encoder. The simulations show that no one estimator algorithm is best for a system that has a large dynamic range of speeds, has large transients, and uses an imperfect (real) encoder.

For this project the fixed-position method is used. This method cannot produce an angular velocity at a fixed rate, which is for example the sample rate of the controller. One can compute the velocity at the moment an encoder pulse is detected. If the joint stops, the velocity will never go to zero because the time between two pulses is infinity. Therefor the elapsed time instead of the time between two pulses is taken. With this strategy the velocity goes asymptotically to zero.

2.7.0.3 Valve System Speed-Up Circuitry

In order to enhance the opening time of the Matrix valves, the manufacturer proposes a speed-up in tension circuitry. With a temporary $24V$ during a period of $2.5ms$ and a remaining $5V$, the opening time of the valve is said to be $1ms$. During practical tests the opening times were in many cases twice as long, in certain ranges of pressure difference over the valves. The opening voltage is therefore increased, but the time during which this voltage is applied is decreased, as such that the valves do not get overheated. Figure 2.29 gives the basic electronic scheme of the speed-up circuitry. A complete scheme can be found in appendix C.2. The micro-controller commands the valves via discrete $5V$ on/off signals. These signals directly activate mosfet Q1 in order to apply $5V$ over the valve. A timing unit ensure the switching of mosfet Q2 and Q3 in order to apply temporally an increased voltage. Whenever the micro-controller commands the valve to close, by disabling mosfet Q1, the discharge path is connected to the increased supply source via diode D2. This provides a fast discharge of the electromagnetic energy of the valve, which results in a faster closing time. Several experiments, see (415), resulted in an opening and closing time of about $1ms$. An increased opening voltage of $36V$ is being applied during

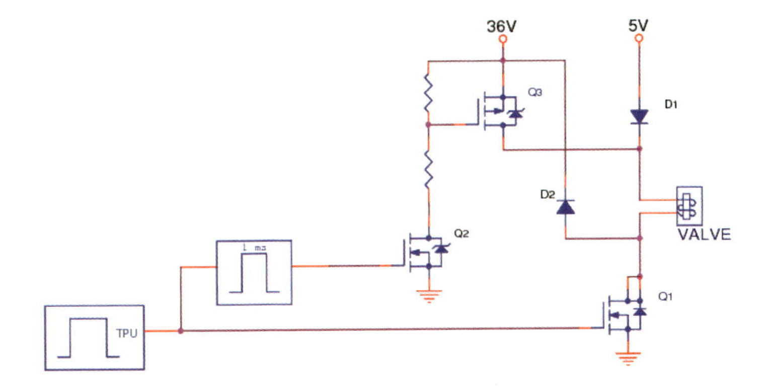

Fig. 2.29 Essential scheme of valve speed-up circuitry.

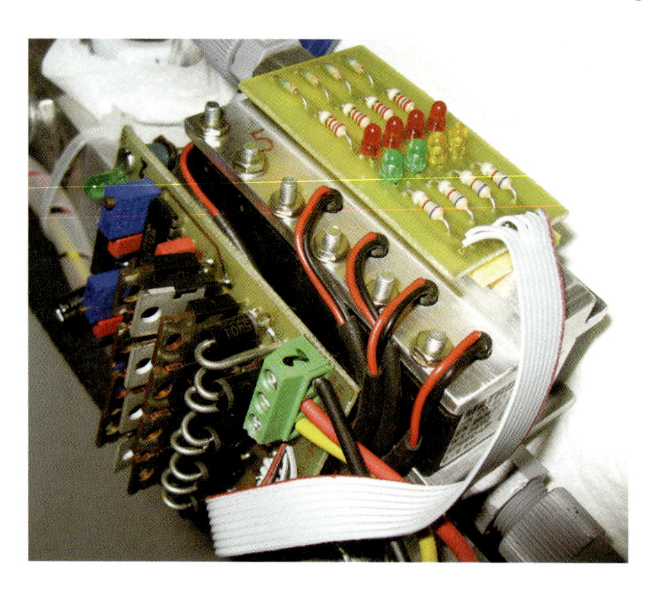

Fig. 2.30 Valve speed-up circuitry.

1*ms*. Figure 2.30 gives a photograph of the speed-up circuitry with its valve island. Four circuits, such as in 2.29, are provided. Two circuits control separately the two inlet valves and two more control the exhaust valves. Hereby three valves are controlled simultaneously by one circuit. Each circuit has 2 LED's: a red one to show the 36*V*, a green to show the 5*V* for an inlet valve and yellow for an outlet valve.

2.7.0.4 Joint Micro-controller Unit

The joint micro-controller units are necessary to capture sensor data and control the valve actions. Part of the control architecture is also locally implemented in these micro-controllers. Therefore a 16*bit* processor was chosen over an 8*bit* and 32*bit* processor. The former is not suited for arithmetics, while the latter is an overkill for the fairly simple local feedback control implementation. The chosen micro-controller is the MC68HC916Y3 of Motorola. This controller has a 16*Mhz* clock rate and an internal 100*kB* flash EEPROM. A separate timer processor unit (TPU) can process sensor information, such as encoder reading, and control outputs without disturbing the CPU.

The basic scheme of the micro-controller board is depicted in figure 2.31. A complete electronic scheme of this board is given in appendix C.1. The basic task of the micro-controller consists of reading the pressure, registering encoder signals, controlling the on/off valves of the two valve islands and communicating with the central PC. The pressure is read via the SPI interface of the micro-controller and the valves are commanded through the TPU output. The TPU is also used to handle the encoder information to provide angle and velocity values.

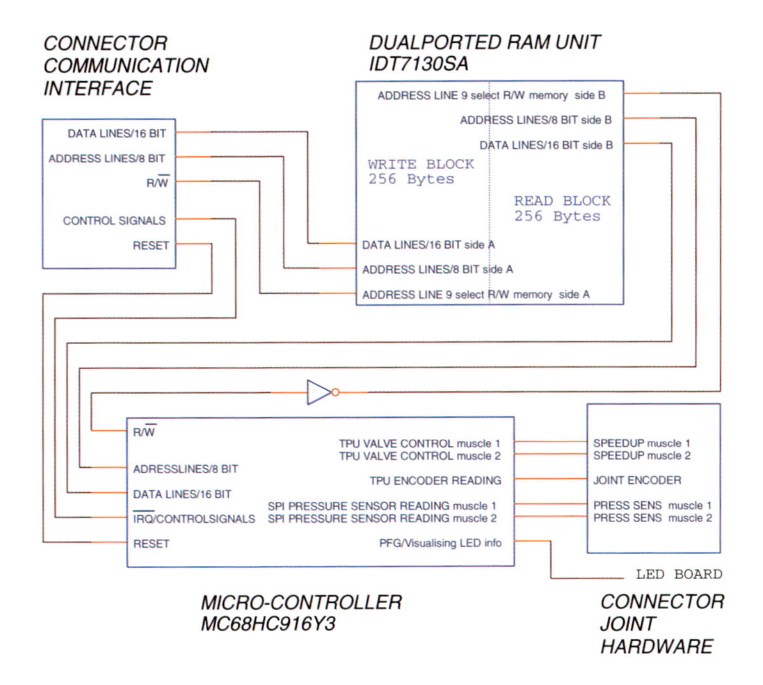

Fig. 2.31 Essential scheme of micro-controller board.

The micro-controller board provides a quasi real-time local control of the robot joints. It performs the local control loop and communicates with a central PC at a refresh rate of $2000Hz$. In order to ensure a real-time operation, the $16bit$ parallel communication lines are buffered via a dual ported RAM structure. The memory of this structure is physically divided into an input and output section of $256bytes$ each, by applying the external $r/\overline{w}$ signal to the higher address lines of the dual ported RAM unit. Additionally, several control lines are linked with the IRQ input/output interface of the micro-controller. The communication interface (see 2.7.1.1) uses these control lines to master the communication protocol and to reset the different micro-controllers. This communication interface is also used to load a program in the micro-controllers. Figure 2.32 shows a picture of the micro-controller board with its dual ported RAM communication interface.

2.7.1 Complete Electronic Hardware

Figure 2.33 gives an overview of the complete electronic hardware. The central PC hosts the program (see section 2.8) and performs the calculations for a large portion of the control scheme. The PC exchanges data with the different dual ported RAM units of the low-level control boards through a data exchange agent which is implemented on an extra micro-controller. This controller distributes the serial USB 2.0 bulk data transfer, originating from the PC, over the several 16 bit parallel data lines going to the dual ported RAM units of the local micro-controllers, and the other way

Fig. 2.32 Micro-controller board.

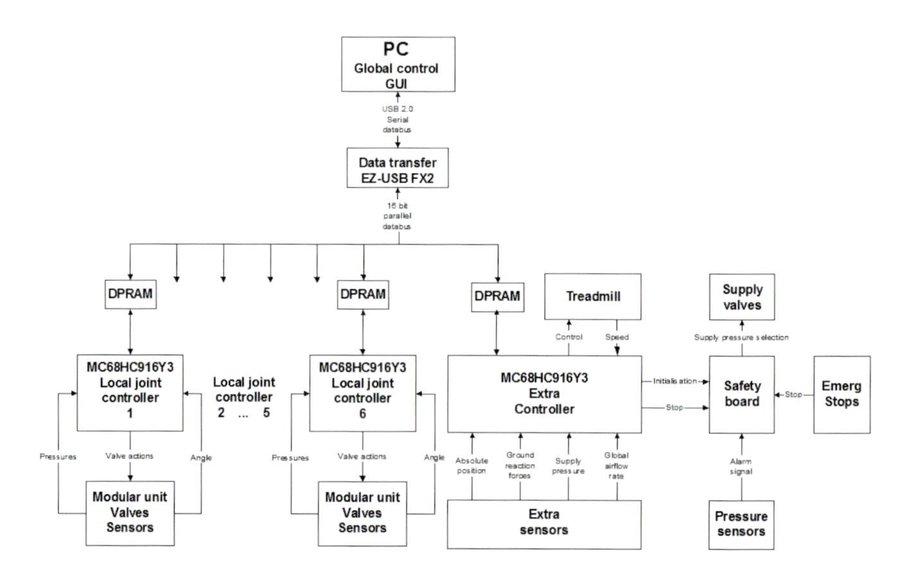

Fig. 2.33 Schematic overview of the robot electronics.

around. Besides the 6 micro-controllers, of which each of them masters a modular unit, an extra controller is provided to read additional sensor information and control the supply valves via a safety board. Extra sensor informs about absolute robot position, ground reaction forces, foot switches, air consumption, supply pressure level, speed of treadmill and control signal of treadmill. The safety board controls the supply valves and depressurizes the supply tubes whenever a muscle pressure sensor gives an overload pressure alarm signal, or whenever an emergency bottom is activated. In the next sections detailed information about this global electronic scheme is given.

2.7.1.1 Communication Hardware and Protocol

Since extensive calculations are required due to the model based control algorithms, a central PC is used. Therefore a fast communication line between PC and robot hardware is provided. A fast communication line could be an extension of the internal PC bus by means of a parallel data communication, but this kind of communication is only suitable for short distance applications. For larger distances (several meters) serial communication protocols are preferable. The most popular serial protocol in the past was the RS232 protocol. This is suitable for slow data transfer (20 to $115Kbit/s$). Nowadays, several other serial protocols, used to branch to computers, have much higher data transfer rates: USB (up to $480Mbit/s$), FireWire (standard IEEE-1394: $400Mbit/s$ and IEEE-1349b: $3.2Gbit/s$) and Ethernet connections (up to 1 Gbit/s). Since USB is a widely used standard, which is available on all modern computers, and since a micro-controller was found, which incorporates a USB 2.0 interface, USB was chosen as communication protocol. Over time the USB standard has evolved from USB 1.1 (1.5 or $12Mbit/s$) to the current USB 2.0 (up to $480Mbit/s$). For normal control operation, the communication line should only transfer pressures and angle information, but in the experimental setup much more information such as control parameter values, valve actions and several status information is transferred. A total set of $226bytes$ are transferred in bulk. Therefore the fastest USB 2.0 protocol is preferred in order to have a high sampling rate.

Since the local Motorola controllers (6 joint controllers+1 extra controller) have a 16 bit parallel communication bus via the dual ported RAM units, the serial USB bulk data block has to be divided into 7 blocks of 16 bit parallel data. Therefore an

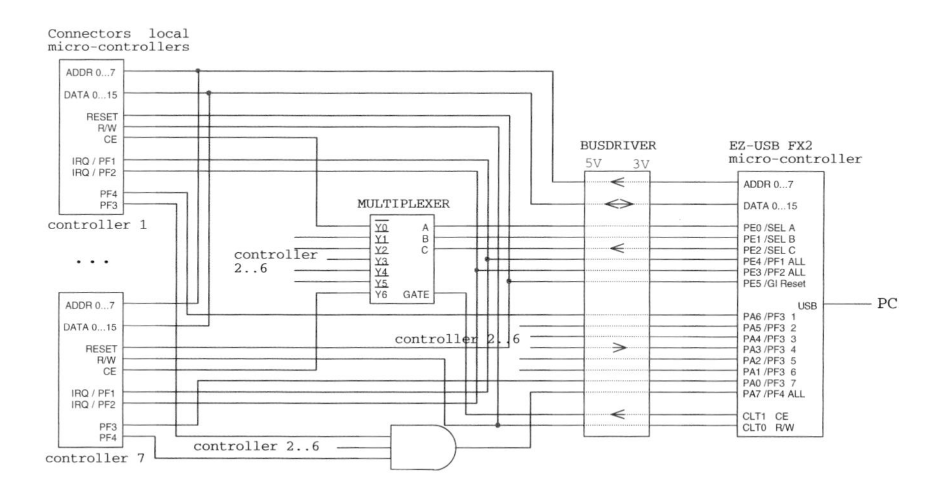

Fig. 2.34 Communication interface overview scheme.

extra micro-controller, EZ-USB FX2 from Cypress Semiconductors, is provided to act as data transfer agent only. This controller runs at $48Mhz$ and is able to transfer the serial data block of $226bytes$ to the peripheral $16bit$ data bus in less than $50\mu s$. Additional to the Cypress development board, an electronic interface has been designed to connect the peripheral bus of the Cypress micro-controller to the different dual ported RAM units. An overview of the communication interface is given in figure 2.34. More information about the electronics can be found in appendix C.7. This interface mainly converts the different voltage levels of address and data lines and connects the Cypress controller, which is the bus master, to the interrupt driven ports PF of the several Motorola micro-controllers. Through the first three pins on port PE, the Cypress controller selects a specific slave micro-controller via a multiplexer. It can generate common interrupts on pins PF1 and PF2 of the different micro-controllers and command a global reset of these controllers, such that a software reset of the complete robot can be ordered by the PC. In the other direction each slave controller can communicate separately or all together, via an AND gate, with the pins of port PA of the Cypress bus master. All these lines are used to exchange communication acknowledgement signals. A photograph of the complete communication interface is given in figure 2.35.

Due to the use of a Windows operating system the refresh rate for the control calculations, implemented on the PC with high priority, is currently set to $2000Hz$, which is the same as the refresh rate of the local micro-controller units. The timing of the communication refresh rate is controlled by the USB Cypress micro-controller. The local micro-controllers ensure low-level, quasi real-time control of the joints. In order to prevent control disturbance of missed torque calculations by the central PC, the incoming data of the local units are buffered via the dual ported RAM hardware. So whenever the central PC does not succeed to perform the

Fig. 2.35 EZ-USB FX2 communication interface.

necessary calculations within the specific sampling time, the local control units use the previously sent data, which are stored in the dual ported RAM structure. One should also remark in the context of this refresh rate, that the delay time of the valves is about $1ms$, which suggests that the communication frequency of $2000Hz$ is high enough.

2.7.1.2 Extra Sensor Implementation and Safety Board

Besides the 6 micro-controller boards, another micro-controller board is provided. This micro-controller is responsible for handling additional sensor information and control of a safety board and treadmill. The controller board is the same as for the joint controllers (2.7.0.4), except that the connections for input and output differ. The TPU of this controller reads three additional encoder signals which are of the same type as for the joints. The encoders measure the horizontal and vertical position of the hip point, which moves together with the guiding XY-frame, and measure the absolute rotation of the upper body. These signals fully determine the absolute position of the robot since it can only move in the sagittal plane. Two extra sensors, air flow and reference pressure sensor, are positioned in the pressure regulating circuit. The standard analogue signals of these sensors are transformed with the same electronic scheme as for the pressure sensor inside the muscles (2.7.0.1). So they are captured by the SPI interface of the extra micro-controller. The flow sensor is needed to have an indication of the air consumption, which becomes crucial when dealing with experiments regarding exploitation of the natural dynamics. This sensor is a SD6000 flow meter from IFM Electronic and measures airflows in a range from 4 to $1250Nl/min$. It has a built in accumulator which gives total air consumption. A reference pressure sensor is required to calibrate all 12 pressure sensors inside the muscles, whenever the robot is initialized. This reference sensor is a PN2024 gauge pressure sensor also from IMF electronics. It measures in a range from -1 to $10bar$ gauge pressure with accuracy smaller then $\pm 0.6\%$ of the range.

Four additional force sensors (THA-250-Q of Transducer Techniques) measure the ground reaction forces in the foot. Each foot requires two such force sensors, one in the front and one at the rear, in order to calculate the ZMP position as a function of the real position of the robot. This information is required to evaluate dynamic stability of the robot and can be used to create a ZMP feedback structure to compensate for errors of the model based trajectory generation. To detect if the foot is standing on the ground or not, 4 on/off switches are mounted in the feet. More details about the electronics placed in the feet can be found in appendix C.4 and in picture 2.21.

2.7.2 Safety Board

The safety board (figure 2.36) consists of electronic hardware, which commands the three valves of the supply pressure regulating unit (figure 2.22). This board handles all the alarm signals, originating from the pressure sensors inside the muscles and

Fig. 2.36 Safety board and supply valves.

several emergency stops. Whenever an alarm signal is activated, the supply valves of the two pressure regulating pneumatic circuits are closed, while the depressurizing valve is opened in order to deflate the complete robot. Opening or closing of the supply valves in the pressure regulating circuits can be commanded by the 7th micro-controller, if the valve commands are not overruled by the electronic hardware during an emergency case. Since this controller is attached to the PC via the USB and dual ported RAM communication structure, selection of the proper supply pressure circuit and depressurization of the robot can be commanded by the central control and GUI. The complete electronic scheme of the safety board can be found in appendix C.6.

2.7.3 Control of Treadmill

The treadmill is powered by a 3 phase synchronous AC Motor controlled by the ACS 350 frequency inverter from ABB. This motor drive contains a vector control to provide enough torque at low rotation speeds. The steering signal for the inverter, coming from the robot, adapts the speed of the treadmill $\tilde{v}_{\mathbf{treadmill}}$ to the speed of the robot $\tilde{v}_{robot}$ so that the hip of the robot $\mathbf{X}_{hip}$ stays in the middle of the treadmill $\tilde{\mathbf{X}}$. The controller consists of a feedforward part which is the desired speed of the

Fig. 2.37 ACS 350 Frequency inverter, AC motor and electronic board with opto-couplers for the treadmill.

robot and a PI feedback part:

$$\tilde{v}_{\mathbf{treadmill}} = \tilde{v}_{robot} - K_p\left(\mathbf{X}_{hip} - \tilde{\mathbf{X}}\right) - K_i \sum\left(\mathbf{X}_{hip} - \tilde{\mathbf{X}}\right) \tag{2.17}$$

The feedback parameters K_p and K_i are tuned manually. The X and Y position of the hip are measured by two linear encoders attached to the rails of the guiding mechanism. The steering signal and the measured rotation speed of the motor are treated by a separate electrical board which can be seen in figure 2.37. This board contains opto-couplers so in case of a fault like an overvoltage on one side, the other side is not corrupted, in particular to protect the low voltage electronics of the robot. This board is also connected to the emergency buttons: if an emergency button is pressed the treadmill stops automatically. The electronics for this subsystem are provided in appendix C.5.

2.8 Interface Program

The interface program, written in Visual C++, has two functions and each has its own thread. A thread in computer science is short for a thread of execution. Threads are a way for a program to split itself into two or more simultaneously (or pseudo-simultaneously) running tasks. The first one is to manage the input/output USB data stream and contains the necessary functions to control the robot. The other one is the interface with the user to enable him to survey the robot. The built Graphical User Interface (GUI), consisting of text and widgets to represent the information and actions available to the user, can be seen in figure 2.38. Because data exchange between both threads is needed and only one thread at a time can be allowed to modify data, "critical sections" are used. To do this the CCriticalSection object of the Microsoft Foundation Class (MFC) is used. Two critical sections were built: one receive critical section and one send critical section.

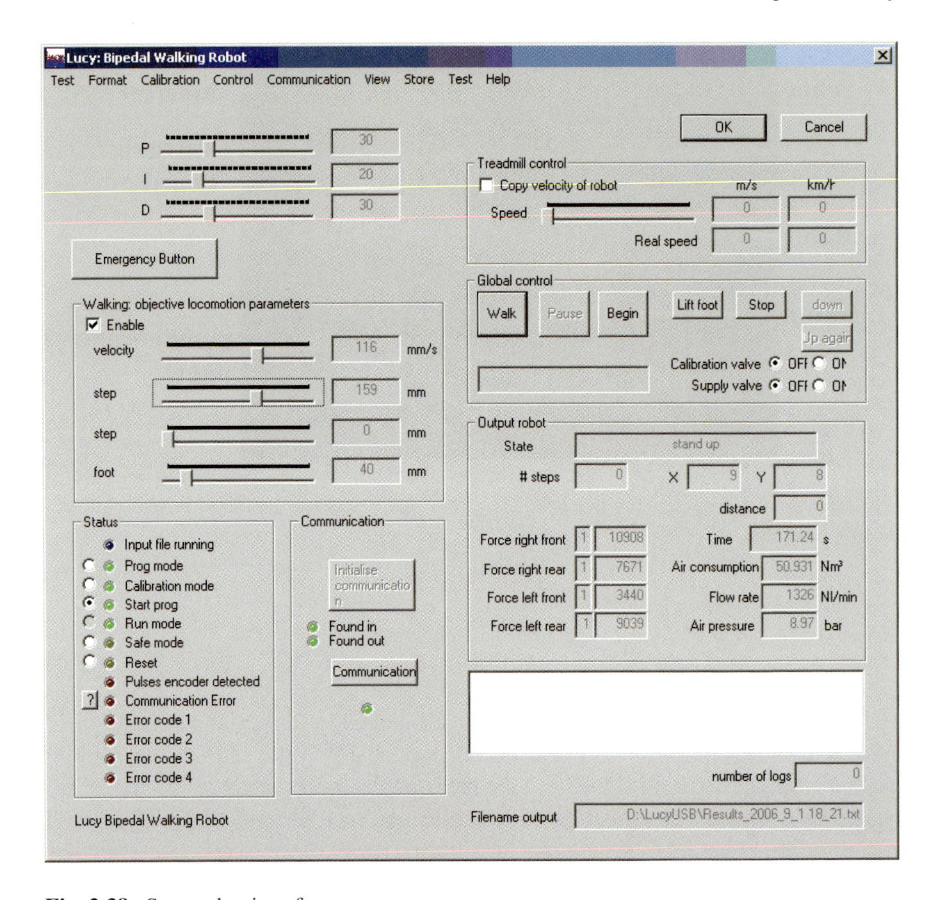

Fig. 2.38 Screenshot interface program.

The first group box *Communication* is the starting point to initialize the communication between the robot and computer. First the micro-controllers are programmed with the low-level controllers. Afterwards the pressure sensors and encoders have to be calibrated. Then the control of the robot can begin. By pushing *Begin* in *Global control* the robot will go from squat position to stand-up position. *Walk* is used to start walking. In *Walking: objective locomotion parameters* the speed, step length, foot lift can be chosen by sliders. *Stop* can be used to stop the treadmill and both feet are placed next to each other. To control the speed of the treadmill two options are available. The user either can control by hand the speed of the robot or if the check box is selected, the speed of the treadmill will be adapted to the speed of the robot so the robot stays in the middle of the treadmill. In *Output robot* some essential information is shown such as the state of the robot, number of steps and so one.

In case of malfunction of the robot different test programs can be executed to facilitate the search for the failure. Test programs for the valves, encoders, pressure sensors, LEDs,... are available.

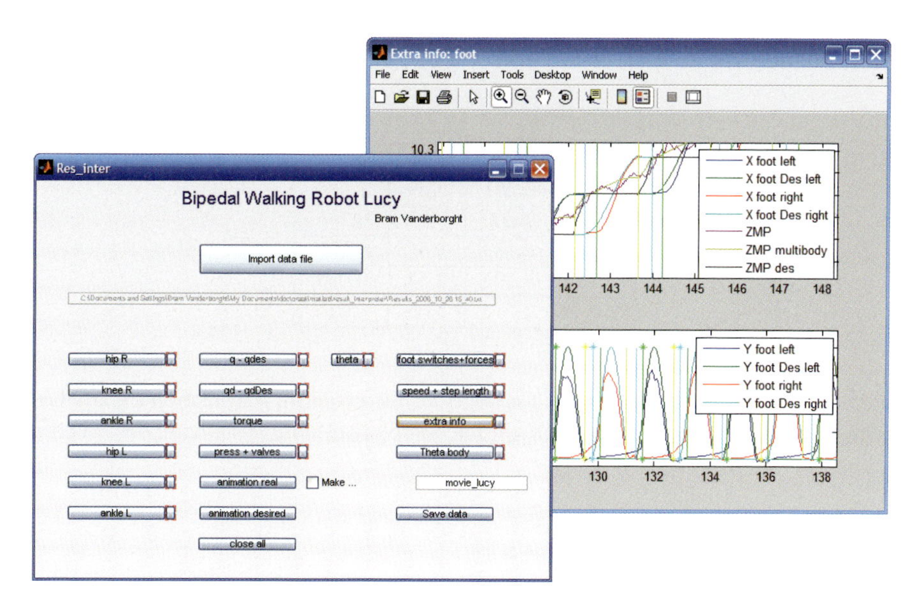

Fig. 2.39 Screenshot Matlab program to view data results.

The data is exported to a *.txt*-file. It is not possible to view in realtime all the data graphically because this will require too much computer power. Every $20ms$ 131 parameters such as time, real and desired angles, velocities, pressure levels, valve actions, torques,... are stored. This creates a datafile of about $2.7MB$ per minute. A result viewer is written in Matlab to analyze this huge data stream afterwards and the main window can be seen in figure 2.39. Matlab is used because of the many built-in functions for graphs.

2.9 Virtual "Lucy"

Besides a real biped a virtual "Lucy" has been built, mainly because it is much easier and faster to test a robot in simulation than for real. The simulator is used to debug control programs and evaluate them before implementing them in a real biped. Matlab was too slow to execute the simulation, but is a powerful tool for engineering purposes. Therefore the complete simulator is written in C++ in a so called MEX-file. Because this is compiled, it is executed very fast.

To have a realistic model both the mechanics of the robot and the thermodynamical processes in the muscles are combined in one set of differential equations. Reported hardware limitations such as valve delays and sampling times, observed on the real robot, where taken into account in the simulation model, as well as some parameter estimation errors. Beside the expected inaccuracies due to the discrete pneumatic valve control system, and parameter estimation and modelling errors for the feedforward trajectory control system, the changing dynamics due to the different phase transitions in the walking motion (e.g. from stance to lift off and at impact)

Table 2.3 Inertial parameters of the robot

i	l_i (m)	J_iG_i (m)	m_i (kg)	I_i (kgm^2)
1	0.45	0.260	3.61	0.060
2	0.45	0.261	3.69	0.062
3	0.45	0.200	10.3	0.145
4	0.45	0.189	3.66	0.060
5	0.45	0.192	3.53	0.058
6	0.30	0.073	2.05	0.016

also might jeopardize the dynamic stability of the robot. In this section the mathematical model for the different parts of the complete robotic system are described.

2.9.1 Mechanics

The biped model during a single support phase is depicted in figure 2.40 and the figure shows the definition of the chosen Lagrange coordinates. These coordinates are the absolute angles of each link of the robot, apart from the stance foot, and are measured with respect to the horizontal axis:

$$\mathbf{q} = \left[\theta_1\ \theta_2\ \theta_3\ \theta_4\ \theta_5\ \theta_6\right]^T \tag{2.18}$$

G_i is the COG of each link, and m_i and I_i are respectively the link mass and the link inertia about G_i. J_i represents the rotation axis between two connected links. The inertial and geometrical parameters of the simulation model are summarized in table 2.3 with l_i the length of link i.

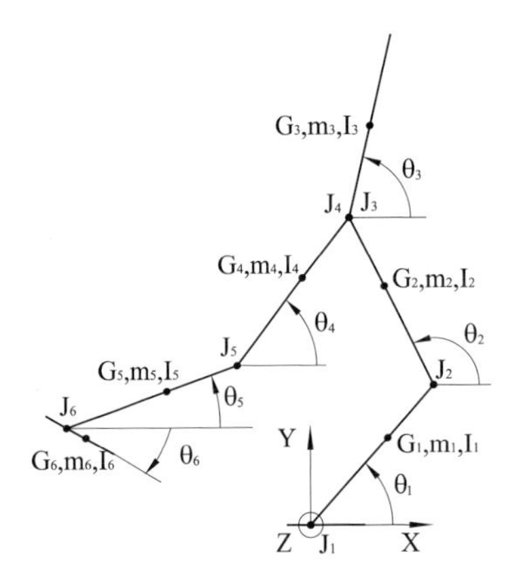

Fig. 2.40 Model of the biped in single support.

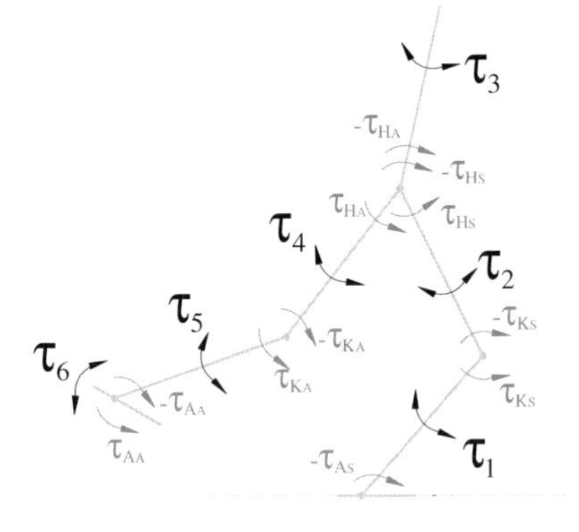

Fig. 2.41 Definition of net torques and joint torques.

The mechanical part of the simulation model contains three different phases: a single support phase, a double support phase and an instantaneous impact phase. During single support, the robot's equations of motion are used (These dynamic equations are written as (376)):

$$D(\mathbf{q})\ddot{\mathbf{q}}+C(\mathbf{q},\dot{\mathbf{q}})\dot{\mathbf{q}}+G(\mathbf{q})=\tau \tag{2.19}$$

with $D(\mathbf{q})$ the generalized inertia matrix, $C(\mathbf{q},\dot{\mathbf{q}})$ the centrifugal/coriolis matrix, $G(\mathbf{q})$ the gravitational torque/force vector. The torque vector τ contains the net torques acting on each link of the robot since the equations of motion are written in absolute coordinates (see figure 2.41):

$$\tau=\begin{bmatrix}\tau_1\\ \tau_2\\ \tau_3\\ \tau_4\\ \tau_5\\ \tau_6\end{bmatrix}=\begin{bmatrix}\tau_{K_s}-\tau_{A_s}\\ \tau_{H_s}-\tau_{K_s}\\ -\tau_{H_s}-\tau_{H_a}\\ \tau_{H_a}-\tau_{K_a}\\ \tau_{K_a}-\tau_{A_a}\\ \tau_{A_a}\end{bmatrix} \tag{2.20}$$

The H, K and A stands for "Hip", "Knee" and "Ankle" respectively, a stands for "air", and s for "stance". Expression (2.20) gives the relations between the net torques and the applied joint torques.

The derivation of the full dynamic model of Lucy can be found in appendix B.

Immediately after impact of the swing leg, three geometrical constraints are imposed on the motion of the system. The three constraints are summarized as follows:

$$l_1cos(\theta_1)+l_2cos(\theta_2)-l_2cos(\theta_4)-l_1cos(\theta_5)-\lambda_{real}=0 \tag{2.21a}$$
$$l_1sin(\theta_1)+l_2sin(\theta_2)-l_2sin(\theta_4)-l_1sin(\theta_5)-\delta_{real}=0 \tag{2.21b}$$
$$\theta_6-C^{te}=0 \tag{2.21c}$$

with λ_{real} and δ_{real} the actual horizontal and vertical position of the front ankle point at touch down. The third constraint expresses that the swing foot stays on the ground, with θ_6 being constant. This constant Cte equals zero for level ground walking. The number of DOF during double support is reduced to 3, but the same 6 Lagrange coordinates (2.18) are used. The equations of motion of single support are adapted with the three geometrical constraints as follows (201):

$$D(\mathbf{q})\ddot{\mathbf{q}}+C(\mathbf{q},\dot{\mathbf{q}})\dot{\mathbf{q}}+G(\mathbf{q})=\tau+J^T(\mathbf{q})\Lambda \tag{2.22}$$

with $J(\mathbf{q})$ the Jacobian matrix, which is calculated by taking the derivative of the constraint equations with respect to the generalized Lagrange coordinates:

$$J(\mathbf{q})=\begin{bmatrix}-l_1sin(\theta_1) & -l_2sin(\theta_2) & 0 & l_2sin(\theta_4) & l_1sin(\theta_5) & 0\\ l_1cos(\theta_1) & l_2cos(\theta_2) & 0 & -l_2cos(\theta_4) & -l_1cos(\theta_5) & 0\\ 0 & 0 & 0 & 0 & 0 & 1\end{bmatrix} \tag{2.23}$$

and Λ the vector of Lagrange multipliers:

$$\Lambda = \left[\lambda_1\,\lambda_2\,\lambda_3\right]^T \tag{2.24}$$

The inertial parameters of the swing foot are taken into account, while the influence of the supporting foot is neglected, since this foot is not moving. The origin of the coordinate system is positioned at the supporting ankle point during single support and at the rear ankle point during double support, which is physically the same point. Each time a transition from double support to single support occurs, the origin of the coordinate system is shifted. In order to have a realistic simulation, an impact phase at touch-down of the swing leg is considered. This impact phase is modelled as an inelastic impulsive impact of the front foot.

2.9.1.1 Single Support Phase

The simulation kernel integrates first order differential equations only. Since the equations of motion (2.19) are of second order, these equations have to be transformed into a first order formulation. This can be done by introducing ω for the angular velocity vector:

$$\omega = \dot{\mathbf{q}} = \left[\omega_1\,\omega_2\,\omega_3\,\omega_4\,\omega_5\,\omega_6\right]^T \tag{2.25}$$

The equations of motion (2.19) are then rewritten as:

$$\begin{cases} \dot{\omega} = D(\mathbf{q})^{-1}\left[\tau - C(\mathbf{q},\omega) - G(\mathbf{q})\right] \\ \dot{\mathbf{q}} = \omega \end{cases} \tag{2.26}$$

Note that the inertia matrix $D(\mathbf{q})$ is symmetric and positive definite and can be inverted. Equations (2.26) represent a set of 12 first order differential equations for which the torques τ depend on the angular positions $\mathbf{q}$ and the pressure values in the muscles of all joints (4.38).

During the simulation process, several conditions need to be observed to check for phase transitions. Whenever the ankle of the swing foot hits the ground, an impact phase will occur, followed by a double support phase, i.e. if the foot does not bounce. If the coordinates of the front foot are given by:

$$X_{A_F} = l_1\cos\theta_1 + l_2\cos\theta_2 - l_2\cos\theta_4 - l_1\cos\theta_5 \tag{2.27a}$$

$$Y_{A_F} = l_1\sin\theta_1 + l_2\sin\theta_2 - l_2\sin\theta_4 - l_1\sin\theta_5 \tag{2.27b}$$

than the condition for phase transition is formulated as:

$$Y_{A_F} < Y_{gr}\left(X_{AF}\right) \tag{2.28}$$

With $Y_{gr}(X)$ representing the specific shape of the ground. In this work simulations only consider walking on flat terrain, thus $Y_{gr}(X) = 0$. Note that an approximation is made by expressing this condition at the ankle joint, neither including the foot dimensions, nor taking into account specific shapes of obstacles which could obstruct the walking motion.

One of the difficulties of controlling legged robots is the unilateral nature of this foot/ground contact. The vertical acceleration of the global COG, $\ddot{Y}_G$, has to be higher than $-g$, otherwise the sign of the total ground reaction force will switch and the robot starts a flight phase which is not foreseen in the programmed control algorithm. Thus a necessary condition for foot/ground contact is:

$$R_y = m_{tot}\left(\ddot{Y}_G + g\right) > 0 \tag{2.29}$$

With the positive direction of the vertical defined upwards. Furthermore, the ZMP position (3.4) has to stay within the physical boundaries of the foot, otherwise the robot starts to tip over while rotating around one of the supporting foot edges:

$$-l_{6B} < -\frac{\tau_A}{m_{tot}\left(\ddot{Y}_G + g\right)} < l_{6F} \tag{2.30}$$

This situation is undesirable and is described by totally different equations of motion, so the simulation should be stopped at this point. It is furthermore assumed that the stance foot of the robot does not slip, meaning that friction between the foot sole and the ground is sufficiently high.

2.9.1.2 Double Support Phase

The equations of motion for the double support phase (4.3) represent 6 equations in 9 unknowns: 6 unknowns for $\ddot{\mathbf{q}}$ and 3 for the Lagrange multipliers Λ. This should be solved by additionally using the three constraint equations (2.21), which constitute a total set of differential algebraic equations (DAE). In order to transform this into a set of ordinary differential equations (ODE), the second derivative of the kinematic constraint equation with respect to time is used (201):

$$J(\mathbf{q})\ddot{\mathbf{q}} + \dot{J}(\mathbf{q})\dot{\mathbf{q}} = 0 \tag{2.31}$$

Combining (4.3) and (2.31) results in:

$$\begin{bmatrix} D(\mathbf{q}) & J^T(\mathbf{q}) \\ J(\mathbf{q}) & 0 \end{bmatrix} \begin{bmatrix} \ddot{\mathbf{q}} \\ \Lambda \end{bmatrix} = \begin{bmatrix} \tau - C(\mathbf{q},\dot{\mathbf{q}})\dot{\mathbf{q}} - G(\mathbf{q}) \\ -\dot{J}(\mathbf{q})\dot{\mathbf{q}} \end{bmatrix} \tag{2.32}$$

Equations (2.32) are then solved for the 9 unknowns. After introducing ω, the following set of 12 first ODE is formed, which have to be integrated numerically:

$$\begin{cases} \dot{\omega} = \mathbf{f}(\mathbf{q},\omega) \\ \dot{\mathbf{q}} = \omega \end{cases} \tag{2.33}$$

with $\mathbf{f}$ being a result of solving (2.32).

When describing the equations of motion with dependent coordinates and Lagrange multipliers, the forces associated with the constraints can be calculated in a straightforward way. In this case, the ground reaction force $\bar{\mathbf{R}}_{\mathbf{F}}$ of the front foot (see figure (3.5)) is linked with the two first constraints of (2.21) by Lagrange multipliers λ_1 and λ_2. The constraint equations can be written in such a way that the horizontal and vertical components of the ground reaction force acting at the front ankle point are found as:

$$R_F^x = \lambda_1 \tag{2.34a}$$

$$R_F^y = \lambda_2 \tag{2.34b}$$

Writing the linear momentum theorem with respect to the global COG allows one to calculate the total ground reaction forces:

$$R_{tot}^x = m_{tot}\ddot{X}_G \tag{2.35a}$$

$$R_{tot}^y = m_{tot}\left(\ddot{Y}_G + g\right) \tag{2.35b}$$

with m_{tot} the total mass of the robot, $\ddot{X}_G$ and $\ddot{Y}_G$ the horizontal and vertical acceleration of the global COG, which can be calculated with equations (B.3) of appendix B. Combining (2.34) with (2.35) allows one to find the ground reaction force acting at the rear ankle point:

$$R_R^x = R_{tot}^x - R_F^x = m_{tot}\ddot{X}_G - \lambda_1 \tag{2.36a}$$

$$R_R^y = R_{tot}^y - R_F^y = m_{tot}\left(\ddot{Y}_G + g\right) - \lambda_2 \tag{2.36b}$$

When the vertical component of the ground reaction force acting at the rear foot (2.36b) becomes negative, the rear foot is lifted of the ground and the double support phase ends. Apart from the rear foot ground reaction force, the vertical component of the front foot ground reaction force is checked if it becomes negative during the double support phase. If so, the simulation should be terminated, since this means that the robot tends to move in the opposite direction, apart from eventual bouncing effects just after impact. Based on the values of the vertical ground reaction forces of the feet, the ZMP position during double support is obtained with equation (3.6).

2.9.1.3 Impact Phase

After the single support phase, an impact occurs when the swing foot touches the ground. This impact causes jumps of the joint angular velocities. The values of these changes in velocity become the starting conditions for the numerical integrator of the next double support phase. The touch-down of the front foot is modeled as an inelastic impulsive impact at the ankle joint. Ignoring the impact on the foot itself, then only the two first equations (2.21a) and (2.21b) are taken into account.

The relation between front foot ankle joint velocity and angular velocities of each link, apart from the feet, is given by:

$$\dot{\mathbf{q}}_{\mathbf{A_F}} = J\dot{\mathbf{q}} \tag{2.37}$$

with

$$\mathbf{q}_{\mathbf{A_F}} = \begin{bmatrix} X_{A_F} \\ Y_{A_F} \end{bmatrix} \tag{2.38}$$

and the Jacobian matrix J:

$$J(\mathbf{q}) = \begin{bmatrix} -l_1 sin(\theta_1) & -l_2 sin(\theta_2) & 0 & l_2 sin(\theta_4) & l_1 sin(\theta_5) \\ l_1 cos(\theta_1) & l_2 cos(\theta_2) & 0 & -l_2 cos(\theta_4) & -l_1 cos(\theta_5) \end{bmatrix} \tag{2.39}$$

Since the Jacobian matrix is non-square it can not be inverted. Zheng and Hemami (465) derived the following expression, which calculates the angular velocity jumps $\Delta\dot{\mathbf{q}}$ using the dynamic model of the robot (2.19):

$$\Delta\dot{\mathbf{q}} = D^{-1}J^T\left(JD^{-1}J^T\right)^{-1}\Delta\dot{\mathbf{q}}_{\mathbf{A_F}} \tag{2.40}$$

with:

$$\Delta\dot{\mathbf{q}}_{\mathbf{A_F}} = \begin{bmatrix} -\dot{X}^-_{A_F} \\ -\dot{Y}^-_{A_F} \end{bmatrix} \tag{2.41}$$

$\dot{X}^-_{A_F}$ and $\dot{Y}^-_{A_F}$ are the horizontal and vertical velocity of the front foot ankle point just before impact. D is the generalized inertia matrix of equation (2.19). It is assumed that the robot configuration and applied torques remain unchanged during the infinitesimal short impact phase.

2.9.2 Thermodynamics

The thermodynamic processes which take place in the antagonistic muscle setup of each joint are described by four first order differential equations. Two equations determine the pressure changes in both muscles of the antagonistic setup and the remaining two describe conservation of mass in the respective muscle volumes. Additionally to these differential equations the perfect gas law is used to determine temperature values.

The pressure inside a muscle is influenced by its volume changes resulting from a variation of the joint angle and by the air flows through the valves which have been activated by the bang-bang pressure controller. Assuming a polytropic thermodynamic process, and assuming that the compressed air inside each muscle behaves as a perfect gas, the first law of thermodynamics, while neglecting the fluid's kinetic and potential energy, can be written for each muscle of the antagonistic setup in the following differential form (appendix A):

$$\dot{p}_i = \frac{n}{V_i}\left(rT^{sup}_{air}\dot{m}^{in}_{air_i} - rT_{air_i}\dot{m}^{ex}_{air_i} - (P_{atm} + p_i)\dot{V}_i\right) \tag{2.42}$$

The total orifice flow through opened inlet valves ($\dot{m}^{in}_{air_i}$) or exhaust valves ($\dot{m}^{ex}_{air_i}$) can be calculated with the following equations which represents a normalized approximation of a valve orifice flow defined by the International Standard ISO6358 (195):

$$\dot{m}_{air} = CP_u\rho_0\sqrt{\frac{293}{T^u_{air}}}\sqrt{1-\left(\frac{P_d/P_u-b}{1-b}\right)^2} \qquad \text{if} \quad \frac{P_d}{P_u} \geq b \tag{2.43}$$

$$\dot{m}_{air} = CP_u\rho_0\sqrt{\frac{293}{T^u_{air}}} \qquad \text{if} \quad \frac{P_d}{P_u} \leq b \tag{2.44}$$

C and b are two flow constants characterizing the valve, ρ_0 the air density at standard conditions. The constant C is associated with the amount of air flowing through the valve orifice, while b represents the critical pressure ratio at which orifice air flows become maximal. Both coefficients have been experimentally determined for the used Matrix valves, which resulted in $C = 22\,\text{Std.l/min/bar}$ and $b = 0.16$ (416). P_u and P_d are the upstream and downstream absolute pressures, while T^u_{air} is the upstream temperature. When choking occurs, equation (2.44) is valid, otherwise equation (2.43) is used.

Once the actions (opening and closing of the valves) for the different inlet and exhaust valves are known, all the air flows can be calculated in order to be substituted in (2.42). The temperature in the muscle T_{air_i} is calculated with the perfect gas law:

$$T_{air_i} = \frac{P_i V_i}{m_{air_i} r} \tag{2.45}$$

The total air mass m_{air_i} is given by integration of the net mass flow entering muscle i:

$$\dot{m}_{air_i} = \dot{m}^{in}_{air_i} - \dot{m}^{ex}_{air_i} \tag{2.46}$$

The volumes and their time derivatives are given by kinematical expressions as a function of the joint angle and joint angular velocity. These functions are determined with the volume fitting function (2.4) and the link between contraction and joint angle, represented by the kinematic expression (2.10) of the pull rod system. The link at torque level between the mechanical equations of motion and these thermodynamic differential equation systems is provided by equation (4.38) which characterizes joint torque as a function of pressures and joint angle.

2.9.3 Complete Simulation Model

In figure 2.42 an overview is given of the complete simulation model. The kernel of this simulator is based on three equation blocks, as depicted in the center of the figure. The 12 first order differential equations (2.26) or (2.33) describe the motion during single support and double support respectively, with addition of the constraint equations for double support. The thermodynamics of each joint are characterized by four first order differential equations on pressure (2.42) and air mass (2.46). This gives a set of 24 differential equations for the thermodynamic differential equation block. Finally, the 12 thermodynamic state equations (2.45) complete the set.

The antagonistic muscle model block creates the link between the mechanics and the thermodynamics by calculating the torque for each joint (j) with the

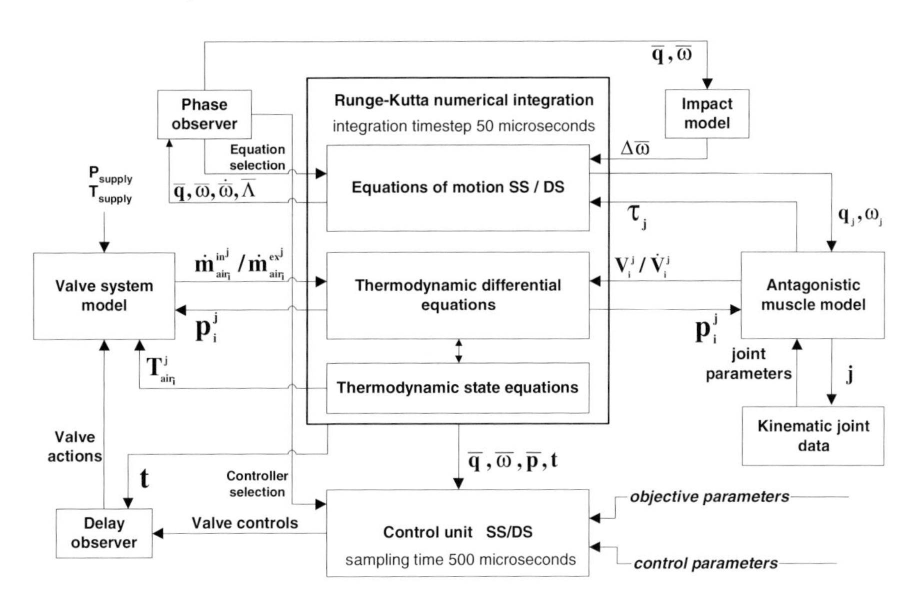

Fig. 2.42 Structure of the complete simulation model.

pressure information of the thermodynamic block. Therefore it needs angle information from the integrated equations of motion. This information allows to calculate the contraction of each muscle (i) within the antagonistic setup (2.10), while using the kinematic data of the pull-rod mechanism of the specific joint. With the contraction values, the linear forces (2.3) of the two muscles can be calculated in order to determine the applied torque with equation (4.38). Additionally, to determine the pressure changes in the thermodynamic differential equation block, muscle volume and volume changes are calculated with (2.4). For the volume changes angular velocity information is required from the integrated equations of motion.

The valve system block determines the air mass flow rates (2.43 or 2.44) for each muscle, depending on the actual pressure and temperature in the muscle and the action taken by the valves. This action is determined by the valve control signals of the control unit. These signals pass through the delay observer, which requires the time instant of the integrator to determine whether the valve may be switched or not. Hereby a valve delay of 1 ms is used.

Finally, the phase observer calculates the vertical ground reaction forces (2.29 or 2.34b, 2.36b) and the position of the front foot (2.27) to check whether the robot is in a single support phase or a double support phase. At touch-down of the front foot, this module commands the impact module to calculate the velocity jumps (2.40). The phase observer requires angles, angular velocities and accelerations and the Lagrange multipliers to determine the ground reaction forces.

The differential equations are numerically integrated using a $4th$ order Runge-Kutta method with an integration time step of $50\mu s$, which is ten times less than the sample time of the control unit. In order to evaluate robustness of the controller with respect to parameter estimation, two systematic errors are introduced. Firstly,

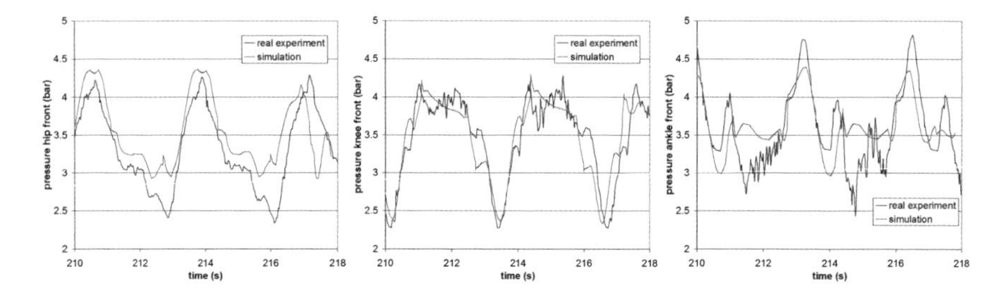

Fig. 2.43 Comparison between real experiment and simulation of pressure course in front hip, knee and ankle muscle of left leg.

the inverse dynamics control unit calculates with deviations on the inertial parameters: 5% for center of gravity and mass and 10% for the inertia of each link. These deviations are applied by increasing the inertial parameters with the respective deviation. Secondly, the reported $\pm 5\,\%$ for the hysteresis on both force functions of the antagonistic set-up is taken into account. In particular this is achieved by adding 5% to the estimated force for one muscle and subtracting the same deviation for the other muscle before calculating the applied joint torque with (4.38). Both muscles of an antagonistic setup, after all move in opposite directions.

In figure 2.43 a comparison is made between simulations and experiments performed on the real robot Lucy for the pressure course in the front hip, knee and ankle muscle of the left leg. The walking conditions (trajectory generator, joint trajectory tracking controller and objective locomotion parameters) are the same as in section 4.2.2. The pressure courses of the real experiments are the ones of figures 4.31, 4.33 and 4.35. The figures give a good approximation of reality and confirms that this hybrid simulator can be used to evaluate control architectures.

2.9.4 Use of Middleware

A major problem of the interface program (section 2.8) and simulator (section 2.9) is that they are two separated programs. An approach as in OpenHRP instead of two seperate programs is recommended for future software developments. In OpenHRP (221) there is unification of the controllers for both the simulation and the real counterpart, this leads to more efficient development of the controllers and the developed code is more reliable. Moreover, as most of the software for research robots, the interface program and simulator are developed independently of the others, mainly driven by the specific application and objective of a pneumatic biped. Further development of the software by other researchers or re-using parts of software will be difficult because it is custom-made. The need among many researchers is arising for a more standardized (e.g. IEEE standard) approach to tackle the problem. The benefits are obvious: development software can be ported so the software development can be limited to the new components and control methods that are not implemented so far. The maintainability will be easier and so on. So called middleware is often used to realize compatibility between separately developed software modules.

Different initiatives have been taken: BABEL (125), Miro (Middleware for Robots) (408), OpenHRP (Open Architecture Humanoid Robotics Platform) (221), YARP (yet another robot platform) (279), MARIE (Mobile and Autonomous Robotics Integration Environment) (102), Player (419), ORCA (63), OROCOS (Open robot control software) (67), (68), MCA2 (Modular Controller Architecture Version 2) (14) and CoRoBA (Controlling Robot with CORBA) (101). But as correctly stated by Hirukawa: "Every-one agree that software should be modularized for recycling and we should have a common architecture, problem is no one agree on how to do it." In June 2006 Microsoft launched Microsoft Robotics Studio (381) because Microsoft sees great potential in robotics. It includes a visual programming language, a 3-D virtual simulation environment and a runtime framework for interfacing with all kinds of hardware that makes it easy to create and debug robot applications. Bill Gates even compares this software with Microsoft BASIC, which was one of the key catalysts for the software and hardware innovations that made the PC revolution possible (134). He states that the robots are in a situation similar with the computer industry during the mid-1970s with a lack of computer standards or platforms. The complexity of this problem is however big. Robotic projects involve very different areas with very different needs: artificial intelligence, control systems, data acquisition, networking, etc, which requires the collaboration of very different people and the integration of a variety of software and hardware components (125).

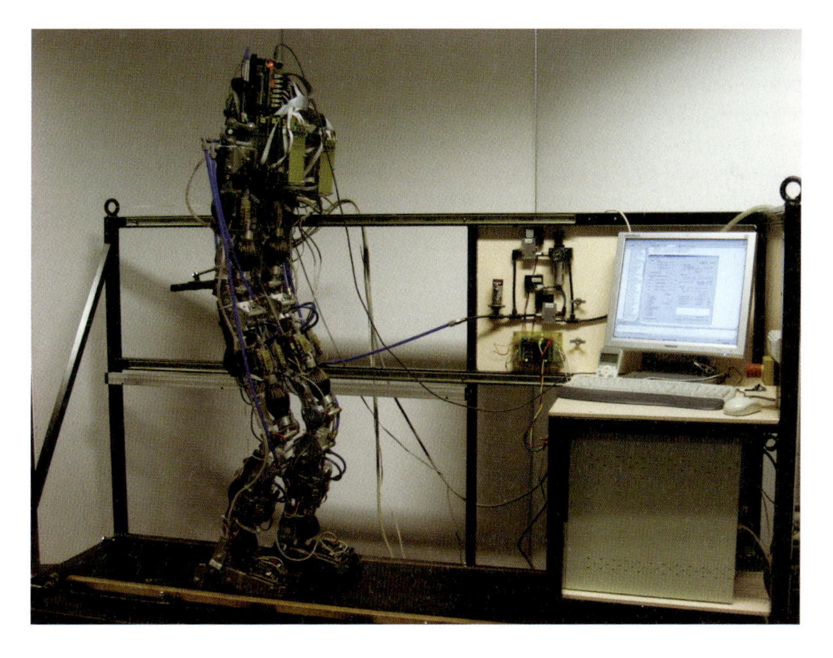

Fig. 2.44 Biped Lucy on treadmill with central computer.

2.10 Conclusion

In this chapter the construction of the biped Lucy is presented. The robot is equipped with 12 pleated pneumatic artificial muscles to power 6 DOF: the hip, knee and ankle of both legs. Initially the compressor or supply tank and PC are not placed on the robot. Building an autonomous robot is not the major concern, the main focus of the research is to investigate how well the pneumatic artificial muscles perform in bipedal locomotion. The muscles however are strong enough to carry an additional payload. The robot Lucy weighs $33kg$ and is $150cm$ tall. The motion of Lucy is restricted to the sagittal plane in order to avoid unnecessary complexity regarding control and design. A guiding mechanism prevents the robot from falling sidewards. Because the rails of the guiding mechanism have limited length a treadmill is used to enable walking for longer distances. The speed of the treadmill is controlled so that the robot stays in the middle. The complete setup is shown in figure 2.44. Key elements in the design phase were modularity and flexibility such as to have the ability to make changes to the robot configuration during the experimental process. This resulted in nearly the same configuration for each structural element such as lower leg, upper leg and body. The same type of modularity is also incorporated in the control hardware. Every joint has its own 16-bit micro-controller (MC68HC916Y3 made by Motorola) which incorporates a low-level pressure controller and collects sensor information from the Agilent HEDM6540 incremental encoder for determining the joint position and velocity and two pressure sensors inside each muscle of the antagonistic setup. The encoder and pressure signals are registered with a separate subprocessor, TPU, on the micro-controller in order not to load the CPU whilst reading their values. An additional micro-controller is used to detect ground contact, ground forces and absolute position of the body. The high-level control is implemented on a PC. All the micro-controller units communicate with this central, Windows operated PC by a USB 2.0 high speed serial bus. As such, the complete biped is controlled at a sample rate of $2000Hz$. The timing of the communication refresh rate is controlled by the EZ-USB FX2 Cypress micro-controller. The local micro-controllers ensure low-level, quasi real-time, control of the joints, and in order to prevent control disturbance of missed torque calculations by the central PC, the incoming data of the local units are buffered in the dual ported RAM. So whenever the central PC does not succeed to perform the necessary calculations within the sampling time, the local control units use the previously sent data. One should also remark that the delay time of the valves is about $1ms$, which suggests that the communication frequency of $2000Hz$ is high enough.

Most parts of the system are highly non-linear: force/torque-contraction of the muscles, the thermodynamic processes in the muscles, the mechanics of the system, difference between single and double support phase, stance and swing leg, impacts. So it is difficult to make a stability/convergence analysis. In order to evaluate control strategies before implementing them in the real biped, while taking into account all these non-linearities, a hybrid simulator was created. This model combines formulations of the thermodynamic processes, taking place in the muscles, with the standard Lagrange representation of the robot dynamics. The simulation model allows to simulate single and double support phases and also impacts are modelled when the swing foot touches the ground. These phase transitions after all have a strong influence on the system. Reported hardware limitations such as valve delays and sampling times, observed on the real robot, are taken into account in the simulation model, as well as some expected parameter estimation errors.

Chapter 3
Trajectory Generator

The challenge for legged robots, especially bipeds, is to maintain postural stability while walking around, whatever the state of the surface the robot is walking on. A possible overall control structure required to steer a biped is shown in schematic overview, depicting several essential control blocks (figure 3.1). A task manager commands the robot to execute a particular task at a specific moment. Depending on the current global robot position and information about its direct environment, a gait planner produces specific objectives for the global robot motion. According to these objectives, while taking into account the biped's configuration, a joint trajectory generator calculates desired trajectories for each joint of the robot. Finally, a tracking controller determines the necessary control actions to be carried out by the different joint actuator units in order to track the trajectories. A joint trajectory generator generally calculates trajectories which incorporate global dynamic postural stability e.g. based on ZMP (435) placement. However since this feedforward ZMP placement is based on estimated robot parameters and approximated dynamics, an extra feedback loop controlling the ZMP, should be provided. This control block commands deviations for the trajectory controller and/or tracking controller, based on ground reaction force measurements in the feet and global orientation information of the robot.

The task manager and gait planner is not within the scope of this project. Well-known research in this field is performed by for example Kuffner. He developed navigation strategies for humanoids through complex environments while using

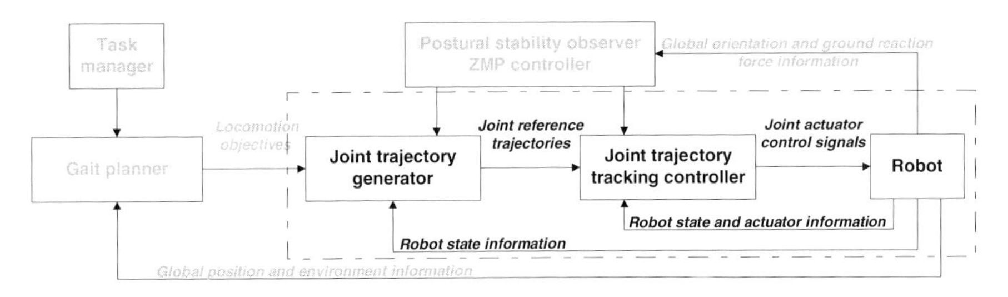

Fig. 3.1 Global control scheme for walking robot.

B. Vanderborght: Bipedal Walking by the Pneumatic Robot Lucy, STAR 63, pp. 93–142.
springerlink.com

their full capabilities. For indoor environments, this includes dealing with furniture, walls, stairs, doors, and possible (movable) objects on the floor (282; 380). For outdoor environments, this includes the ability to navigate on rough terrain and uneven surfaces (88). Yoshida et al. (460; 461) developed a humanoid motion planner to manipulate objects in complex environments. First the kinematic and geometric motion planner generates the trajectory for both the humanoid body as the carried object. Afterwards the path is reshaped to provide dynamically feasible robot motions without collisions. Part of the their research was performed using the path planning software KineoWorks^{TM} (11). On the humanoid platform Johnnie (103) and the biped robot BARt-UH (363) a vision guided path planning and obstacle avoidance was implemented. Gutmann et al. (147) presented a three level architecture for the navigation of QRIO where the motion capabilities of the robot are represented by a collection of different behavior modules as an ordinary walk, stair climb, crawl, sidewards walk and so on.

The current research and the topic for this project focusses on the two control blocks depicted with boldface in figure 3.1: the trajectory generator and the joint trajectory tracking controller. The trajectory generator will be the topic of this chapter, the joint trajectory tracking controller is described in the next chapter. Some remarks on the ZMP controller or "stabilizer", as referred to in (163), is also given in the next chapter.

Two different trajectory generators for Lucy will be described. Both use the Zero Moment Point as stability criterium and this chapter starts with an explanation of this concept. There are several possibilities to calculate trajectories for walking robots. In section 3.2 an overview is given were the methods are divided in three main categories depending on their main underlying working principle. The first proposed trajectory generator is based on the inverted pendulum approximation, which models the robot as a single point mass. The trajectory generator allows the step length, intermediate foot lift and velocity to be chosen for each step while keeping the zero moment point in the ankle point during the single support phase and it provides a smooth transition of the ZMP from the rear ankle point to the front ankle point during the double support phase. This method is good for low walking speeds, but at higher speeds the real and desired ZMP will differ. The main reason is that the complete multibody distributed masses are not taken into account. A second trajectory generator copes with this problem by taking the full multibody model into account. The trajectory generator is a kind of servo tracking controller which tracks a ZMP reference path. It will be shown that future information is needed and a preview control method is introduced. This method is also used for another research project which the author participated in during his 6 weeks research stay at JRL in Tsukuba, Japan. The goal was to generate trajectories to dynamically step over large obstacles by the humanoid robot HRP-2.

3.1 Dynamic Balance

There are two main kinds of control regarding stability of legged robots: static and dynamic balancing.

Statically balanced robots keep the center of mass within the polygon of support in order to maintain postural stability (374; 276), this is sufficient when the robot moves slow enough so all the inertial forces are negligible. The support polygon of the robot during single support phase is the area of the supporting foot and during the double support phase it is the polygon created by the convex boundary around the two feet. A support area does not exist in the case when both the feet are off the ground (running or jumping) or when the contact area has degenerated to a point or a line (this, however, means that the rigid foot rotates about an axis or point and that the mechanism as a whole is pivoting) (436). Contrary to statically balanced robots, for dynamically balanced robots the inertial effects have to be taken into account in the different control strategies. However the boundary between static and dynamic balancing is very loose and often dynamic gaits are referred to as not statically balanced at all times (349). The robot is certainly dynamically balanced when phases can be distinguished were the vertical projection of the COG is outside the support area during walking. For bipeds having single-point feet (for example RABBIT (330)), purely static balance during motion is also impossible. Problem of the term "dynamic balance" is that within the robotics community one does not agree about the definition and that consequently also many different concepts exist to judge if the robot's dynamic stability is guaranteed or not. The most popular criteria is the zero moment point (ZMP), introduced by Vukobratović (434).

The ZMP can be referred to as "*an overall indicator of the mechanism's behaviour, and is the point where the influence of all forces acting on the mechanism can be replaced by one single force*" (435). Or as interpreted by Dasgupta and Nakamura (107): *The ZMP is defined as that point on the ground at which the net moment of the inertial forces and the gravity forces has no component along the horizontal axis*. For a better comprehension, the ZMP formulation is given here for a planar robot system.

During the single support phase, the ZMP concept is about avoiding tipping over of the stance foot. After all, it is important to be able to use the total supporting foot area in order to influence the robot's behaviour. In figure 3.2, all the forces, inertial and ground reaction forces, which act on the foot are depicted. The influence of the dynamics of the complete robot on the foot are replaced by the torque $\bar{\tau}_{\mathbf{A}}$ (exerted by the ankle actuator) and the force $\bar{\mathbf{F}}_{\mathbf{A}}$, acting at the ankle point A. The total resultant of the ground reaction force $\bar{\mathbf{R}}$ works at point P and gravity acts on the foot in the center of gravity G_f. Note that for the sake of clearness the discussion is restricted to a 2D problem representation with the foot aligned to the horizontal ground. In the vertical direction, the ground reaction force $\bar{\mathbf{R}}$ compensates the vertical component of $\bar{\mathbf{F}}_{\mathbf{A}}$ and the weight of the foot $\mathbf{m_f}\bar{\mathbf{g}}$. The horizontal component of $\bar{\mathbf{R}}$, generated by friction forces, only compensates the horizontal component of $\bar{\mathbf{F}}_{\mathbf{A}}$. Note that, besides the robot stability criteria on rotation, friction between foot sole and ground has to be sufficient in order to have a non-slipping foot condition. To prevent the foot from rotating around one of its edges, the ground reacting forces will also counteract the moment induced by gravity and the inertial forces:

$$\overline{\mathbf{OP}} \times \bar{\mathbf{R}} + \overline{\mathbf{OG_f}} \times \mathbf{m_f}\bar{\mathbf{g}} + \overline{\mathbf{OA}} \times \bar{\mathbf{F}}_{\mathbf{A}} + \bar{\tau}_{\mathbf{A}} = 0 \tag{3.1}$$

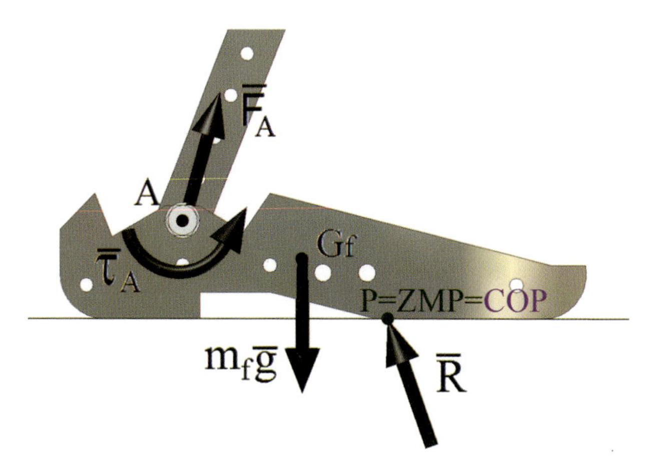

Fig. 3.2 Forces acting on a foot.

Writing equation 3.1 with respect to point P, the ground reaction force $\bar{\mathbf{R}}$ disappears from the equation. So with respect to this point the moment of the inertial and gravitational forces acting on the robot has to be zero. This explains the name of point P, zero moment point, and clarifies the equality between ZMP and COP, center of pressure. The center of pressure is defined as the distance-weighted average location of the individual pressures on the foot (331), thus the point P where the resultant $\bar{\mathbf{R}}$ of the ground reaction forces acts. The ZMP and COP are frequently mixed up in the legged robotics community, the ZMP can be seen as defined from the robot dynamic's point of view, while the COP is determined by the ground reaction forces. Whenever, the moment generated the ankle actuator $\bar{\tau}_{\mathbf{A}}$ is too large for the unilateral ground reaction force $\bar{\mathbf{R}}$ to compensate, the force $\bar{\mathbf{R}}$ will act on one of the foot edges, while an uncompensated part of the force moment will cause the robot to start tipping over (figure 3.3). This means that, in this undesirable situation, the COP is located at the foot edge, but that the ZMP actually, doesn't exist anymore. In this context, Goswami et al. (141) defined the foot-rotation index (FRI). The FRI is the point on the foot/ground contact surface, within or outside the support area where the net ground reaction force would have to act to keep the foot stationary. This point coincides with the ZMP when the foot is stationary, and diverges from the ZMP for non-zero angular foot accelerations.

The criteria above can judge wether or not the contact is kept, without solving the equations of motions when the robot moves on a flat plane under the assumption of sufficient friction. Humanoids however can move in an arbitrary terrain (225) and are able to use their arms to grasp a handrail (150), push, pull or lean on an object (151) or crawl (222). Hirukawa et al. presented a more universal stability criterion for the contact with the environment of legged robots (162). The proposed method checks if the sum of the gravity and the inertia wrench applied to the COG of the robot is inside the polyhedral convex cone of the contact wrench between the feet of a robot and its environment, which is proposed to be the stability criterion. The method has the advantage that it also takes into account the use of hands and it can

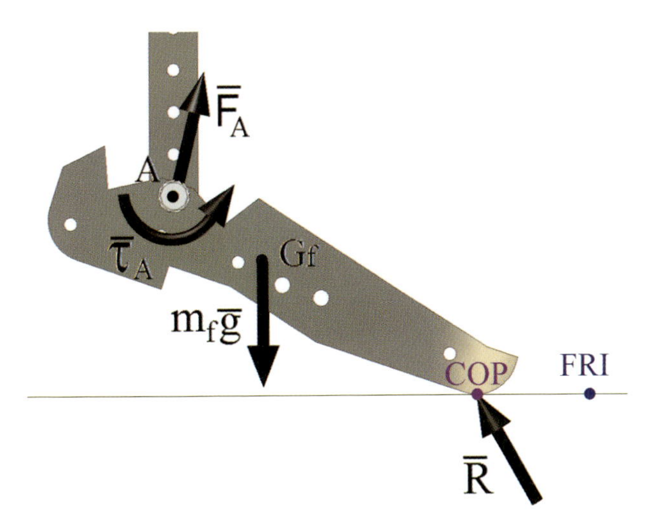

Fig. 3.3 Forces acting on the foot when the foot starts turning.

determine if the foot contact is sufficiently weakly stable incorporating the effect of friction. However, it is proved in the paper that the proposed criterium is equivalent to check if the ZMP is inside the support polygon of the feet when the robot walks on a horizontal floor with sufficient friction. Both these assumptions are fulfilled and Lucy has no arms, the ZMP criterion is maintained. As stated by the authors the ZMP can be drawn on a plane which is very convenient.

In a further development of the control strategy in this chapter an approximation is made by neglecting the weight of the foot and the height of the ankle point. In figure 3.4 the origin is placed at the ankle point A and τ_A is the applied ankle

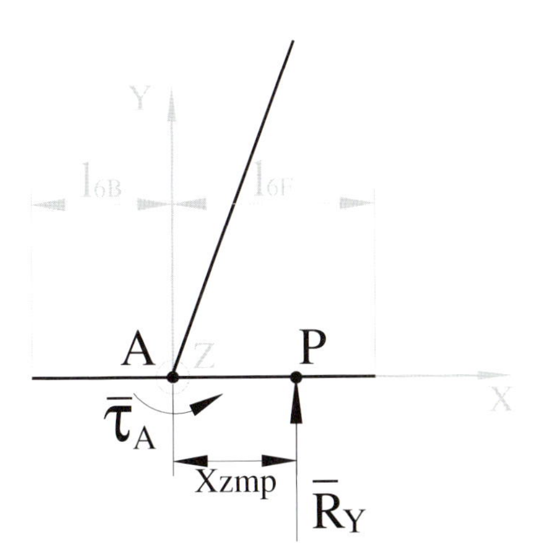

Fig. 3.4 Calculation of the ZMP.

torque in the ankle joint of the supporting foot, during the single support phase. The horizontal ZMP position, X_{zmp}, is then defined as:

$$R_y X_{zmp} + \tau_A = 0 \tag{3.2}$$

$$R_y = m_{tot}\left(\ddot{Y}_G + g\right) > 0 \tag{3.3}$$

with m_{tot} the total mass of the robot and $\ddot{Y}_G$ the vertical acceleration of the global COG, which can be calculated with equation (B.3b) of appendix C, representing the vertical position of the COG. Equation (3.3) is a result of the vertical component of the linear momentum theorem expressed for the global COG of the robot. Combining (3.2) and (3.3) gives:

$$X_{zmp} = \frac{-\tau_A}{m_{tot}\left(\ddot{Y}_G + g\right)} \tag{3.4}$$

with the necessity to keep the foot on the ground

$$\ddot{Y}_G > -g \tag{3.5}$$

To ensure dynamic stability during the single support phase $\left|X_{zmp}\right|$ has to be respectively smaller than the distances l_{6B} and l_{6F}, which are the respective distances from the heel and from the toe to the ankle point.

For the double support phase, instead of calculating the ZMP with the inertial and gravitational forces, the ground reaction forces are used to calculate the COP. In figure 3.5 the robot is depicted during a double support phase. At the front foot (F_F) the ground reaction force $\bar{\mathbf{R}}_{\mathbf{F}}$ is acting, and at the rear foot $\bar{\mathbf{R}}_{\mathbf{R}}$. In the absence of ankle torques, the total reaction forces on both feet act at the ankle points. The COP, or ZMP, location P is then found as:

$$X_{zmp} = \frac{R_F^y X_{A_F}}{R_F^y + R_R^y} \tag{3.6}$$

with X_{A_F} the distance between both ankle points during double support. Contrary to the single support phase, the ZMP stability margin is generally much larger and does not imply the same critical situation towards postural stability. During double support, the ZMP will have to be shifted from the rear to the front foot by gradually changing the “weight” of the robot from the back to the front. The ZMP will be located at the front foot when the rear foot is about to be lifted to start the next single support phase.

3.2 Trajectory Generator

Biped dynamics are high-order and nonlinear and therefore difficult to understand. There are different ways to generate biped locomotion control. “Natural dynamics-based control”, “soft computing” and “model-based trajectory generation” are possible groups of strategies.

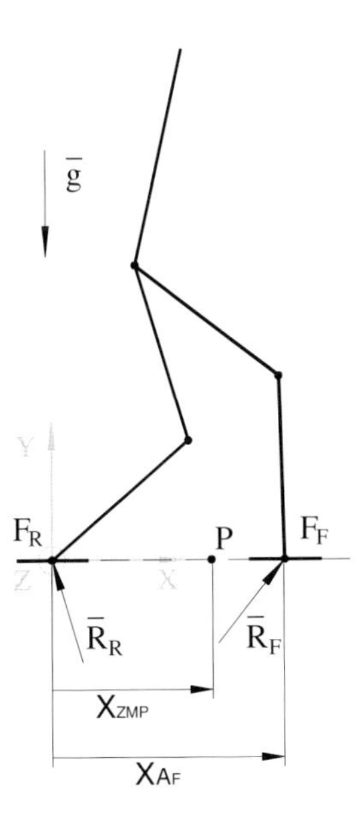

Fig. 3.5 ZMP during a double support phase.

3.2.1 *Natural Dynamics-Based Control*

The first group encompasses passive walking robots, where the idea is to put the intelligence not in the control of the robot but in the mechanical design. These robots are carefully designed mechanical systems with tuned natural dynamics so they can achieve a dynamic gait despite the lack of any control. By adding additional control like changing the duration of the swing of the leg, the stability is increased (395). Wisse et al. increased the stability of Mike by putting the swing leg fast enough in front of the stance leg (445). Schuitema et al. (362) used reinforcement learning to control the torque applied in the hip to the upper legs for the Meta biped.

Also the virtual model control is counted in this category. The robot Spring Flamingo is able to walk without any generation of trajectories and uses an intuitive control scheme. Virtual components are placed at strategic locations within the robot or between the robot and the environment (334) (337) to control the pitch, height and speed of the robot. The virtual forces applied by the springs are mapped to physical torques at each of the robots joints. The resulting reaction forces on the body exactly mimic the virtual forces created by the virtual elements.

These methods give very nice results regarding energy consumption and the walking motion looks very natural. However, they lack the versatility of trajectory-controlled robots. A major disadvantage is that there is not a strategy to build and control such a robot, but mainly depends on the experience of the researcher. Many parameters have to be tuned by hand, so from a practical point of view the size and weight of the robot must be adapted to this. Lucy is too heavy and big for such trial-and-error experiments and probably also too fragile to survive a big fall.

3.2.2 Soft Computing

Soft computing consists of neural networks, fuzzy logic techniques, genetic algorithms, etc or combinations of these methods (204). They are characterized by the fact that a model of the robot is not needed and are very tolerant against imprecision and uncertainty.

Park constructed a fuzzy-logic controller for the trunk to control the ZMP to stabilize the robot (319). Jha et al. (204) made a gait generator using a fuzzy logic controller whose rule base was optimized off-line, using a genetic algorithm. Genetic algorithms are also used to minimize the consumed energy by finding the optimal locations of the center of mass of the links (238).

There is definite experimental evidence in lower vertebrates, and suggestive evidence in higher mammals, that pattern generators can be found in spinal cords and are used to coordinate movement between multiple limbs (400). This biological inspiration has led to the development of Central Pattern Generators (CPGs). A widely used oscillator was first proposed by Matsuoka (1987) and is based on the mutual inhibition of two artificial neurons that generate a periodic signal as output.

These systems can realize robust bipedal locomotion control in terms of external disturbances and energy consumption and are not dependent on precise modelling (119). A drawback of the CPG approach is that most of the time these CPGs have to be tailor made for a specific application, and there are very few methodologies to construct a CPG for generating an arbitrary periodic signal (350).

Taga successfully applied a CPG controller for an 8-link simulated planar biped model (389). A pair of CPGs, modelled by an ANN (artificial neural network), controlled the muscles of the trunk and the left and right hip, knee and ankle joints. Once the model had been trained, it not only produced level gait under normal conditions, but it also adapted to environmental perturbations such as uneven terrain or increased carrying load. Taga also demonstrated that the speed of walking could be controlled by a single parameter which drove the neural oscillators, and the step cycle could be entrained by a rhythmic input to the oscillators (387) (388). Miyakoshi et al. extended Taga's work from 2D to 3D and also simplified the CPG control mechanism (288).

In (289) it is shown that a humanoid robot can step and walk using simple sinusoidal desired joint trajectories with their phase adjusted by a coupled oscillator model. The control approach was successfully applied in a hydraulic humanoid robot developed by SARCOS and the small humanoid QRIO. Also HOAP2 is able to walk using a system of coupled nonlinear oscillators that are used as programmable central pattern generators (350). RunBot has a sensor-driven controller built with biologically inspired sensor- and motor-neuron models (135). This way RunBot can reach a relative speed of 3.5 leg lengths per second ($0.8m/s$) after only a few minutes of online learning. An active upper-body component was added to walk on terrains with different slopes up to $7.5°$ (269). Nakanishi et al. proposed a learning method for biped locomotion from human demonstration (296). Other research in this field was performed by Aoi and Tsuchiya (37), Komatsu and Usui (244) and Matsubara et al. (272).

To use such a strategy in a real robot, it is preferred that the learning has been done in simulation first. The success greatly depends on the quality of the model of the real world used in simulation. Also when one wants to use the strategy in another robot, the whole learning algorithm has to be redone.

3.2.3 Model-Based Trajectory Generation

This group obtains trajectories for the different joints based on a model of the robot and a joint trajectory tracking controller has to follow the desired path. This can be further classified in "trajectory replaying" and "realtime generation" as proposed by Sugihara (383), or roughly speaking off-line and on-line techniques.

The first category prepares a joint-motion trajectory in advance, and applies it to the real robot with little on-line modification. Mita et al. (285) recorded human data and applied a tracking control of the human gait trajectories. Unfortunately, measuring the angle trajectories during human walking for a wide range of step lengths and walking speeds is difficult and time consuming (81). Moreover a humanoid robot does not necessarily have the same kinematical and dynamical properties (e.g. link dimensions, number of DOF, number of actuators, etc.) as a human individual, such that the recorded data from humans have to be processed to fit the robot's specifications. Zarrugh et al. (463) also investigated the walking pattern for a biped robot by recording human kinematic data.

Numerous off-line techniques mainly focus on the aspect of optimization of a certain criterion, such as e.g. energy consumption. Since computation time is not an issue in this case, numerical optimization techniques have been developed in order to obtain energy optimal trajectories. Chevallereau and Aoustin focused on optimal cyclic gaits for a walking and running biped robot without actuated ankles (91). The coefficients of the polynomial functions were chosen to optimize various criteria (maximal forward velocity, minimal torque, and minimal energy) and to insure cyclic motion of the biped. Polynomials were also used by Beletskii et al. (51) and Channon et al. (85), Cabodevilla et al. (76) used Fourier series, Ono and Liu (315) used Hermite polynomial functions.

Denk et al. developed a systematic method for generating databases of walking primitives for humanoid robots allowing step length adaptation, direction changes and stepping over obstacles (112). It was demonstrated that walking primitives can be computed efficiently by optimal control techniques using direct collocation methods. To be ever applicable in a real world, off-line trajectory generators are not very useful.

On-line techniques, on the other hand, generate joint trajectories in real-time, while using actual robot feedback information. This method executes planning and control in a unified way. Although realtime generation is more promising than trajectory replaying from the viewpoint of high mobility, they commonly suffer from a large amount of computations which have to be performed in real time.

This class can on its turn be subdivided into two categories (212). The first category uses precise knowledge of dynamic parameters of a robot e.g. mass, location of center of mass and inertia of each link to prepare walking patterns. Therefore, it mainly relies on the accuracy of the model data. This method is sometimes called the ZMP based approach since they often use the ZMP for pattern generation and walking control.

In (181) a method was developed to plan a walking pattern consisting of a foot trajectory and a hip trajectory represented by 3rd order periodic spline functions. They changed the hip trajectory iteratively to obtain a smooth trajectory $x_h(t)$ with the largest stability margin by defining different values for x_{sd} and x_{ed} which denote the distances along the x-axis from the hip to the ankle of the stance foot at the beginning and the end of the double support phase.

In (422) the trajectory planner generates motion patterns based on two specific concepts, being the use of objective locomotion parameters, and exploiting the natural upper body dynamics by manipulating the angular momentum equation. The trajectories of the leg links, represented by 6th order polynomials, are planned in such a way that the upper body motion is naturally steered. Natural motion of the upper body is defined here as the motion generated by an underactuated system, i.e. without ankle torque. By using the angular momentum equation in an adequate way, the motion of the leg links can be defined such that the upper body motion is indirectly controlled on the position, the velocity and the acceleration level. Since the upper body performs this motion naturally, the resulting ankle actuator action is limited. It is restricted to cover the minor differences between a polynomial tracking function and the natural trajectory, and the compensation for non-modelled external disturbances. This limited action avoids problems concerning ZMP and foot rotation. An interesting aspects of this method is that they are based on fast converging iteration loops, requiring only a limited computation time. A disadvantage of his work is that for each set of objective locomotion parameters, the user has to provide the vertical position, velocity and acceleration of the hip at the start and end of the single support phase.

The second category uses limited knowledge of dynamics e.g. location of total center of mass, total angular momentum, etc. Since the controller knows little about the system structure, this approach much relies on a feedback control. The fundament of this method is often the inverted pendulum approach. The first trajectory generator that is developed is based on this concept, so a more elaborated overview is given in the next section.

3.3 Requirements

The trajectory generator for Lucy has to comply with a number of requirements:

- Dynamically balanced trajectories.
- Gait generator must be planar.
- Use of objective locomotion parameters being horizontal velocity v, step length λ and intermediate foot lift κ without the need to provide additional parameters.
- These objective parameters must be allowed to change during the locomotion process, in order to make motion on irregular terrain possible.
- Must contain a double support phase.

- The method should always give a solution in real time.
- Gait generator must be planar.
- Angle position, speed and accelerations of different joint trajectories continuous.

In section 3.1 is showed that dynamic balancing is more interesting than static balancing because higher walking speeds can be attained. It is chosen to use the ZMP as stability criterion. This point is easy to measure by ground force sensors.

Using objective locomotion parameters is an elegant way of characterizing steps of a motion pattern. They are calculated by a high level path planning control unit to perform a certain predefined task which is beyond the scope of this work. It should be possible to change these parameters while walking to adapt to a continuously changing environment and to be able to walk on irregular terrain. The user interface (section 2.8) allows to change the objective locomotion parameters by sliders, this means that the method should always give a result in real time.

The objective locomotion parameter "step height" is not used in the proposed locomotion generator because it is impossible to emulate the climbing of stairs with a treadmill. Moreover theoretical analysis reveals that dynamic gaits are not energy efficient for stair climbing (370). More information about climbing stairs can be found in (370; 291).

Some authors only consider the single support phase and study trajectory generators with instantaneous double support phase. This is interesting from an academic point of view. For controlling real robots, a double support phase is very important for improving the smoothness of the biped locomotion. As correctly remarked by Shih and Gruver (371) a double support phase is needed to enable the robot to start and stop its motion.

Only gait generation in the sagittal plane is considered because Lucy is a planar walker. For more information about controlling the robot in the frontal plane the reader is referred to (139; 154; 93; 92).

It is preferred for Lucy that besides joint position and velocity, also the accelerations are continuous because the joint trajectory tracking controller (chapter 4) use this to calculate the desired pressures for the muscles. Large pressure discontinuities are found to destabilize the system due to the slow dynamics of the valves (see section 4.1.4).

3.4 Trajectory Generation Based on Inverted Pendulum Mode

Kajita et al. (218) suggested the Linear Inverted Pendulum Mode (LIPM). The LIPM assumes a concentrated mass at the torso and neglects all other mass distributions. Kajita considered applying constraint control so that the body of the robot moves on a particular straight line and rotates at a constant angular velocity. This makes the dynamics of the center of mass completely linear. By doing this there is no ankle torque needed and the ZMP (437) stays in the ankle. The ankle torque is used to control the horizontal motion of the body to include the effect of the masses of the legs and to cope with external disturbances. An extension to three dimensions is presented in (211) and is called the Three-Dimensional Linear Inverted Pendulum Mode (3D-LIPM). This method was tested on the 12 DOF biped HRP-2L.

A walking pattern generated by the 3D-LIPM tends to have singularities and excessive knee joint torque problem (high knee joint angular velocities) since it requires planar constraint of the waist. In (291) a more natural and more efficient walking pattern using mostly stretched knees was generated by introducing a parametric constraint surface. This means that the motion of the center of gravity follows a desired surface which is designed by considering a landing position and a movable space of legs. Human walking also contains almost stretched knees.

At the Waseda University, the WABIAN-2 robot was developed. It has two 7-DOF legs, a 2-DOF waist and a 2-DOF torso. An algorithm was developed that enables the robot to stretch its knees in steady walking, avoiding singularities by using the waist motion (394). It was shown that the required knee joint torques of the stretched walking are much lower than those of the conventional walking with bent knees (306) and that the energy consumption of the knee actuators was lower (308), (307). Sekiguchi et al. developed a walking strategy based on LIPM that changes the leg motion direction by using proper ankle control around the singularities (366).

Because the LIPM method doesn't take the mass of the legs into account, errors occur between the position of the real ZMP and the desired ZMP. Park and Kim (320) proposed the gravity-compensated inverted pendulum mode (GCIPM). Instead of taking only one mass as in the LIPM method the robot model consists of two masses: one mass is for both the base link and the supporting leg, and the other is for the free leg. Using this technique, they developed an on-line trajectory generation method to increase the stability robustness of locomotion, based on the ZMP equation and the sensed information of the ZMP (323). This strategy was further refined in (322), where it was expanded to be used during the double support phase.

In (297) a two-mass inverted pendulum model was proposed with one mass for the lower part and another mass for the upper body of the humanoid robot.

The Multiple Masses Inverted Pendulum Mode (MMIPM) models the robot with several masses (136) (31). The user prescribes the foot motion of the swinging leg and the remaining trajectories of the robot are then calculated iteratively. Simulations and real measurements of the ZMP by force sensors show that this leads to a higher gait stability with respect to the ZMP, which is logical because the model is more accurate.

For LIPM no angular momentum has been generated since it assumes that the COG is a mass point and that the ground force vector always passes through the COG of the system. In (250) the LIPM method is extended so that angular momentum can be induced by the ground reaction force. This method is called the Angular Momentum inducing inverted Pendulum Model (AMPM). The LIPM is enhanced in two ways. The ZMP is allowed to move over the ground and the ground vector does not have to be parallel to the vector between the ZMP and the COG, as far as its horizontal component is linearly dependent on the COG position. This method is the base to counteract the large amount of angular momentum induced by strong external perturbations applied to the body during gait motion (246).

3.4.1 Objective Locomotion Parameters Based Inverted Pendulum Mode (OLPIPM) Trajectory Generator

The LIPM method generates a stable gait. The foot placement is controlled so the velocity of the next single support phase can be controlled. Consequently this method is not applicable if the foot must be placed on specified locations. This is needed to walk for example on stepping-stones or in an area with obstacles (212). The goal of this new trajectory generator is that the objective locomotion parameters, which are intermediate foot lift, desired speed and step length can be chosen from one step to another. The method is based on the inverted pendulum model so the ZMP stays in the ankle point during single support phase. Consequently this phase the motion is passive. This has as consequence that the necessary control actions have to take place during double support phase to realize the desired objective locomotion parameters of the next step. This is achieved by controlling the accelerations during the double support phase. Here is the difference with the LIPM method where the velocity during double support phase is a constant and the acceleration is zero. By making the position, velocity and acceleration continuous when switching between single and double support phases there is a smooth transition of the ZMP from rear ankle point to front ankle point.

Many trajectory generators based on the inverted pendulum principle have a body of the robot that moves on a particular straight line. However, humans almost stretch the knee of the stance leg and studies on bipeds show that walking with stretched knees reduces the energy consumption and torque level of the knee actuator (253). For planar walking bipeds it is impossible to walk with stretched knees and have a double support phase without the use of the toes and heel of the foot because at impact the robot comes in a singular state were both legs are completely stretched. To be able to walk with explicit use of the toes and heel requires a special design of the feet. Lucy doesn't have these specially designed feet. To solve this problem trajectories are calculated with almost stretched knees. During single support the distance between ankle and hip of the stance leg is kept constant. The proposed controller can however also be used for completely stretched knees.

Figure 3.6 shows 3 states of the robot and definitions of parameters used. The shown states are: start position of single support phase, end position of single support phase which is also the start position of the double support phase and finally the end of the double support phase which is also the beginning of the single support phase of the next step.

3.4.1.1 Hip Motion during Single Support Phase

The hip motion during single support is that of an inverted pendulum. A passive motion is desired so no ankle torque is considered, consequently the ZMP stays in the ankle joint. The equation of motion for small angles for this model can be written as:

$$mL^2\ddot{\theta} - mgLsin(\theta) = 0 \tag{3.7}$$

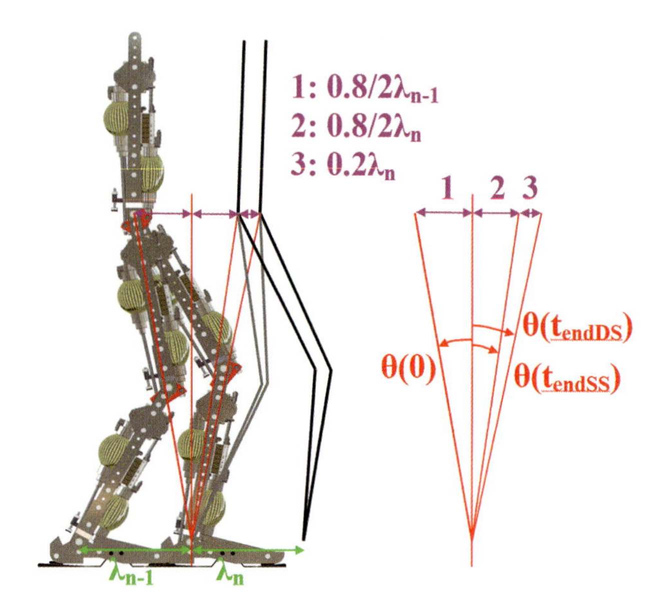

Fig. 3.6 Definition of parameters used (θ positive for clockwise rotations).

This can be simplified for small angles θ:

$$mL^2\ddot{\theta} - mgL\theta = 0 \tag{3.8}$$

With L the length from the contact point to the COG (here we have taken the hip) and m the mass of the complete robot.

The step length λ_{n-1} is calculated out of the measured joint angles at impact of the previous swing phase, while λ_n is given by the user. The duration of the single support phase will be chosen as 80% of the total step duration, corresponding to its duration in human walking at low speeds (152). The start angle $\theta_{SS}(0)$ and end angle $\theta_{SS}(t_{end_{SS}})$ are:

$$\theta(0) = -asin(\frac{0.8\lambda_{n-1}}{2L}) \tag{3.9}$$

$$\theta(t_{end_{SS}}) = asin(\frac{0.8\lambda_n}{2L}) \tag{3.10}$$

with the duration of the single support phase $t_{end_{SS}}$:

$$t_{end_{SS}} = \frac{0.8\lambda_{n-1} + 0.8\lambda_n}{2v} \tag{3.11}$$

v is the mean horizontal hip velocity. The trajectory of the hip can be calculated out of equation (3.8) with a desired $\theta(0)$ and $\dot{\theta}(0)$ as initial condition:

$$\theta_{SS}(t) = \frac{\theta(0) + T_c\dot{\theta}(0)}{2} e^{\frac{t}{T_c}} + \frac{\theta(0) - T_c\dot{\theta}(0)}{2} e^{\frac{-t}{T_c}} \tag{3.12}$$

with:

$$T_c = \sqrt{\frac{L}{g}} \tag{3.13}$$

At $t_{end_{SS}}$, the end condition (3.10) has to be reached. This is possible by choosing the start velocity $\dot{\theta}(0)$ as:

$$\dot{\theta}(0) = \frac{2\theta(t_{end_{SS}}) - \theta(0)e^{\frac{t_{end_{SS}}}{T_c}} - \theta(0)e^{\frac{-t_{end_{SS}}}{T_c}}}{T_c e^{\frac{t_{end_{SS}}}{T_c}} - T_c e^{\frac{-t_{end_{SS}}}{T_c}}} \tag{3.14}$$

The velocity and acceleration of θ are found by deriving equation (3.12):

$$\dot{\theta}_{SS}(t) = \frac{\theta(0) + T_c\dot{\theta}(0)}{2T_c} e^{\frac{t}{T_c}} + \frac{\theta(0) - T_c\dot{\theta}(0)}{-2T_c} e^{\frac{-t}{T_c}} \tag{3.15}$$

$$\ddot{\theta}_{SS}(t) = \frac{\theta(0) + T_c\dot{\theta}(0)}{2T_c^2} e^{\frac{t}{T_c}} + \frac{\theta(0) - T_c\dot{\theta}(0)}{2T_c^2} e^{\frac{-t}{T_c}} \tag{3.16}$$

The motion of the hip, in Cartesian coordinates, is given by:

$$\begin{aligned}
x_{hip}(t) &= Lsin\big(\theta_{SS}(t)\big) \\
\dot{x}_{hip}(t) &= Lcos\big(\theta_{SS}(t)\big)\dot{\theta}_{SS}(t) \\
\ddot{x}_{hip}(t) &= Lcos\big(\theta_{SS}(t)\big)\ddot{\theta}_{SS}(t) - Lsin\big(\theta_{SS}(t)\big)\dot{\theta}_{SS}(t)^2 \\
y_{hip}(t) &= Lcos\big(\theta_{SS}(t)\big) \\
\dot{y}_{hip}(t) &= -Lsin\big(\theta_{SS}(t)\big)\dot{\theta}_{SS}(t) \\
\ddot{y}_{hip}(t) &= -Lsin\big(\theta_{SS}(t)\big)\ddot{\theta}_{SS}(t) - Lcos\big(\theta_{SS}(t)\big)\dot{\theta}_{SS}(t)^2
\end{aligned} \tag{3.17}$$

This means that the coordinate system is located in the ankle joint of the stance foot. At the end of the double support phase the coordinate system is shifted to the next stance foot.

3.4.1.2 Hip Motion during Double Support Phase

The motion of the hip during the double support phase has to connect the old and the new single support phases while position, velocity and acceleration have to be continuous. The boundary conditions are:

$\mathbf{X}(t_{start_{DS}})_n = \mathbf{X}(t_{end_{SS}})_n$ and $\mathbf{X}(t_{end_{DS}})_n = \mathbf{X}(t_{start_{SS}})_{n+1}$
with $\mathbf{X} = [x, \dot{x}, \ddot{x}, y, \dot{y}, \ddot{y}]^T$.

The most evident method is to use 5th order polynomials with the boundary conditions determining the coefficients. A problem arises for the ZMP when

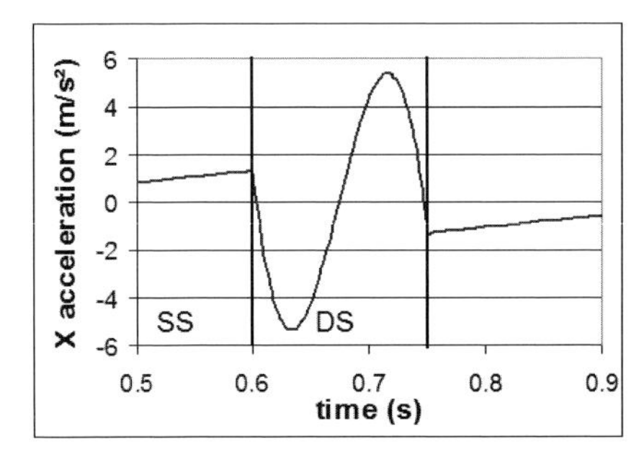

Fig. 3.7 x-acceleration.

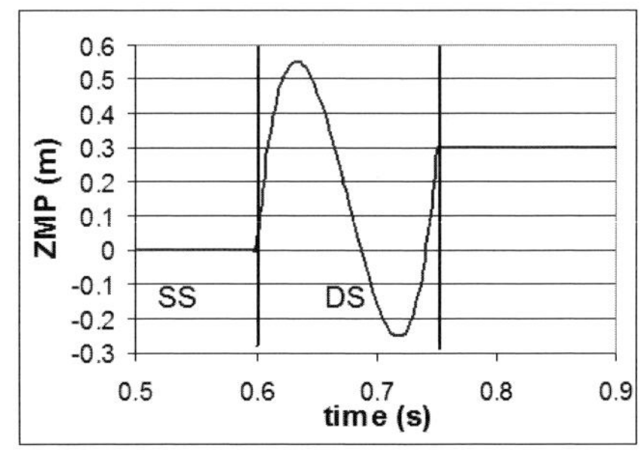

Fig. 3.8 ZMP.

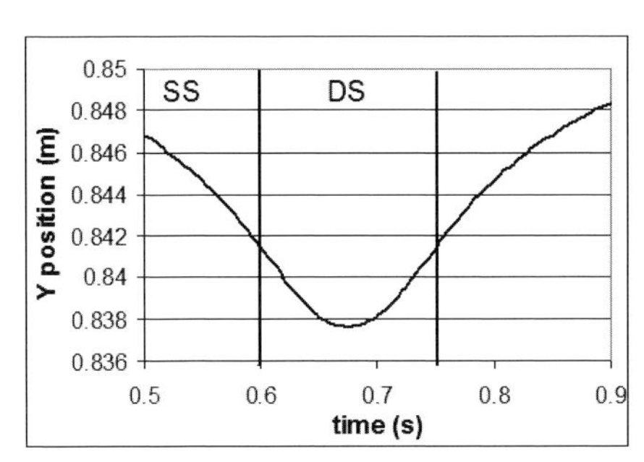

Fig. 3.9 y-position.

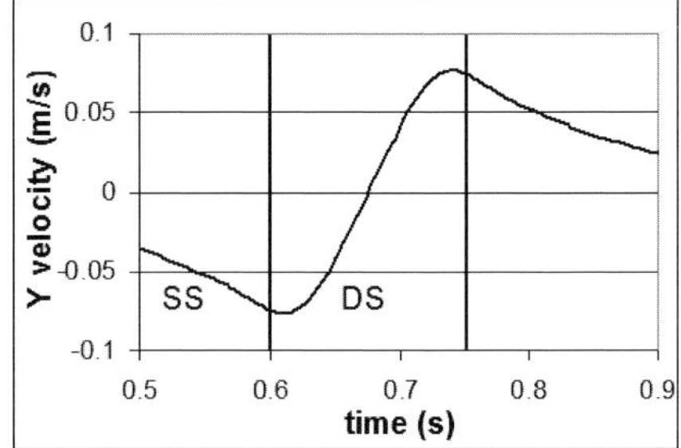

Fig. 3.10 y-velocity.

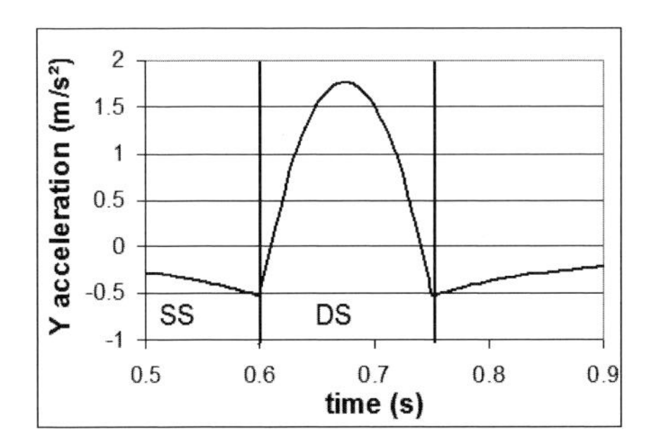

Fig. 3.11 y-acceleration.

with $\lambda = 0.3m$; $v = 0.4m/s$

implementing this strategy, as is illustrated in figures 3.7-3.8. The x-acceleration (figure 3.7) oscillates severely in the time interval, meaning that the zero moment point (figure 3.8) goes forward and backward very violently. This is not the case for the y-position so we keep this strategy for the y-direction as can be seen in figures 3.9-3.11, showing the y-position, velocity and acceleration.

For the x-position another strategy is used to shift the ZMP from the rear ankle to the front ankle smoothly. Since the evolution of the ZMP is mainly dependent on the x-acceleration, the strategy is built up from this second order derivative. The idea is to use the opposite of the acceleration curve of θ in the single support phase during the double support phase. So equation (3.16) is taken with a minus sign. By choosing this particular θ-acceleration, the acceleration is at once continuous and also the velocity when there is no change in desired speed. One could also choose for example a linear change of $\ddot{\theta}_{DS}$ between the end of the single support phase and the start of the next single support phase, but then more calculations will be needed to fulfill the boundary conditions. So this gives[1]:

$$\ddot{\theta}_{DS}(t) = -\left(\frac{\theta(0)+T_c\dot{\theta}(0)}{2T_c^2}e^{\frac{t}{T_c}} + \frac{\theta(0)-T_c\dot{\theta}(0)}{2T_c^2}e^{\frac{-t}{T_c}}\right) \tag{3.18}$$

Because the distance between the two feet remains constant (λ_n) during this double support phase instead of going from λ_{n-1} to λ_n equation (3.9) is

$$\theta(0) = -asin(\frac{0.8\lambda_n}{2L}) \tag{3.19}$$

Equation (3.10) remains the same and equation (3.11) to calculate $t_{end_{SS}}$ has to be changed because the double support phase is shorter in time than the single support phase.

$$t_{end_{DS}} = \frac{0.2\lambda_n}{v} \tag{3.20}$$

With these values it is possible to use equation (3.14) to calculate $\dot{\theta}(0)$.
Integrating (3.18) gives the velocity:

$$\begin{aligned}\dot{\theta}_{DS}(t) = &-\left(\frac{\theta(0)+T_c\dot{\theta}(0)}{2T_c}e^{\frac{t}{T_c}} + \frac{\theta(0)-T_c\dot{\theta}(0)}{-2T_c}e^{\frac{-t}{T_c}}\right)\\ &+\dot{\theta}(t_{end_{SS}}) + \frac{\theta(0)+T_c\dot{\theta}(0)}{2T_c} + \frac{\theta(0)-T_c\dot{\theta}(0)}{-2T_c}\end{aligned} \tag{3.21}$$

The last 3 terms are introduced to guarantee the continuity ($\dot{\theta}_{DS}(0) = \dot{\theta}_{SS}(t_{end_{SS}})$) because by definition θ remains the same as in single support phase. Integrating (3.21) gives the position:

$$\begin{aligned}\theta_{DS}(t) = &-\left(\frac{\theta(0)+T_c\dot{\theta}(0)}{2}e^{\frac{t}{T_c}} + \frac{\theta(0)-T_c\dot{\theta}(0)}{2}e^{\frac{-t}{T_c}}\right)\\ &+\left(\dot{\theta}(t_{end_{SS}}) + \frac{\theta(0)+T_c\dot{\theta}(0)}{2T_c} + \frac{\theta(0)-T_c\dot{\theta}(0)}{-2T_c}\right)t\\ &+\theta(t_{end_{SS}}) + \frac{\theta(0)+T_c\dot{\theta}(0)}{2} + \frac{\theta(0)-T_c\dot{\theta}(0)}{2}\end{aligned} \tag{3.22}$$

If we do this for a constant speed v, the x-velocity (figure 3.13) and x-acceleration (figure 3.14) at start and end are continuous, but not the position (figure 3.12). This is due to the higher mean velocity during double support phase as can be seen in figure 3.13, which comes from the last 3 terms of (3.21) that were introduced to guarantee the continuity of $\dot{\theta}_{DS}(t)$ at $t = 0$. This can be solved by decreasing the time of the double support phase by a factor ζ.

[1] For better readability of the equations the start time of the double support phase is set to zero, so $t_{start_{DS}} = 0$.

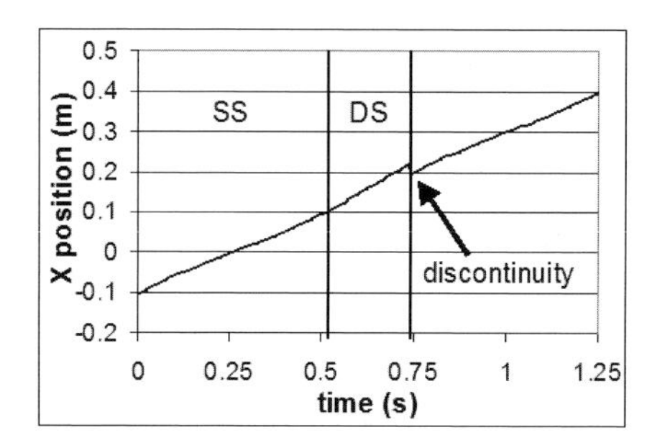

Fig. 3.12 x-position with $\lambda = 0.3m$, $v = 0.4m/s$.

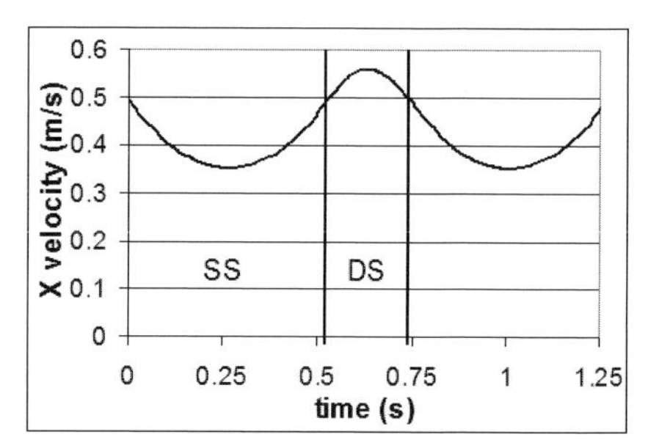

Fig. 3.13 x-velocity with $\lambda = 0.3m$, $v = 0.4m/s$.

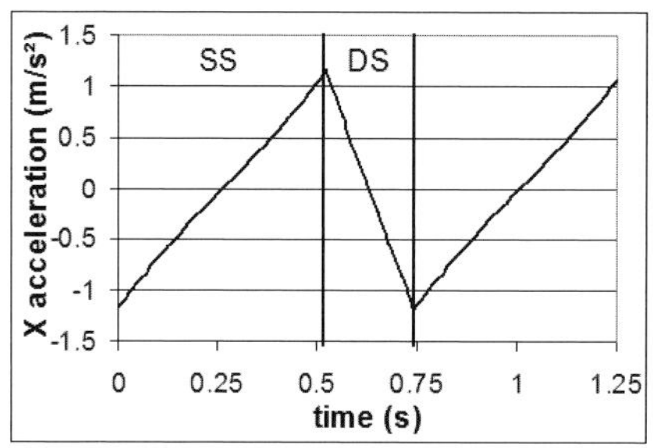

Fig. 3.14 x-acceleration with $\lambda = 0.3m$, $v = 0.4m/s$.

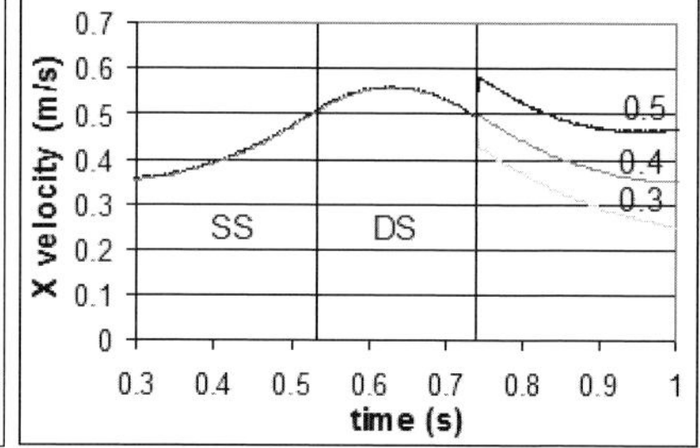

Fig. 3.15 Discontinuity of velocity, a change of velocity from $v_1 = 0.4m/s$ to $v_2 = 0.3, 0.4, 0.5m/s$ with $\lambda = 0.3m$.

$$t_{end_{DS}}^{adjusted} = \frac{t_{end_{DS}}}{\zeta} \tag{3.23}$$

To find the adjusted $t_{start_{DS}}^{adjusted}$ one has to replace $t = \zeta t^*$ in equation (3.18) and again integrate twice, just as in equations (3.21) and (3.22). At $t^* = t_{end_{DS}}/\zeta$ θ has to be the start position of the next single support phase or $\theta_{DS}(t_{end_{DS}}) = asin\left(\frac{0.8\lambda_n}{2L} + \frac{0.2\lambda_n}{L}\right)$. This gives the following equation $a + \frac{b}{\zeta} + \frac{c}{\zeta^2} = 0$ with:

$$\begin{aligned}
a &= \theta(t_{end_{SS}}) - asin\left(\frac{0.8\lambda_n}{2L} + \frac{0.2\lambda_n}{L}\right) \\
b &= \dot{\theta}(t_{end_{SS}})t_{end_{DS}} \\
c &= \left(\frac{\theta(0) + T_c\dot{\theta}(0)}{2} + \frac{\theta(0) - T_c\dot{\theta}(0)}{2}\right) \\
&- \left(\frac{\theta(0) + T_c\dot{\theta}(0)}{2} e^{\frac{t_{end_{DS}}}{T_c}} + \frac{\theta(0) - T_c\dot{\theta}(0)}{2} e^{\frac{-t_{end_{DS}}}{T_c}}\right) \\
&+ \left(\frac{\theta(0) + T_c\dot{\theta}(0)}{2T_c} + \frac{\theta(0) - T_c\dot{\theta}(0)}{-2T_c}\right) t_{end_{DS}}
\end{aligned} \tag{3.24}$$

So we get a quadratic equation $a\zeta^2 + b\zeta + c = 0$ with 2 solutions of which the negative solution can be neglected. This makes also the position at the end of the double support phase continuous.

The single support phase is intended to be passive. When one wants to change the velocity from one step to another an extra acceleration or deceleration $\ddot{\theta}_{DS}(t)^{extra}$ is needed during the double support phase to attain the necessary velocity for the next single support phase, as can be seen in figure 3.15. Because the acceleration at $t = t_{start_{DS}}$ and $t = t_{end_{DS}}$ is fixed by the pendulum motion and independent of the velocity, the extra acceleration at these points has to be zero. So a quadratic acceleration is chosen with at the start and end position zero acceleration and for which the integral is the desired velocity change. The extra acceleration is:

$$\ddot{\theta}_{DS}(t)^{extra} = kt^2 + lt + m \tag{3.25}$$

with boundary conditions:

$$\begin{aligned} \ddot{\theta}_{DS}(0)^{extra} &= 0 \\ \ddot{\theta}_{DS}(t_{end_{DS}}/2)^{extra} &= a^{extra} \\ \ddot{\theta}_{DS}(t_{end_{DS}})^{extra} &= 0 \end{aligned} \tag{3.26}$$

These boundary conditions yield:

$$\begin{aligned} k &= \frac{-4a^{extra}}{t_{end_{DS}}^2} \\ l &= \frac{4a^{extra}}{t_{end_{DS}}} \\ m &= 0 \end{aligned} \tag{3.27}$$

As a result the extra terms for position, velocity and acceleration become:

$$\ddot{\theta}_{DS}(t)^{extra} = \frac{-4a^{extra}}{t_{end_{DS}}^2} t^2 + \frac{4a^{extra}}{t_{end_{DS}}} t \tag{3.28a}$$

$$\dot{\theta}_{DS}(t)^{extra} = \frac{-4a^{extra}}{t_{end_{DS}}^2} \frac{t^3}{3} + \frac{4a^{extra}}{t_{end_{DS}}} \frac{t^2}{2} \tag{3.28b}$$

$$\theta_{DS}(t)^{extra} = \frac{-4a^{extra}}{t_{end_{DS}}^2} \frac{t^4}{12} + \frac{4a^{extra}}{t_{end_{DS}}} \frac{t^3}{6} \tag{3.28c}$$

with $\theta_{DS}(0)^{extra} = 0$ and $\dot{\theta}_{DS}(0)^{extra} = 0$. a^{extra} is calculated out of the extra velocity change needed by using (3.28b):

$$a^{extra} = \frac{3(\dot{\theta}_{nextSS}(0) - \dot{\theta}(0))}{2t_{end_{DS}}} \tag{3.29}$$

$\dot{\theta}_{nextSS}(0)$ is calculated out of equation (3.14) for the next step. Additionally the term $0.5T_{DS}(\dot{\theta}_{nextSS}(0)-\dot{\theta}(0))$ has to be added in the b-term of equation (3.24). This term comes from (3.28c) and (3.29) at $t_{end_{DS}}$.

So $\theta_{DS}(t)$, $\dot{\theta}_{DS}(t)$ and $\ddot{\theta}_{DS}(t)$ during double support become:

$$\begin{aligned}\theta_{DS}(t) = &-\left(\frac{\theta(0)+T_c\dot{\theta}(0)}{2\zeta^2}e^{\frac{\zeta t}{T_c}}+\frac{\theta(0)-T_c\dot{\theta}(0)}{2\zeta^2}e^{\frac{-\zeta t}{T_c}}\right)\\ &+\left(\dot{\theta}(t_{start_{DS}})+\frac{\theta(0)+T_c\dot{\theta}(0)}{2\zeta T_c}+\frac{\theta(0)-T_c\dot{\theta}(0)}{-2\zeta T_c}\right)t\\ &+\theta(t_{start_{DS}})+\frac{\theta_{DS}(0)+T_c\dot{\theta}(0)}{2\zeta^2}+\frac{\theta(0)-T_c\dot{\theta}(0)}{2\zeta^2}\\ &+\frac{-4a^{extra}}{t_{end_{DS}}^2}\frac{t^4}{12}+\frac{4a^{extra}}{t_{end_{DS}}}\frac{t^3}{6}\end{aligned} \tag{3.30}$$

$$\begin{aligned}\dot{\theta}_{DS}(t) = &-\left(\frac{\theta(0)+T_c\dot{\theta}(0)}{2\zeta T_c}e^{\frac{\zeta t}{T_c}}+\frac{\theta(0)-T_c\dot{\theta}(0)}{-2\zeta T_c}e^{\frac{-\zeta t}{T_c}}\right)\\ &+\dot{\theta}(t_{start_{DS}})+\frac{\theta(0)+T_c\dot{\theta}(0)}{2\zeta T_c}+\frac{\theta(0)-T_c\dot{\theta}(0)}{-2\zeta T_c}\\ &+\frac{-4a^{extra}}{t_{end_{DS}}^2}\frac{t^3}{3}+\frac{4a^{extra}}{t_{end_{DS}}}\frac{t^2}{2}\end{aligned} \tag{3.31}$$

$$\begin{aligned}\ddot{\theta}_{DS}(t) = &-\left(\frac{\theta(0)+T_c\dot{\theta}(0)}{2T_c^2}e^{\frac{\zeta t}{T_c}}+\frac{\theta(0)-T_c\dot{\theta}(0)}{2T_c^2}e^{\frac{-\zeta t}{T_c}}\right)\\ &+\frac{-4a^{extra}}{t_{end_{DS}}^2}t^2+\frac{4a^{extra}}{t_{end_{DS}}}t\end{aligned} \tag{3.32}$$

With $t_{end_{DS}}$ at once the adjusted $t_{end_{DS}}^{adjusted}$.

This strategy makes it possible to change the desired step length and velocity from one step to another without discontinuities when switching between single and double support phase. Figures 3.16-3.19 show a change in desired velocity v. The ZMP is calculated out of the motion of the concentrated mass. One can see the smooth transition of the zero moment point, shifting from the rear foot to the front foot. When the desired velocity increases (line 3) the ZMP is behind the ZMP when the velocity is kept constant (line 2), due to the higher forces in the rear ankle to increase the speed. The opposite happens when the desired velocity of the next step decreases (line 1). Figures 3.20-3.23 show a change in desired step length λ. Again the ZMP shifts very smoothly from the rear ankle point to the front ankle point.

Since the developed strategy does not contain any iterations and consists of very straightforward calculations, it always gives results and is very fast, suitable for realtime use.

3.4.1.3 Foot Motion during Single Support Phase

Two sixth order polynomial functions for the leg links of the swing leg are established, which connect the initial, intermediate and final boundary values for the

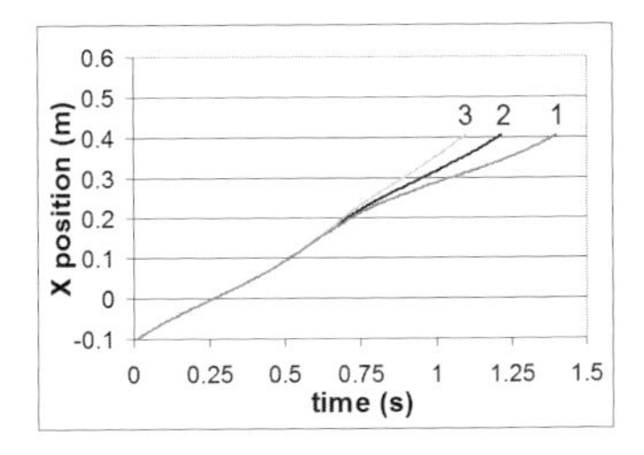

Fig. 3.16 x-position of hip.

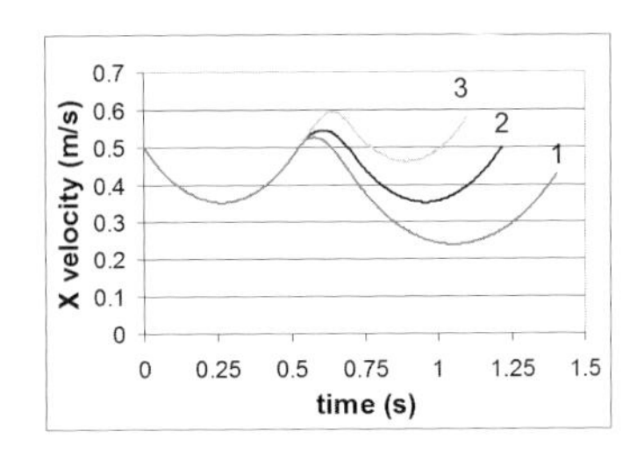

Fig. 3.17 x-velocity of hip.

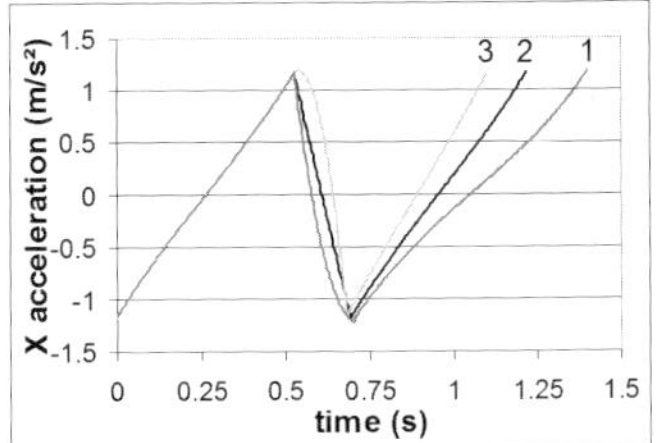

Fig. 3.18 x-acceleration of hip.

Fig. 3.19 ZMP

Single, double and single support phase with parameters:
1: $\lambda_1 = 0.3m\ \lambda_2 = 0.3m$; $v_1 = 0.4m/s\ v_2 = 0.3m/s$
2: $\lambda_1 = 0.3m\ \lambda_2 = 0.3m$; $v_1 = 0.4m/s\ v_2 = 0.4m/s$
3: $\lambda_1 = 0.3m\ \lambda_2 = 0.3m$; $v_1 = 0.4m/s\ v_2 = 0.5m/s$

swing foot motion. The intermediate condition at $t = t_{end_{SS}}/2$ is used to lift the foot, with height κ, whenever an obstacle has to be avoided during the swing phase. At the start of the single support phase the boundary conditions are: $x_{foot}(0) = -\lambda_{n-1}$ and $\dot{x}_{foot}(0) = \ddot{x}_{foot}(0) = y_{foot}(0) = \dot{y}_{foot}(0) = \ddot{y}_{foot}(0) = 0$ At the end of the single support the boundary conditions are: $x_{foot}(t_{end_{SS}}) = \lambda_n$ and $y_{foot}(t_{end_{SS}}) = 0$, the velocity and acceleration can be chosen freely, here both are taken zero. This special kinematic requirement at touch-down was introduced by Beletskii et al. (51), who described it as the softness of gait. According to Blajer and Schiehlen (57) the impacts due to collision of the legs with the ground create destabilizing effects on the walking cycle, and should therefore be avoided. However, Chevallereau and Aoustin (90) stated that in most cases high joint torques are needed in order to achieve this specific requirement, especially when walking at high speeds. This seems logical since one deliberately has to slow down the dynamics in order to avoid the impact. If necessary, it is however not a problem to choose a non-zero touchdown velocity in this strategy.

3.4.2 Calculation of the Joint Trajectories

Out of the desired trajectory for the hip (x_{hip}, $\dot{x}_{hip}$, $\ddot{x}_{hip}$, y_{hip}, $\dot{y}_{hip}$ and $\ddot{y}_{hip}$) and the motion of both feet (x_{foot}, $\dot{x}_{foot}$, $\ddot{x}_{foot}$, y_{foot}, $\dot{y}_{foot}$ and $\ddot{y}_{foot}$) it is

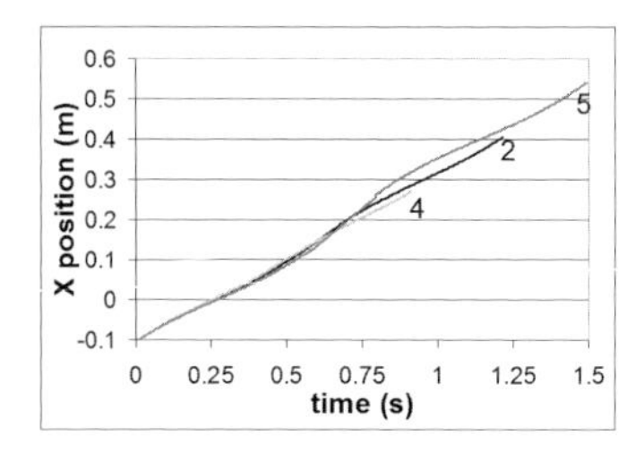

Fig. 3.20 x-position of hip.

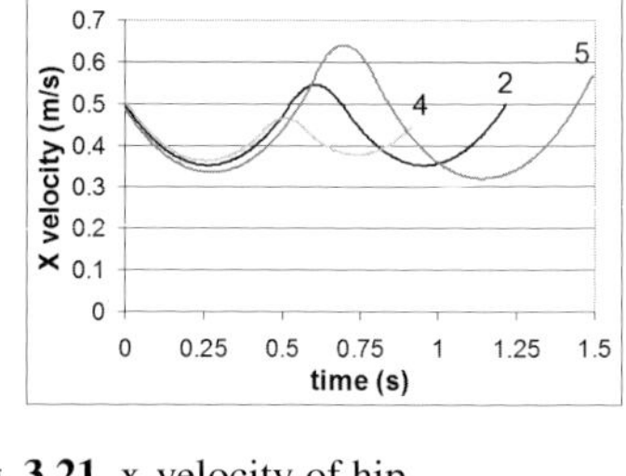

Fig. 3.21 x-velocity of hip.

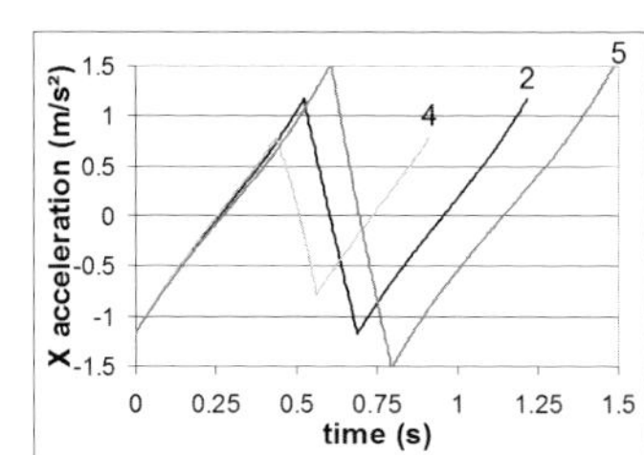

Fig. 3.22 x-acceleration of hip.

Fig. 3.23 ZMP

Single, double and single support phase with parameters:
4: $\lambda_1 = 0.3m\ \lambda_2 = 0.2m$; $v_1 = 0.4m/s\ v_2 = 0.4m/s$
2: $\lambda_1 = 0.3m\ \lambda_2 = 0.3m$; $v_1 = 0.4m/s\ v_2 = 0.4m/s$
5: $\lambda_1 = 0.3m\ \lambda_2 = 0.4m$; $v_1 = 0.4m/s\ v_2 = 0.4m/s$

straightforward to calculate the desired joint angles, velocities and accelerations using inverse kinematics.

x_{diff} and y_{diff} are respectively horizontal and vertical distances between hip and foot. D is the distance between those two points.

$$x_{diff} = x_{hip} - x_{foot} \tag{3.33}$$

$$y_{diff} = y_{hip} - y_{foot} \tag{3.34}$$

$$D = \sqrt{x_{diff}^2 + y_{diff}^2} \tag{3.35}$$

The angle γ between horizontal and line between hip and ankle is:

$$\gamma = atan\frac{y_{diff}}{x_{diff}} \qquad x_{diff} > 0 \tag{3.36}$$

$$\gamma = \pi/2 \qquad x_{diff} = 0 \tag{3.37}$$

$$\gamma = atan\frac{y_{diff}}{x_{diff}} + \pi \qquad x_{diff} < 0 \tag{3.38}$$

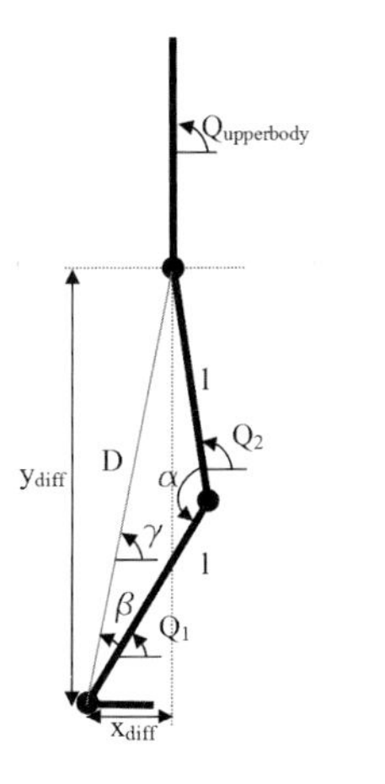

Fig. 3.24 Used parameters for calculating the inverse kinematics.

By using the "law of cosines" it is possible to find α and the "law of sines" for β:

$$\alpha = acos(\frac{-D^2+2l^2}{2l^2}) \tag{3.39}$$

$$\beta = asin(\frac{lsin(\alpha)}{D}) \tag{3.40}$$

Using α en β it is easy to calculate the absolute angles Q_1 and Q_2. The absolute angles Q are measured with respect to the horizontal axis.

$$Q_1 = \gamma - \beta \qquad \text{absolute angle lower leg} \tag{3.41}$$

$$Q_2 = Q_1 + (\pi - \alpha) \qquad \text{absolute angle upper leg} \tag{3.42}$$

With those absolute angles forward kinematics can be used to find x_{diff} and y_{diff}.

$$x_{hip} - x_{foot} = lcos(Q_1) + lcos(Q_2) \tag{3.43}$$

$$y_{hip} - y_{foot} = lsin(Q_1) + lsin(Q_2) \tag{3.44}$$

When the derivative is taken, it is possible to find $\dot{Q}_1$ and $\dot{Q}_2$ by taking the inverse.

$$\dot{x}_{hip} - \dot{x}_{foot} = -lsin(Q_1)\dot{Q}_1 - lsin(Q_2)\dot{Q}_2 \tag{3.45}$$

$$\dot{y}_{hip} - \dot{y}_{foot} = lcos(Q_1)\dot{Q}_1 + lcos(Q_2)\dot{Q}_2 \tag{3.46}$$

$$\dot{Q}_1 = \frac{(\dot{x}_{hip} - \dot{x}_{foot})cos(Q_2) + (\dot{y}_{hip} - \dot{y}_{foot})sin(Q_2)}{lsin(Q_2 - Q_1)} \tag{3.47}$$

$$\dot{Q}_2 = -\frac{(\dot{x}_{hip} - \dot{x}_{foot})cos(Q_1) + (\dot{y}_{hip} - \dot{y}_{foot})sin(Q_1)}{lsin(Q_2 - Q_1)} \tag{3.48}$$

When the derivative is taken a second time, $\ddot{Q}_1$ and $\ddot{Q}_2$ can be found in an analogous way.

$$\ddot{x}_{hip} - \ddot{x}_{foot} = - lcos(Q_1)\dot{Q}_1^2 - lsin(Q_1)\ddot{Q}_1 \quad (3.49)$$
$$- lcos(Q_2)\dot{Q}_2^2 - lsin(Q_2)\ddot{Q}_2 \quad (3.50)$$
$$\ddot{y}_{hip} - \ddot{y}_{foot} = - lsin(Q_1)\dot{Q}_1^2 + lcos(Q_1)\ddot{Q}_1 \quad (3.51)$$
$$- lsin(Q_2)\dot{Q}_2^2 + lcos(Q_2)\ddot{Q}_2 \quad (3.52)$$

$$A = (\ddot{x}_{hip} - \ddot{x}_{foot}) + lcos(Q_1)\dot{Q}_1^2 + lcos(Q_2)\dot{Q}_2^2 \quad (3.53)$$
$$B = (\ddot{y}_{hip} - \ddot{y}_{foot}) + lsin(Q_1)\dot{Q}_1^2 + lsin(Q_2)\dot{Q}_2^2 \quad (3.54)$$
$$D = -l^2 sin(Q_1)cos(Q_2) + l^2 sin(Q_2)cos(Q_1) \quad (3.55)$$

If $Q_1 \neq Q_2$, then $D \neq 0$:

$$\ddot{Q}_1 = \frac{Alcos(Q_2) + Blsin(Q_2)}{D} \quad (3.56)$$
$$\ddot{Q}_2 = \frac{-Blsin(Q_1) - Alcos(Q_1)}{D} \quad (3.57)$$

The trajectories for the upper body and feet are:

$$Q_{upperbody} = \pi/2 \quad (3.58)$$
$$Q_{foot} = 0 \quad (3.59)$$

So the upper body is always kept vertical and the foot is kept parallel with the ground. In practice, it is generally desired that the posture of the trunk is kept nearly stationary, in an upright position. This would allow the robot to carry objects in a stable manner, or to get scenery information with vision cameras in the head (321).

3.4.3 Influence of the Complete Multibody Model

The trajectory generator calculates joint trajectories with the ZMP located in the ankle, but with the assumption that all the mass of the robot is located in the hip. This is off course not the case because the robot has distributed masses. To have an idea of these effects the proposed trajectories were tested in simulation with a multibody model of the biped Lucy, hereby supposing the trajectories are tracked perfectly. In figure 3.25 one can see the evolution of the desired and multibody ZMP, x-position of both feet and hip with the following sequence of steps:

- Stand-up: $\lambda = 0.0m$; $v = 0.0m/s$ (start position)
- $\lambda = 0.20m$; $v = 0.3m/s$
- $\lambda = 0.30m$; $v = 0.4m/s$

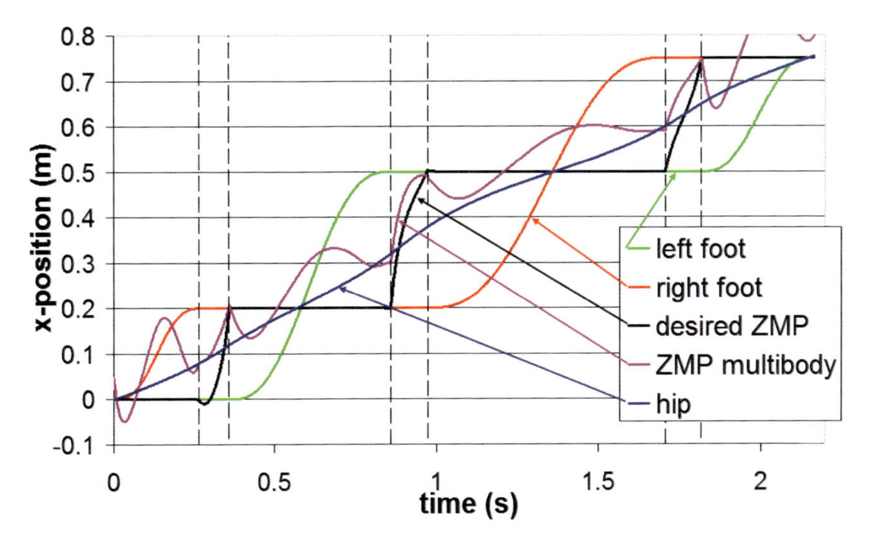

Fig. 3.25 desired and multibody ZMP, x-position of both feet and hip.

- $\lambda = 0.25m$; $v = 0.3m/s$
- $\lambda = 0.00m$; $v = 0.3m/s$
- $\lambda = 0.00m$; $v = 0.00m/s$ (end position)

The vertical lines are the phase transitions when switching between single support and double support. The ZMP deviation between desired and real ZMP is large. This is not a problem as long as the ZMP multibody stays within the support polygon. The higher the walking speed, the higher the dynamics of for example the swinging leg will influence the ZMP multibody. If the mass of the upper body of Lucy was higher compared to the mass of the leg, then the deviation would be smaller. So this method is suitable for low walking speeds. To be able to use this method also at higher walking speeds a stabilizer is needed. A stabilizer is a sensor feed back module which adapts the motion of the robot based on a difference between desired and actual measured ZMP to cancel this difference. In section 4.3.3 an overview of possible stabilizers is given. Beside the compensation for the deviation of ZMP due to the used simplification of the robot model also destabilizing effects due to unexpected disturbances or tracking errors can be handled by a stabilizer. To draw a line up to which speed this method can be used is not unambiguous to say. It depends on the size of the feet and the stability region, the tracking performances and the expected disturbances that will act on the robot.

3.5 Implementation of a Preview Controller of the Zero Moment Point to Generate Trajectories

In the previous section a trajectory generator was built without taking into account the distributed masses of a complete multibody model. For moderate walking speeds without stabilizer this is unnecessary, but for higher walking speeds the disturbances due to for example the swing leg threatens the dynamic stability of the robot. In the trajectory generator proposed in this section the complete multibody model of the robot will be used too, so this will make the strategy suitable for higher walking speeds. The goal is to follow a predefined trajectory of the ZMP using a method proposed by Kajita (212). This is not so straightforward as calculating the ZMP out of the joint trajectories. The controller used here is based on the linear quadratic integral (LQI) technique to derive an optimal controller having state feedback plus integral and preview actions, called preview controller.

3.5.1 Introduction to Preview Controller

When the control signal is solely calculated based on the error signal, it is a feedback problem. When the controller also utilizes a prediction calculated by a model and the desired trajectory it is called a feedback/feedforward control. Those two approaches are very popular. Not so commonly used is a preview control where also future information is available and used.

The preview control method can either be used when the future information of the reference signals or the disturbances are available to greatly improve the performance of transient responses (392). The method is used in a number of engineering problems. The tracking control of a suspension durability test simulator is presented in (280), (281). The objective was to control a hydraulically actuated durability simulator, so that the vehicle responses, previously measured on the test track, can be reproduced in the laboratory. Katayama et al. (229) used this method to make an optimal control tracker for a heat exchanger in the presence of load changes. In (95) the objective of the control system for the rolling stand of the tandem cold mill in the steel-making works is to minimize thickness error of the exit strip and tension variation between stands simultaneously. The entry strip thickness to the stand and the roll gap variation are considered as previewable disturbances, since they can be measured and estimated.

In robotics the preview controller was used by Kajita et al. to build a trajectory generator to track a predefined ZMP trajectory (212). This walking pattern generator is still very successfully used in the humanoid robot HRP-2. Because a preview controller for trajectory generation needs future trajectory information till twice T_{prev} ahead as will be explained in section 3.5.6, the trajectory cannot be changed in this period from the original desired motion. However this is sometimes necessary in order to realize a quick response to a change of input commands. In (298) a walking pattern generation is implemented on HRP-2 that can update the pattern at a short cycle such as $40ms$. Kanzaki et al. (228) implemented the preview control method in the humanoid robot HOAP-2 to generate bracing behaviour against external impact. To reduce the ZMP error at impact the future information of external forces was used. Wieber (442) focused on the compensation of strong perturbations of the dynamics of the robot and proposed a new Linear Model Predictive Control scheme which is an improvement of the original ZMP preview control scheme. The control performance of the mine detection hexapod robot COMET-III, powered by hydraulics, is improved by a preview sliding mode controller. The preview control prevents the flattery delay caused by the strong nonlinear characteristic of the oil pressure system, by using a future target value (303). Verrelst et al. (379) used this preview control scheme as base to step dynamically over large obstacles. More information about this will be given in section 3.7.

This section is organized as follows. In section 3.5.2 the cart-table model is described, this is a convenient way to represent the dynamics of a biped and the corresponding ZMP equations are given. The idea to use this cart-table model with a preview controller to track a desired ZMP trajectory was introduced by Kajita (212) and is described in section 3.5.3. In section 3.5.4 the necessary equations are given to reproduce the trajectory generator. Because not all the necessary equations are given in the work of Kajita (212), they are taken from Katayama et al. (230). Section 3.5.5.1 discusses the methodology used and the influence of some parameters. Section 3.5.6 deals with the deviation of the ZMP of the complete multibody model and how this error can be reduced by using the preview controller for a second time with a discussion of the results.

3.5.2 Cart-Table Model

Kajita et al. (212) proposed to approximate a humanoid robot by a cart-table model which is shown in figure 3.26. A running cart of mass m is placed on a pedestal table with negligible mass. If the COG of a cart at rest is outside of the foot area of the table, the table will fall. However, the ZMP can be positioned inside the stability region by choosing a proper acceleration of the cart. This will keep it upright for a while. Since the moment around the ZMP must be zero, one has:

$$\tau_{ZMP} = mg(x - p_{ZMP}) - m\ddot{x}z_c = 0 \tag{3.60}$$

where z_c is the height of the COG, g is the gravity acceleration. So the COG is supposed to move horizontally on a constant height, while the COG of the previous method moved up and down. Out of equation (3.60) the position of the ZMP can be derived:

$$p_{ZMP} = x - \frac{z_c}{g}\ddot{x} \tag{3.61}$$

These are called ZMP equations in (212).

3.5.3 Trajectory Generation as Servo Tracking Control of ZMP

The goal of the trajectory generator is to design an optimal controller for a biped to track a desired ZMP path p_{ref}. The control scheme is given in figure 3.27. When an ordinary servo control is used the hip will move like in figure 3.28. The hip moves of course too late and the ZMP is not able to track the desired ZMP trajectory. The reason is that the cart has to move before the ZMP changes and that is why also future information is needed as shown in figure 3.29. To employ this a preview

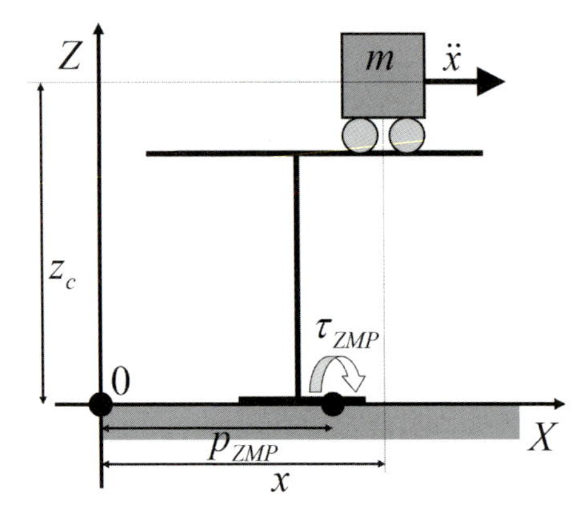

Fig. 3.26 Cart-table model.

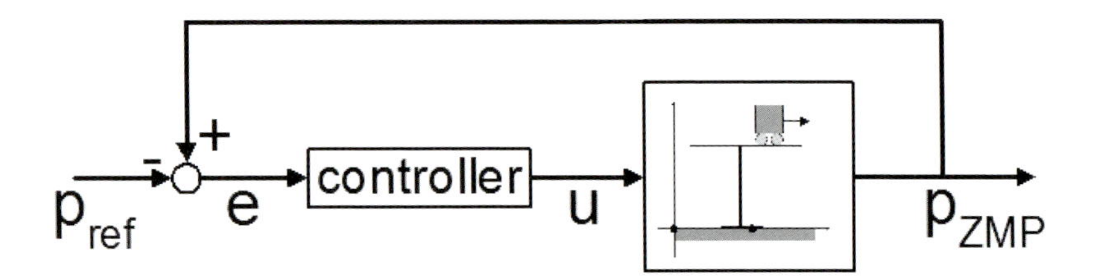

Fig. 3.27 Control scheme to track desired ZMP.

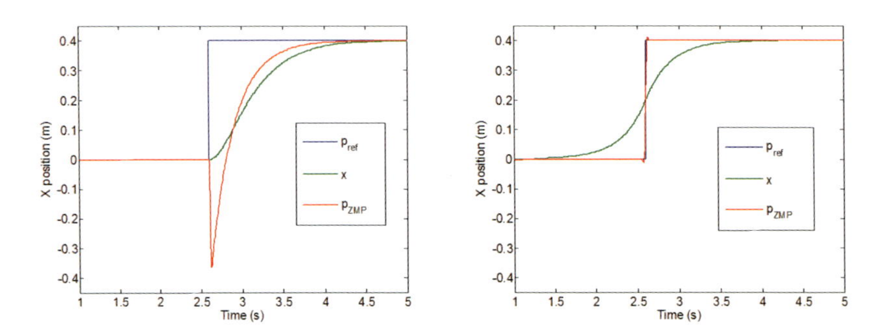

Fig. 3.28 Ordinary servo control.

Fig. 3.29 Preview control.

controller is implemented as proposed by Kajita (212). The concept and naming originates from Sheridan (369) and the LQI optimal controller technique used to solve the problem was developed by Tomizuka and Rosenthal (402) and extended for MIMO systems by Katayama et al. (230).

3.5.4 Trajectory Generation Using Preview Control

If u_x is defined as the jerk (derivative of acceleration) of the cart and used as input value in the dynamical system describing the cart-table model, the following set of equations can be written:

$$\frac{d}{dt}\begin{pmatrix} x \\ \dot{x} \\ \ddot{x} \end{pmatrix} = \begin{pmatrix} 0 & 1 & 0 \\ 0 & 0 & 1 \\ 0 & 0 & 0 \end{pmatrix}\begin{pmatrix} x \\ \dot{x} \\ \ddot{x} \end{pmatrix} + \begin{pmatrix} 0 \\ 0 \\ 1 \end{pmatrix} u_x \tag{3.62a}$$

$$p_{ZMP} = \begin{pmatrix} 1 & 0 & -z_c/g \end{pmatrix}\begin{pmatrix} x \\ \dot{x} \\ \ddot{x} \end{pmatrix} \tag{3.62b}$$

The discretized version of (3.62) with sampling time T is:

$$\begin{pmatrix} x((k+1)T) \\ \dot{x}((k+1)T) \\ \ddot{x}((k+1)T) \end{pmatrix} = \begin{pmatrix} 1 & T & T^2/2 \\ 0 & 1 & T \\ 0 & 0 & 1 \end{pmatrix}\begin{pmatrix} x(kT) \\ \dot{x}(kT) \\ \ddot{x}(kT) \end{pmatrix} + \begin{pmatrix} T^3/6 \\ T^2/2 \\ T \end{pmatrix} u_x(kT) \tag{3.63a}$$

$$p_{ZMP}(kT) = \begin{pmatrix} 1 & 0 & -z_c/g \end{pmatrix}\begin{pmatrix} x(kT) \\ \dot{x}(kT) \\ \ddot{x}(kT) \end{pmatrix} \tag{3.63b}$$

or

$$\mathbf{x}(k+1) = A\mathbf{x}(k) + Bu(k) \tag{3.64a}$$

$$p(k) = C\mathbf{x}(k) \tag{3.64b}$$

with

$$\mathbf{x}(k) = \begin{pmatrix} x(kT) & \dot{x}(kT) & \ddot{x}(kT) \end{pmatrix}^T \quad \text{state vector} \tag{3.65}$$

$$u(k) = u_x(kT) \quad \text{control vector} \tag{3.66}$$

$$p(k) = p_{ZMP}(kT) \quad \text{output vector (to be controlled)} \tag{3.67}$$

and

$$A = \begin{pmatrix} 1 & T & T^2/2 \\ 0 & 1 & T \\ 0 & 0 & 1 \end{pmatrix}$$
$$B = \begin{pmatrix} T^3/6 \\ T^2/2 \\ T \end{pmatrix} \tag{3.68}$$
$$C = \begin{pmatrix} 1 & 0 & -z_c/g \end{pmatrix}$$

$p_{ref}(k)$ is taken as the reference vector of the desired ZMP. This vector is assumed to be an arbitrary bounded signal convergent to a constant vector ($\lim_{k\to\infty}(p_{ref}(k)) = \bar{p}_{ref}$)

and is previewable in the sense that at each instant k, N_L step future values $p_{ref}(k+1)\dots p_{ref}(k+N_L)$ are available for control purpose. The values beyond time $k+N_L$ are hypothetically approximated by the values at $k+N_L$. The product $T*N_L$ is called the preview period T_{prev}. The incremental state vector is $\Delta x(k) = \mathbf{x}(k) - \mathbf{x}(k-1)$, the tracking error $e(k) = p(k) - p_{ref}(k)$, the incremental control vector $\Delta u(k) = u(k) - u(k-1)$ and the incremental demand $\Delta p_{ref}(k) = p_{ref}(k) - p_{ref}(k-1)$. The optimal controller is the one who makes the performance index

$$J = \sum_{i=k}^{\infty}[e^T(i)Q_e e(i) + \Delta x^T(i)Q_x\Delta x(i) + \Delta u^T(i)R\Delta u(i)] \tag{3.69}$$

minimal at each time k. Q_e and R are symmetric positive definite matrices and Q_x is a symmetric non-negative definite matrix. Qe, Qx and R penalize the loss due to tracking error, incremental state and incremental control vector respectively. The physical interpretation of J is to achieve the asymptotic regulation without excessive rate of change in the state and control vectors (230).

From (3.64a), the incremental state is described by

$$\Delta x(k+1) = A\Delta x(k) + B\Delta u(k) \tag{3.70}$$

and the ZMP tracking error from (3.64b) and (3.70):

$$e(k+1) = e(k) + CA\Delta x(k) + CB\Delta u(k) - \Delta p_{ref}(k+1) \tag{3.71}$$

Combining (3.70) and (3.71) gives:

$$\begin{pmatrix} e(k+1) \\ \Delta x(k+1) \end{pmatrix} = \begin{pmatrix} 1 & CA \\ \mathbf{0} & A \end{pmatrix} \begin{pmatrix} e(k) \\ \Delta x(k) \end{pmatrix} + \begin{pmatrix} CB \\ B \end{pmatrix} \Delta u(k) - \begin{pmatrix} 1 \\ \mathbf{0} \end{pmatrix} \Delta p_{ref}(k+1) \tag{3.72}$$

where $\mathbf{0}$ is a 3x1 zero matrix. Or:

$$\mathbf{X}(k+1) = \tilde{A}\mathbf{X}(k) + \tilde{B}\Delta u(k) - \tilde{I}\Delta p_{ref}(k+1) \tag{3.73}$$

With

$$\begin{aligned} \mathbf{X}(k) &= \begin{pmatrix} e(k) \\ \Delta x(k) \end{pmatrix} \\ \tilde{A} &= \begin{pmatrix} 1 & CA \\ \mathbf{0} & A \end{pmatrix} \\ \tilde{B} &= \begin{pmatrix} CB \\ B \end{pmatrix} \\ \tilde{I} &= \begin{pmatrix} 1 \\ \mathbf{0} \end{pmatrix} \end{aligned} \tag{3.74}$$

In (230) the optimal controller $u^{\circ}(k)$ is given by:

$$u^{\circ}(k) = -G_I \sum_{i=0}^{k} e(i) - G_x \mathbf{x}(k) - \sum_{l=1}^{N_L} G_d(l) p_{ref}(k+l) \tag{3.75}$$

where G_I, G_x and $G_d(l)$ are the gains calculated from the weights Qe, Qx and R and the system parameters of equation (3.63). So the preview control is made of three terms, the integral action on the tracking error, the state feedback and the preview action using the future reference.

The gains are:

$$[G_I G_x] = [R + \tilde{B}^T \tilde{P} \tilde{B}]^{-1} \tilde{B}^T \tilde{P} \tilde{A} \tag{3.76a}$$

$$G_d(1) = -G_I \tag{3.76b}$$

$$G_d(l) = [R + \tilde{B}^T \tilde{P} \tilde{B}]^{-1} \tilde{B}^T \tilde{X}(l-1), l = 2, \ldots, N_L \tag{3.76c}$$

Where matrix $\tilde{P}$ is the non-negative definite solution of the algebraic Riccati equation:

$$\tilde{P} = \tilde{A}^T \tilde{P} \tilde{A} - \tilde{A}^T \tilde{P} \tilde{B} [R + \tilde{B}^T \tilde{P} \tilde{B}]^{-1} \tilde{B}^T \tilde{P} \tilde{A} + \tilde{Q} \tag{3.77}$$

Furthermore, the matrices $\tilde{X}(l)$ are given by:

$$\tilde{X}(1) = -\tilde{A_c}^T \tilde{P} \tilde{I} \tag{3.78}$$

$$\tilde{X}(l) = -\tilde{A_c}^T \tilde{X}(l-1), l = 2, \ldots, N_L \tag{3.79}$$

where $\tilde{A_c}$ is the closed-loop matrix defined by

$$\tilde{A_c} = \tilde{A} - \tilde{B} [R + \tilde{B}^T \tilde{P} \tilde{B}]^{-1} \tilde{B}^T \tilde{P} \tilde{A} \tag{3.80}$$

These are the necessary equations to be able to track the desired ZMP trajectory. In the next section it will be explained how they need to be used practically and what is the influence of the different parameters.

3.5.5 Methodology and Influence of Parameters

3.5.5.1 Methodology

First of all the gains for the optimal controller have to be calculated by using equations (3.76a)-(3.76c). For these gains the solution of the Riccati equation $\tilde{P}$ is needed. The Riccati equation is not amenable to elementary techniques in solving differential equations. The matlab function "dlqry" (Linear quadratic regulator design with output weighting for discrete-time systems) is used to calculate the steady-state solution to the associated discrete matrix Riccati equation.

Figure 3.30 shows the gains for the preview action $G_d(l)$ with $T = 0.005s$, $z_c = 0.6m$, $Q_e = 1$, $Q_x = 0$ and $R = 1.10^{-6}$. After $t = 2s$ the gains become very small so the controller doesn't need the information of the far future. $Q_x = 0$ because otherwise it cannot be solved by the matlab function. The gains are independent of the preview period T_{prev}.

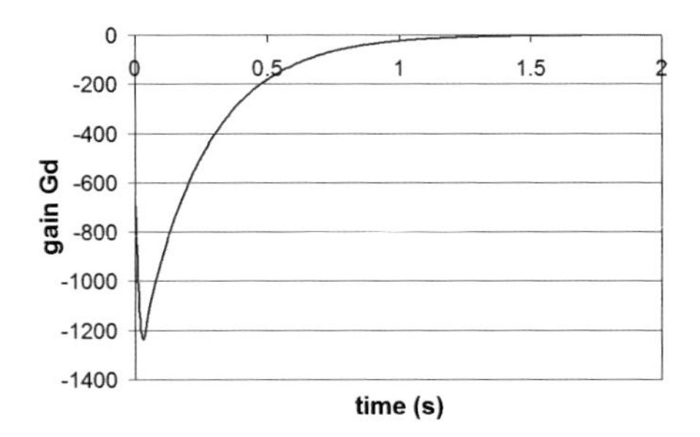

Fig. 3.30 Gains $G_d(l)$ of preview controller.

The imposed trajectory of the ZMP is the one that has to be controlled and is calculated out of the objective locomotion parameters of the desired motion. The objective locomotion parameters used are horizontal velocity v, step length λ (measured between ankle joints) and intermediate foot lift κ.

When a 80%-20% time distribution is taken between single and double support, the phase durations are:

$$T_{SS} = \frac{2(\lambda_{n-1} + \lambda_n)}{5v} \tag{3.81a}$$

$$T_{DS} = \frac{T_{SS}}{4} \tag{3.81b}$$

with λ_{n-1} the length between rear and front ankle joint at the start of the swing motion and λ_n at impact. The trajectory of the ZMP $p_{ref}(k)$ can be freely chosen as long as the ZMP stays in the stability region. In section 3.5.5.4 three different trajectories will be discussed.

At the same time, sixth (for the vertical Z direction) and fifth (for the horizontal X direction) order polynomial function for the trajectories of the swing foot are established, which connect the initial, intermediate and final boundary values for the swing foot motion. The additional order for the Z direction is to include also the foot lift. This intermediate condition at $t = t_{end_{SS}}/2$ is used to lift the foot, with height κ, whenever an obstacle has to be avoided during the swing phase. At the start of the single support phase the boundary conditions are: $x_{foot}(0) = \sum_{l=0}^{n-1} \lambda_l$ and $\dot{x}_{foot}(0) = \ddot{x}_{foot}(0) = z_{foot}(0) = \dot{z}_{foot}(0) = \ddot{z}_{foot}(0) = 0$. At the end of the single support the boundary conditions are: $x_{foot}(T_{SS}) = \sum_{l=0}^{n} \lambda_l$ and $z_{foot}(T_{SS}) = 0$, the velocity and acceleration can be chosen freely, here both are taken zero. So the coordinate system is fixed and does not move with the stance foot as is the case in the previous method.

The gains (G_I, G_x, G_d) and the trajectory of the ZMP (p_{ref}) are used to calculate control vector $u^{\circ}(k)$ (equation (3.75)) and afterwards equations (3.64a) and (3.64b) are taken to obtain the state vector $\mathbf{x}(k)$ and the output vector $p(k)$. The state vector $\mathbf{x}(k)$ describes the x-motion of the COG, the height of the COG of the robot is taken constant and equals z_c. The motion of the hip is the same as the COG with a constant vertical deviation of $0.25m$ so the hip moves at a height of $0.85m$. Out of the desired trajectory for the hip and the motion of both feet it is straightforward to calculate the desired joint angles, velocities and accelerations are shown in 3.4.2.

For the step sequences shown in figures 3.31 - 3.37, 3.40, 3.45 and 3.46, the same objective locomotion parameters as in section 3.4.3 have been taken:

- Stand-up: $\lambda = 0.0m$; $v = 0.0m/s$ (start position)
- $\lambda = 0.20m$; $v = 0.3m/s$
- $\lambda = 0.30m$; $v = 0.4m/s$
- $\lambda = 0.25m$; $v = 0.3m/s$
- $\lambda = 0.00m$; $v = 0.3m/s$
- $\lambda = 0.00m$; $v = 0.00m/s$ (end position)

So the robot starts at rest, takes 4 steps and stops; to show the robot is able to walk at different walking speeds and step lengths.

3.5.5.2 Influence of Preview Period T_{prev}

Figures 3.31-3.34 give the trajectory of the hip and ZMP of the cart-table model and trajectory of both feet for 4 different preview periods $T_{prev} = 0.5s$, $T_{prev} = 1.0s$, $T_{prev} = 1.5s$, $T_{prev} = 2.0s$. For $T_{prev} = 2.0s$ and $T_{prev} = 1.5s$ the imposed ZMP trajectory is followed precisely. When the previewing period is reduced, the trajectory tracking of the ZMP becomes worse. Taking a larger preview window requires more calculations and should be avoided if unnecessary.

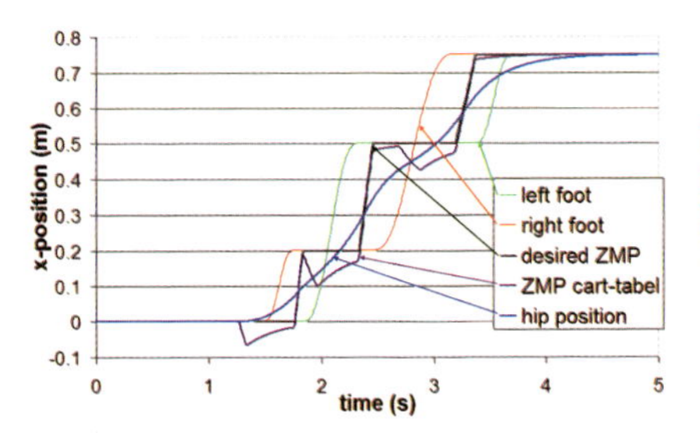

Fig. 3.31 $T_{prev} = 0.5s$.

Fig. 3.32 $T_{prev} = 1.0s$.

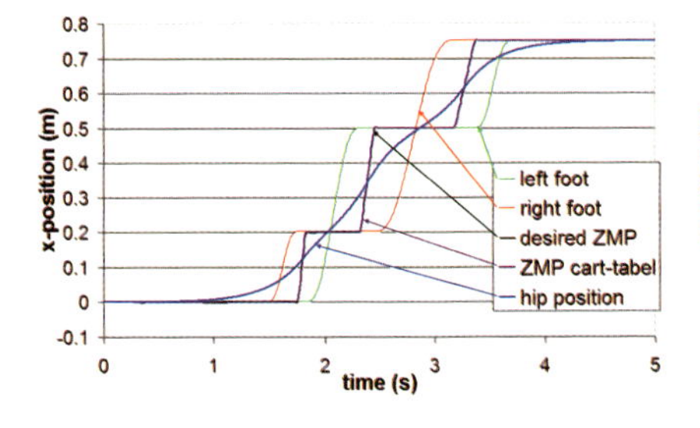

Fig. 3.33 $T_{prev} = 1.5s$.

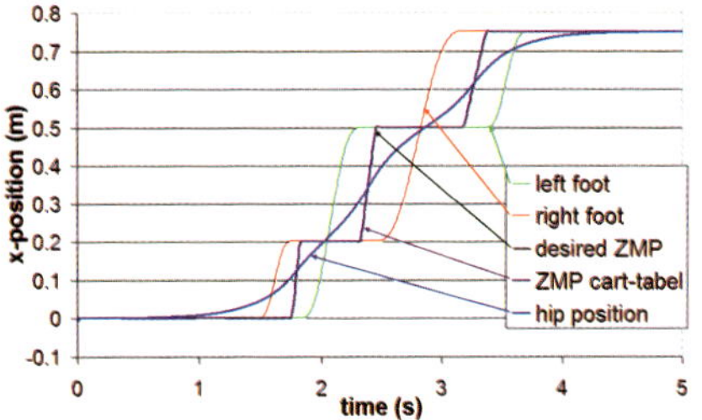

Fig. 3.34 $T_{prev} = 2.0s$.

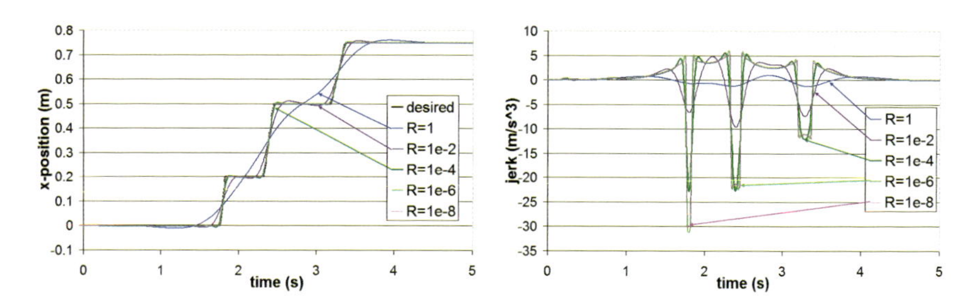

Fig. 3.35 Influence of R on ZMP position ($Q_e = 1$).

Fig. 3.36 Influence of R on jerk u_x ($Q_e = 1$).

3.5.5.3 Influence of R and Q_e

The parameter R penalizes the loss due the to incremental control vector. Figures 3.35-3.36 show 5 different values for R on the ZMP position and the jerk respectively. Taking $R = 0$ is impossible because this matrix has to be positive definite. The higher R, the lower the jerk u_x, but the tracking of the ZMP becomes worse. The parameter Q_e penalizes the tracking error. The higher Q_e, the better the tracking as can be seen in figure 3.37, but the jerk u_x will also increase.

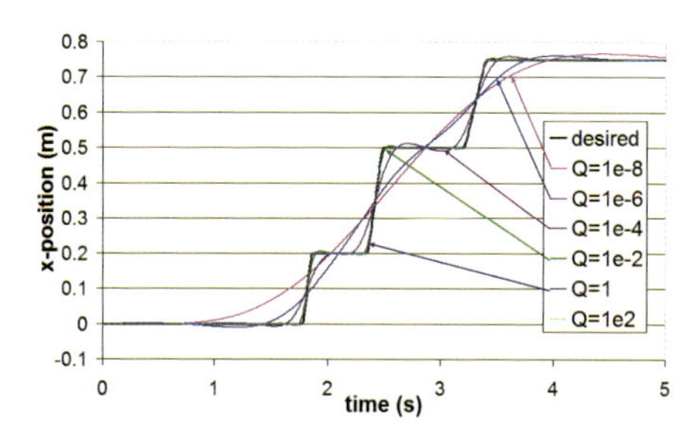

Fig. 3.37 Influence of Q_e on ZMP position ($R = 1.10^{-6}$).

In (213) it is shown that the third component of Q_x ($Q_x[3]$) is enlarged to penalize the horizontal acceleration of the COG because this is more suitable for walking on low friction floors. This of course reduces the ZMP tracking performance but the ZMP is still within the supported foot area. In the same work a slip concerned ZMP was proposed to treat the slip and non-slip condition.

3.5.5.4 Influence of Desired ZMP Trajectory $\mathbf{p_{ref}}(k)$

The preview controller uses a desired trajectory as input so different trajectories can be used. Three different ZMP trajectories are proposed:

- “ZMP trajectory 1” ZMP is in the ankle point during single support phase and makes a step to the ankle point of the next single stance leg in the middle of the double support phase.
- ‘ZMP trajectory 2” ZMP is in the ankle point during single support phase and evolves linear from the rear to the front ankle point during double support phase.
- ‘ZMP trajectory 2” ZMP evolves both during the single and double support phase. In this example, during single support phase, the ZMP shifts from $5cm$ behind the ankle point to $5cm$ in front of the ankle point. The limits are that the ZMP moves from the heel to the tip of the foot during single support phase.

These desired ZMP trajectories are shown in figure 3.38. Figure 3.39 shows the velocity of the hip for those 3 different ZMP trajectories. “ZMP trajectory 1” has the highest velocity peaks, the accelerations of “ZMP trajectory 3” are the smallest. Consequently it is preferable to use “ZMP trajectory 3”, but then the ZMP comes closer the boundary of the support area. This is the reason why for this application “ZMP trajectory 2” is chosen. More research is needed to find an optimal ZMP trajectory. Also the time distribution between single and double support phase should be studied. Here an arbitrary value of 80%-20% is taken; maybe a more optimal distribution can be found, possibly dependant on speed and step length.

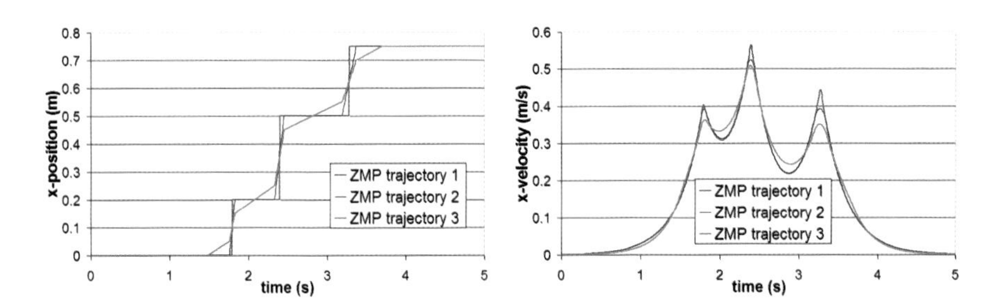

Fig. 3.38 3 different ZMP trajectories.

Fig. 3.39 Velocity for 3 different desired ZMP trajectories.

3.5.6 Complete Multibody Model

When the physical parameters of the complete multibody model of Lucy are used, the ZMP calculated with equation (3.82) deviates from the desired trajectory as can be seen in figure 3.40, stage 1.

$$p_{multibody} = \frac{\sum_{l=1}^{7}\left(m_l\left((\ddot{z}_l+g)x_l-\ddot{x}_l z_l\right)+I_l\ddot{\theta}_l\right)}{\sum_{l=1}^{n} m_l(\ddot{z}_l+g)} \tag{3.82}$$

This deviation is caused by the difference between the simple cart-table model and the complete multibody model consisting of 7 links. In order to solve this, as proposed by Kajita (212), the complete multibody calculated ZMP trajectory is re-feeded into the preview control by means of taking the error between the calculated

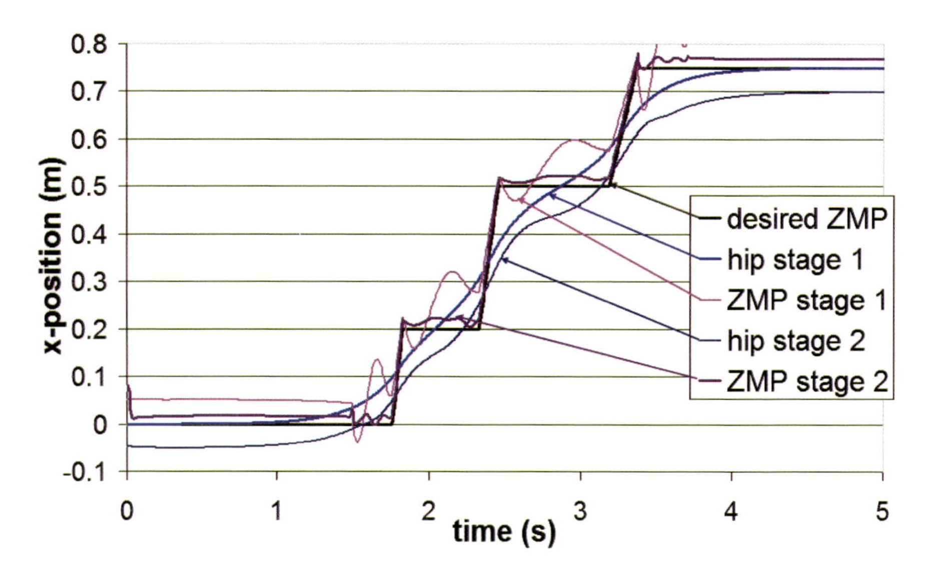

Fig. 3.40 Comparison of X position hip and ZMP (stage 1 after 1st preview controller, stage 2 after 2nd preview controller).

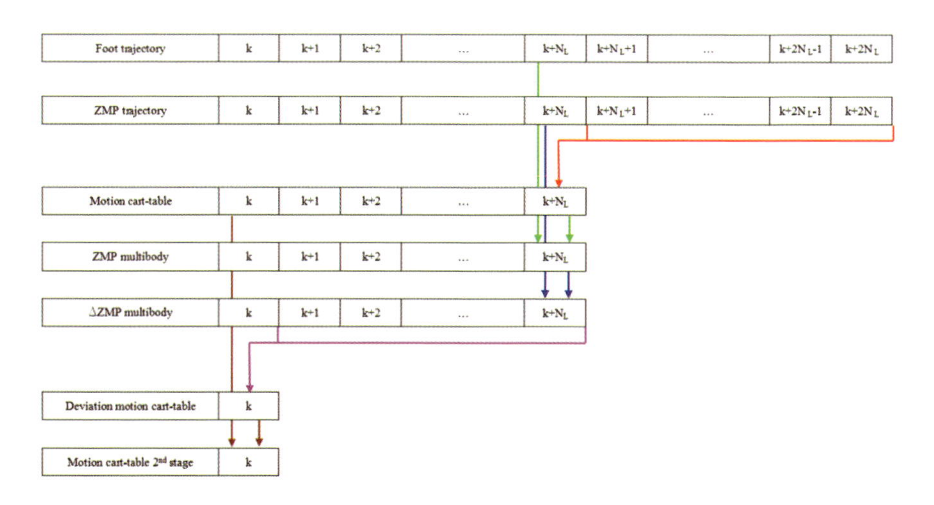

Fig. 3.41 Used buffers for preview controller.

multibody trajectory $p_{multibody}$ and the desired trajectory p_{ref}. This error Δp_{ref2} is again presented as input to a second stage of preview control with the same cart-table model, resulting in deviations of the horizontal motion of the COG $\Delta\mathbf{x}(k)$. So the X position, velocity and acceleration of the COG is adapted such that the ZMP stays closer to the desired trajectory.

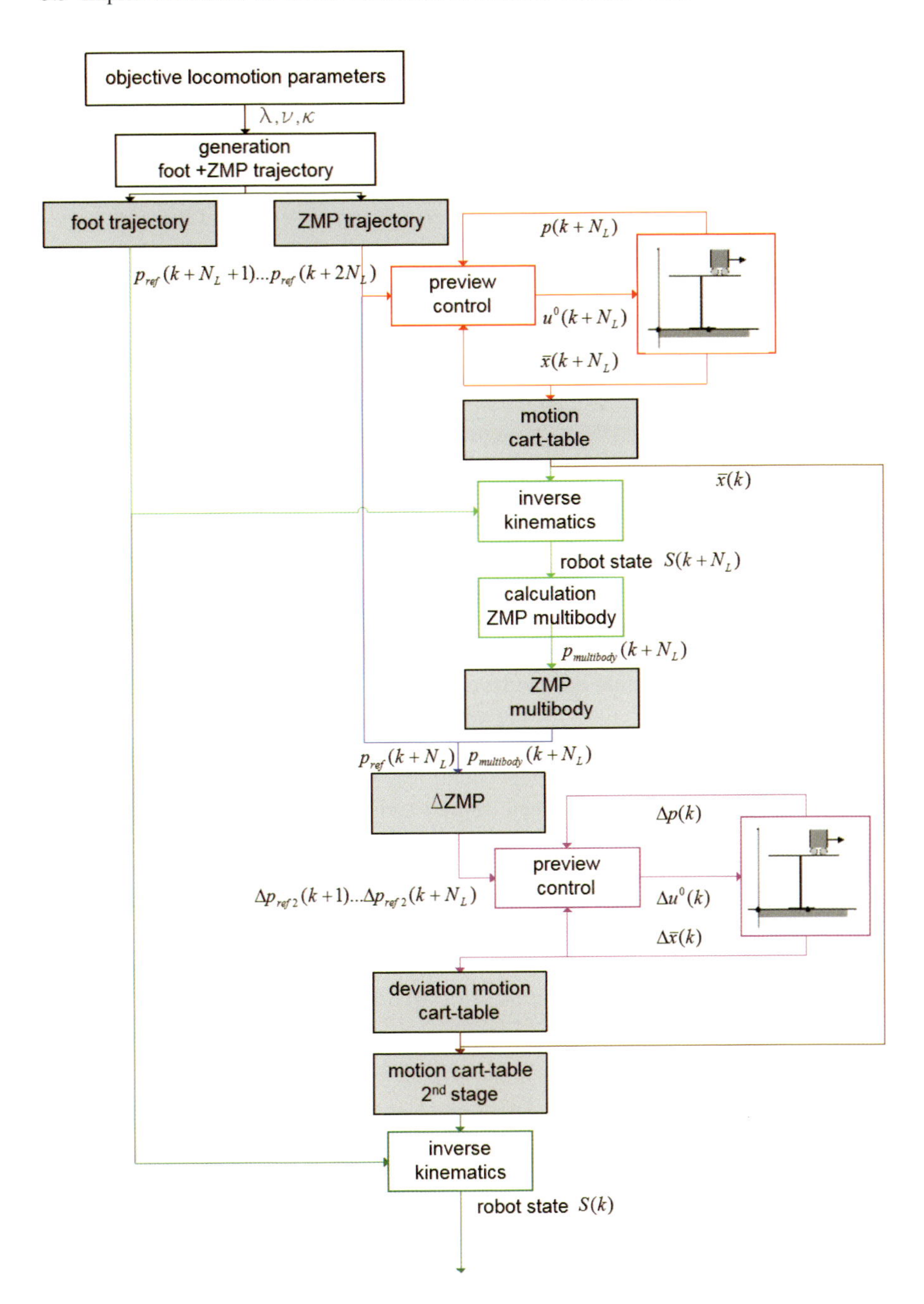

Fig. 3.42 Total scheme of preview controller.

The complete control scheme of the trajectory generator based on preview control of the ZMP can be summarized as shown in figure 3.42. Figure 3.41 depicts the different buffers shown in the grey boxes of figure 3.42. At the sample k, out of the desired objective locomotion parameters, the desired ZMP and foot trajectory at $k+2N_L$ can be constructed. Equation (3.75) requires $p_{ref}(k+N_L+1)...p_{ref}(k+2N_L)$ out of the ZMP trajectory buffer and also uses $p(k+N_L)$ and $\mathbf{x}(k+N_L)$. Using equation (3.64a) it is possible to find the motion of the cart-table. The use of these equations are shown by red lines in both figures. From the motion of the cart-table model, which represents the motion of the COG, and the foot trajectory it is possible to find the robot state with the equations of section 3.4.2. Using equation (3.82) it is possible to calculate $p_{multibody}(k+N_L)$, shown in green. The difference between this ZMP of the multibody and the desired ZMP trajectory ΔZMP ($\Delta p_{ref2}(k+1)...\Delta p_{ref2}(k+N_L)$) is calculated and is shown in blue. This is re-feeded in a seconde stage of the preview controller as shown in pink to obtain the deviation of the cart-table motion $\Delta\mathbf{x}(k)$. Together with the motion of the cart-table obtained after the first stage of the preview controller the final motion of the cart-table is obtained, as shown by brown lines. In a way as was done after the first stage, the final robot stage can be calculated with the foot trajectory, in dark green.

Figure 3.43 and 3.44 shows respectively the velocity and acceleration after the 1st and 2nd preview stage for Lucy. The large adaptation of the hip trajectory to compensate for the swinging leg can be noticed.

Using equation (3.82) the position of $p_{multibody}$ is calculated again out of the final robot stage and also depicted in figure 3.40, stage 2. One can see that after this second round through the preview controller the ZMP is tracked substantially better, however a difference between desired and real ZMP can still be observed. This is mainly due to the low inertia and weight of the body of the robot Lucy. The body of Lucy weighs $10.2kg$, while the total weight of the robot is $33kg$. This also has as consequence that the COG is located under the hip. Consequently the influence on the ZMP of for example the swinging leg is large. This can be solved by using for a third time the preview controller, but then the number of calculations becomes very big as the preview time T_{prev} triples. Another strategy is to increase

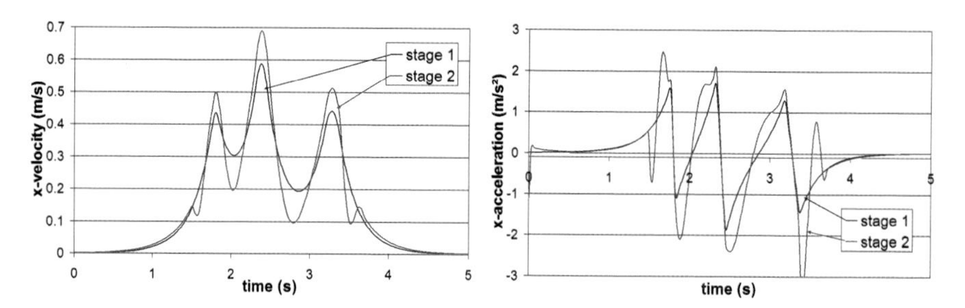

Fig. 3.43 X velocity after 1st and 2nd preview stage

Fig. 3.44 X acceleration after 1st and 2nd preview stage.

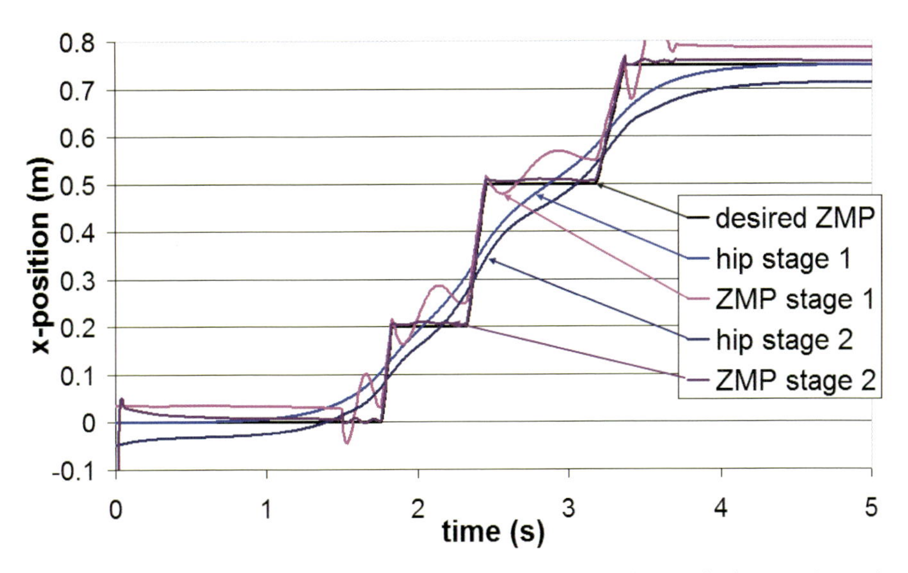

Fig. 3.45 Comparison of X position hip and ZMP with increased upper body mass (stage 1 after 1st preview controller, stage 2 after 2nd preview controller).

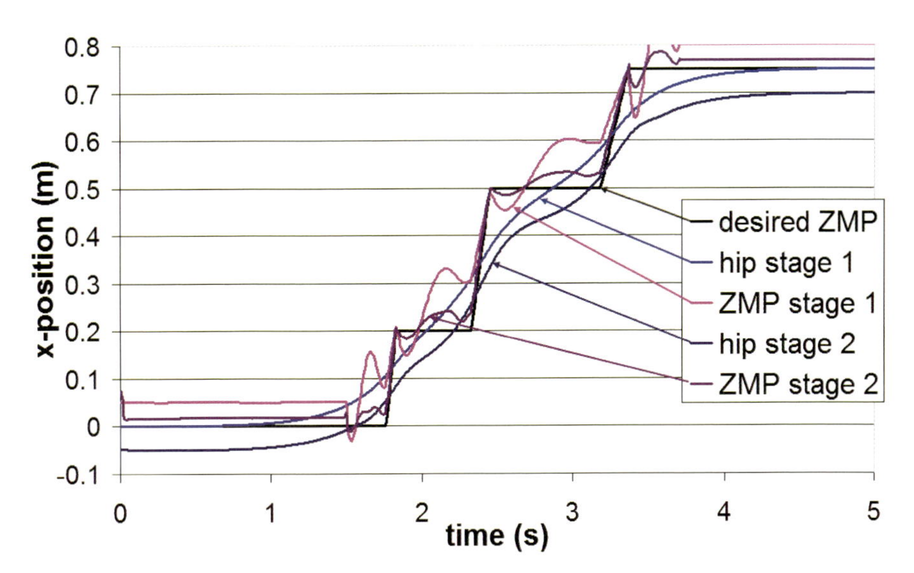

Fig. 3.46 Comparison of X position hip and ZMP with $z_c = z_h = 0.85m$ (stage 1 after 1st preview controller, stage 2 after 2nd preview controller).

the weight of the body of the robot. When the parameters of the biped used in ((183), table 1) are taken, the results are much better because the influence of for example the swinging leg on the ZMP is smaller compared to the one generated by the body (see figure 3.45). Here the mass distribution of the upper body to the complete robot is $43kg/81kg$. The specifications of HRP-2P ((164), table 1) are even better $38.9kg/54.1kg$.

It is also important to take for z_c the actual height of the COG instead of the hip height z_h. For the biped Lucy $z_c = 0.6m$, while $z_h = 0.85m$. If $z_c = 0.85m = z_h$ the deviation between real and desired ZMP course becomes worse as can be seen in 3.46.

3.6 Comparison between OLPIPM en Preview Control Method

The results of the implementation of the OLPIPM and preview control method in the real biped are provided in section 4.2.1, because they are strongly related to the performances of the joint trajectory tracking controller, which is discussed in the next chapter.

The ZMP multibody trajectory tracks better the desired ZMP trajectory when using the preview control method compared to the OLPIPM method. This is logical because the OLPIPM method only uses the simplified inverted pendulum model, while the preview control method also takes into account the complete multibody distributed masses. So this method is more suitable for walking at higher walking speeds. Moreover the desired ZMP trajectory can be constructed as desired because it is an input. More research should be performed to determine what the best desired ZMP trajectory is. Possible questions are for example should the ZMP move in the foot during the single support phase or is it better that the ZMP stays in the ankle point or in the middle of the foot? What is the best time distribution between single and double support phase? Is it dependent on the speed?

The cost of the better performance of the preview control method is that more calculations need to be performed. To calculate the trajectories for the different joints links out of the step sequence of section 3.4.3 the OLPIPM method needs about $0.015s$ while the preview control method requires $4.58s$. The time to calculate the gains is not included because they are constants for a biped. This is the average value for 1000 loops, calculated using Matlab on an $2GHz$ Intel Dual-Core processor. Another disadvantage of the preview control method is that future information ($3.2s$) about the desired ZMP trajectory is needed. In (298) a modified preview controller is proposed that can update the pattern at a short cycle such as $40ms$. If enough computation power is available the preview control method is advised. For robots with less computation power the OLPIPM method gives nice results and if a stabilizer is added probably also higher walking speeds can be achieved.

An important remark when evaluating the different trajectory generators is that the upper body mass of Lucy is too low. The influence of for example the swinging leg on the ZMP has to be canceled by manipulating the hip trajectory or the upper body trajectory. The trajectory generator developed by Vermeulen (421) exploits the natural upper body dynamics by manipulating the angular momentum equation. In (427) this method is implemented in simulation for the biped Lucy. Even when the upper body mass is increased from $10.5kg$ to $18kg$ the upper swings 5° to keep the ZMP in the ankle joint. The OLPIPM method can only work at low walking speeds except a stabilizer is added and using the preview method the deviated motion of the cart-table model after the second preview stage is considerable. In figure 4.19 one can notice that the hip has to move backwards during a short period so the ZMP multibody tracks the desired ZMP trajectory.

3.7 Dynamically Stepping over Large Obstacles by the Humanoid Robot HRP-2

The advantage of the method of preview control is that it has as core a simplified model of the robot, represented by the cart-table model, while the multibody dynamics of the robot are only used during a second control loop. This strategy makes it possible to handle disturbances of for example the swing leg. To further explore the possibilities of preview control, this method has been used for dynamically stepping over large obstacles by the humanoid robot HRP-2[2]. Especially during the stepping over phase the swing leg creates severe disturbances because large and fast step motions are needed. In the cart-table model of the robot it is also presumed that the COG and hip height stays on a constant height. During the stepping over, the hip height has to change in order to have feasible stepping over, causing additional disturbances. It will be shown that the second preview stage can also cope with such a disturbance. In this section a summary of this research is given, for a more detailed description of this problem the reader is referred to (379; 428; 426; 423).

3.7.1 Introduction to Stepping Over

Humanoid robots have the potential to navigate through complex environments such as standard living surrounding of humans. This is mainly due to the legged nature of the robotic system, which allows higher mobility than its wheeled counterpart. For example legged robots have the capacity to negotiate obstacles by stepping over them. Few research has been performed in this field. The elaborated strategy adapts the foot, hip and body trajectories for a collision free and dynamical stable stepping over.

Previous work on stepping over large obstacles, conducted by Guan (144), investigated the feasibility of the stepping over. Hereby focusing on quasi-static stepping over procedures by keeping the projection of the global COG of the robot within the polygon of support. Since the postural stability only takes into account the COG, the motion of the robot has to be slow in order not to induce substantial accelerations and as such not demanding for dynamic stability criteria, e.g. ZMP.

If large obstacles are considered, this quasi-static stepping over motion has a quite unnatural resemblance due to the continuous restricting balancing of the COG

[2] From the end of April 2006 the author Bram Vanderborght participated for 6 weeks in the ongoing research "Dynamically Stepping Over Large Obstacles by the Humanoid Robot HRP-2" which was conducted by Björn Verrelst at the Joint Japanese-French Robotics Laboratory (JRL) located at the National Institute of Advanced Industrial Science and Technology (AIST) in Tsukuba, Japan.

Fig. 3.47 Quasi-static stepping over performed by Guan (144).

(see figure 3.47). Moreover, a large double support phase is required, in order to shift the COG from the rear to the front during the double support phase. This implies kinematical restrictions and consequently limits the dimensions of the obstacles which can be negotiated.

On the contrary, a dynamic stepping over procedure cancels the restriction of the COG balancing and allows a shorter double support phase. A dynamic walking pattern is characterized by postural stability on the ZMP criterion and allows the COG to leave the supporting foot as long as the ZMP stays within the polygon of support. As such the COG can be shifted over the obstacle during one single support phase, which in theory should allow for using an instantaneous double support phase only, if running is not regarded. This results in larger obstacle dimensions which can be negotiated.

3.7.2 Feasibility Unit

First the feasibility unit calculates the step length, step height and foothold positions during the stepping over procedure because the actual leg layout of HRP-2 and the closed kinematic chain during the double support phase makes this phase mainly determine the actual obstacles which can be stepped over. Figure 3.48 shows all the essential parameters which are of concern for these calculations. The obstacle

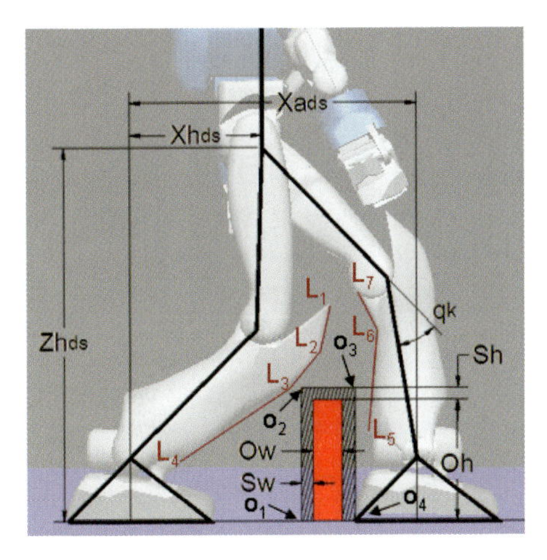

Fig. 3.48 Double support phase feasibility.

is regarded to be rectangular with certain width O_w and height O_h. For the stepping over trajectory planning a safety margin (S_w, S_h) around the obstacle is included, not only to cope with deviations on calculated kinematics due to tracking errors during the actual stepping over, but mainly regarding the uncertainty of the vision system, determining the obstacle dimensions. Since the rear leg is most likely to collide with the obstacle, due to the knee which is directed towards the obstacle, the heel of the front foot behind the obstacle is positioned near the safety boundary around the obstacle at point o_4. The selection of the step length and hip height starts with the normal walking values, while piecewise increasing step length and decreasing hip height until a collision free configuration is found. Hereby taking into account a minimum angle (q_{min}) for the knee angle (q_k) which cannot be exceeded in order to avoid the singular configuration of knee overstretching. When the optimal step length is determined the step sequence to reach the obstacle can be calculated as well as the desired ZMP trajectory.

The calculation of the feasibility takes place at a certain time instant t_{DS} when the hip is at a certain horizontal position $X_{h_{DS}}$. In fact the horizontal trajectory $X_h(t)$ is calculated by the preview controller and consequently not known yet. The value of this parameter $X_{h_{DS}}$ originates from a look-up table containing an estimate for different step lengths created by the pattern generator for normal walking, for which a specific step-time has to be chosen. For the Y direction in the frontal plane, generally $Y_{h_{DS}} = 0$ if the left and right foot are positioned symmetrically with respect to the center waist frame. The height of the hip of course is determined by the feasibility selection itself.

3.7.3 Spline Foot Trajectories

Contrary to regular walking the stepping over large obstacles requires more information to be used for the design of the foot trajectories, in order not to collide with

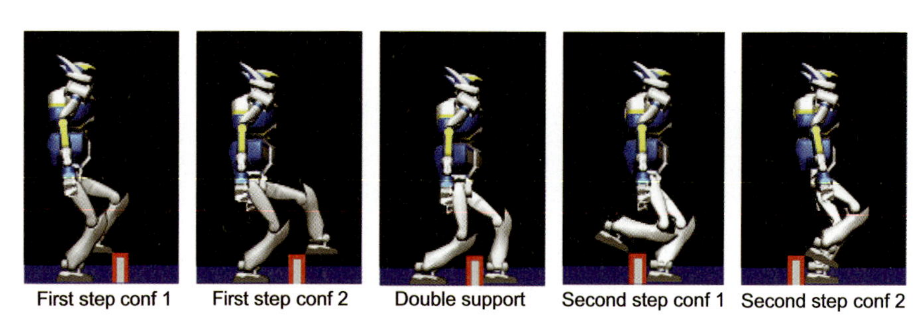

Fig. 3.49 Snapshots of the step over procedure for an obstacle of $15cm$ plus $3cm$ safety boundary zone, showing the two intermediate configurations for both steps involved in the stepping over.

the obstacle. Therefore Clamped Cubic Splines (CCS), for the 3 translations (X,Y,Z) and pitch rotation ω of the foot (ccw+ angle between horizontal and foot sole), are chosen over the more traditional polynomials because these tend to oscillate when different control points are chosen. Clamped Cubic Splines are constructed of piecewise third-order polynomials which pass through a set of control points with a chosen start and end velocity. These boundary values on the velocity are chosen zero to avoid impacts at touch-down and have a smooth transition at lift-off.

Two intermediate control configurations P_1 and P_2 are selected to construct the foot trajectories. For most cases the Y coordinate (horizontal frontal plane axis) of the feet is kept constant, so the focus is set on the sagittal horizontal X and vertical Z coordinate. The two intermediate control configurations P_1 and P_2 are used for both steps of the step over procedure as depicted in the snapshots in figure 3.49, and are determined such that the tip of the foot coincide with point o_2 of the obstacle and the ankle of the foot with point o_3 respectively. Two rotation angles at the intermediate points are chosen to prevent self-collision of the leg and the foot.

3.7.4 Preview Control on ZMP

Because the desired footholds and the ZMP trajectory is known the preview controller can calculate the horizontal waist motion (both in the horizontal X and Y direction) by the first stage of the preview controller. The vertical motion of the waist (Z_h) is a regular 3th order polynomial to change (in general to lower) from the normal walking height to the one ($Z_{h_{DS}}$) at double support over the obstacle determined by the feasibility unit, subsequently it is restored to normal height again during the second step. To clear more space during the double support over the obstacle and consequently allow for larger obstacles to be stepped over, the waist of the robot is rotated. The HRP-2 robot has two degrees of freedom (yaw and pitch) between the waist and upper-body such that the upper-body and the head (with vision system) is oriented towards the walking direction. This motion is achieved with an analogous polynomial structure as for the vertical waist motion. Because the complete motion of the robot is now known the real ZMP motion can be calculated. This will again

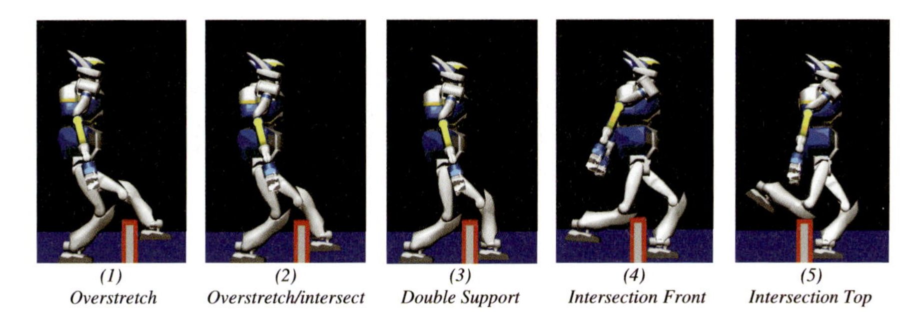

Fig. 3.50 Snapshots of the step over procedure for an obstacle of $25cm$ plus $3cm$ safety boundary zone, showing the difficulties using the basic trajectory planning.

differ from the desired ZMP trajectory as explained in section 3.5.6, also because now the hip height changes during the stepping over. A re-feeding of ΔZMP in a second stage of the preview controller filters out these disturbances as is shown in section 3.7.6, discussing the results.

3.7.5 Trajectory Adaptations for Higher Obstacles

When higher obstacles ($> 20\,cm$) are considered, the basic foot trajectory planning is not sufficient anymore. In this case overstretch during the first step and collision with the rear leg during the second step occurs as depicted in figure 3.50.

3.7.5.1 Avoiding near Knee-overstretch

Changing the foot trajectory to avoid overstretch situation is undesired because the impact at touch-down has to be kept low. By rotating the arms to the rear, the second preview loop will compensate for this COG shift by moving the waist forward in order to keep track of the desired ZMP. This avoids the overstretch situation.

3.7.5.2 Avoiding Intermediate Collisions

The basic trajectory planner has taken into account several intermediate collision free configurations. And, although these intermediate points are selected carefully there is no guarantee that tracking these specific foot and waist trajectories will result in collision free stepping over. Especially when large obstacles are negotiated, due to the complex movement and shape of the leg itself. So, the last required tool is a trajectory adapter which makes small corrections to the planned base trajectories. Two extra changes are performed by the collision free trajectory adapter. First the trajectory of the waist height is altered as such that the knee does not intersect with the safety boundary on top of the obstacle, as depicted in the picture on the right of figure 3.50. Secondly, the foot trajectory is adapted to avoid any other collision as shown in the second picture on the right of figure 3.50. Comparing this picture with the equivalent picture of figure 3.49 (intermediate configuration 1 for stepping over smaller obstacles), it is clear that the foot around configuration 1 has to be lifted much higher to the rear for higher obstacles.

3.7.6 Simulation Results

The simulations are performed using OpenHRP (Open Architecture Humanoid Robotics Platform) (221). OpenHRP is a software platform for humanoid robotics, and consists of a dynamics simulator, view simulator, motion controllers and motion planners of humanoid robots. Figure 3.51 shows the results of a simulation for stepping over an obstacle of $15cm$ (plus $3cm$ safety boundary zone) for the height and $5cm$ (plus 2x$3cm$ safety boundary zone) for the width. The figure gives ZMP and waist position for both the walking direction X and the perpendicular horizontal direction Y. The simulation shows 7 steps, for which normal steps last $0.78s$ and $0.02s$ for single support and double support respectively, while the stepping over steps and both previous and subsequent steps take 1.5 s and 0.04 s respectively.

The stability of the system is given by the position of the ZMP, which is calculated with the complete multibody model of the robot. For normal steps, the desired ZMP position is right underneath the ankle point of the stance foot, while for the actual stepping over steps the desired ZMP is shifted a little to the front as such that it falls in the middle of the foot, increasing the stability. The graphs show both ZMP calculations after first and second preview. A big difference between both can be witnessed, specifically for the stepping over. But it is clear that the second preview loop almost perfectly compensates for the use of the simplified model and is able to cope with the severe dynamic disturbance of the large swing leg motions and waist height variation during the stepping over. Of course, in these simulations imperfections such as compliance in the feet, parameter uncertainties, tracking errors etc. are not taken into account.

Figure 3.52 shows the results of a simulation for stepping over an obstacle of $25cm$ (plus $3cm$ safety boundary zone) for the height and $5cm$ (plus 2x$3cm$ safety boundary zone) for the width. This graph is given to depict the effect of the overstretch and collision free trajectory adaptations. Waist and left foot trajectories are given before and after adaptation. The waist is more to the rear before the adaptation, which induces the overstretch. The foot lift is clearly higher than in figure 3.51 to cope with the high obstacle and the step-length over the obstacle, calculated by the feasibility tool, is larger. Again, the ZMP course shows that overall stability is guaranteed by the second preview loop. Of course, for these calculations the second preview loop was executed several times in order to conduct the overstretch and collision detections for the two adaptation strategies.

3.7.7 Experimental Results

Figure 3.53 shows a photograph sequence of HRP-2 stepping over an obstacle of $15cm$ height and $5cm$ width, again $3cm$ safety boundary around the obstacle is taken

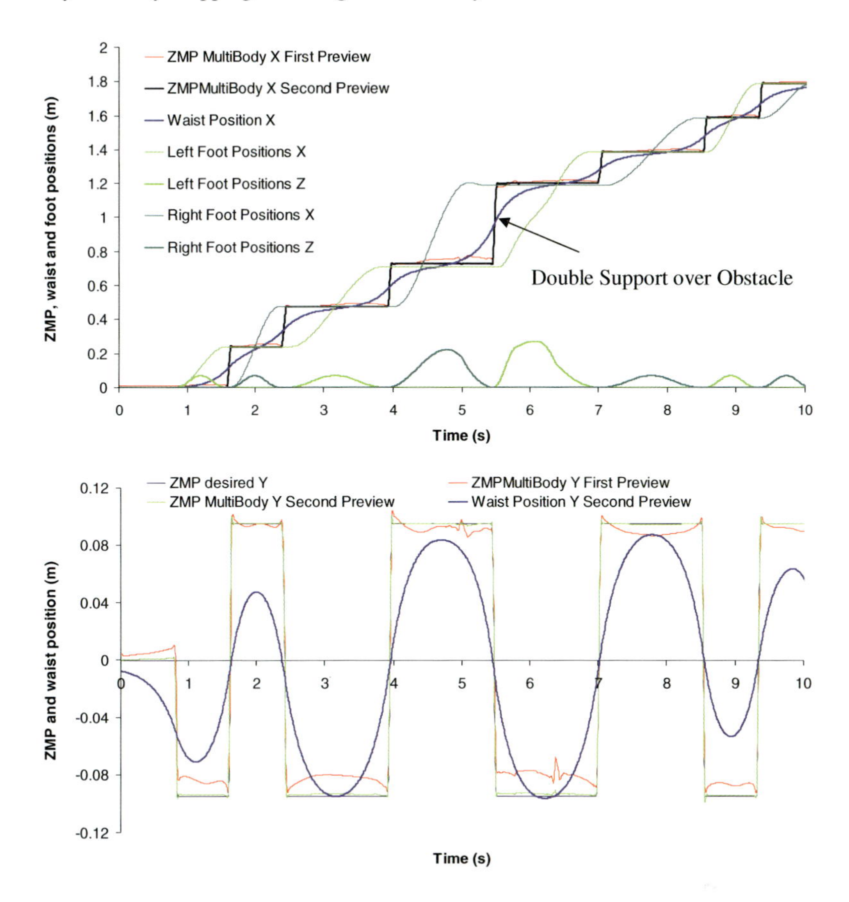

Fig. 3.51 Stepping over an obstacle of 15*cm* plus 3*cm* safety boundary zone: *ZMP* and waist position in both walking (*X*) and perpendicular horizontal (*Y*) direction and horizontal and vertical foot positions.

into account. One can notice the waist rotation and the arm motion to the back. The obstacle limit for real experiments so far is 15*cm* mainly due to two reasons. A first influencing factor is the presence of the extra stabilizing control loop (225; 458). The preview pattern generator takes into account the complete multibody model of the robot but does not include model parameter errors, the compliance of the feet and all kinds of extra external perturbations. Therefore the stabilizer acts on the posture of the robot trying to match the real measured ZMP with the desired one. This feedback loop controls hip motions and consequently the stance leg configuration, but it also adapts the swing leg according to the changing hip. Even if near overstretch situations are carefully avoided by the step over planner, the stabilizer tends to induce overstretch again, mainly because the performance of the feedback control of the stabilizer is poor in near stretched positions. Therefore the step over planner needs to apply more severe boundaries to avoid overstretch. Another issue is the speed and torque limitation of the motors, which reduces the tracking performance

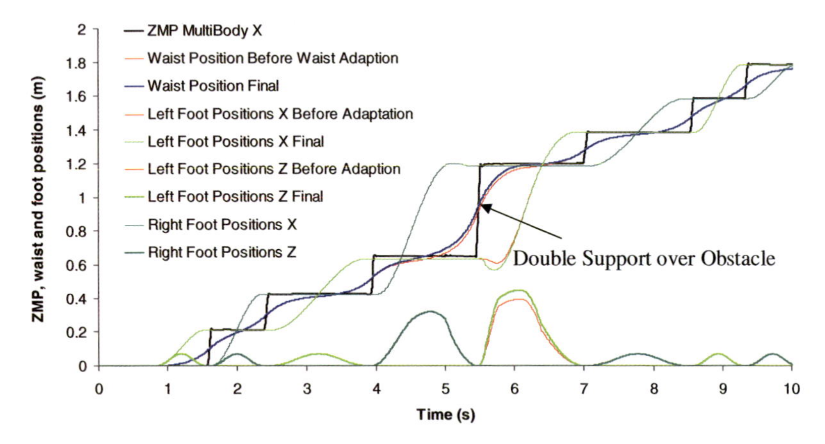

Fig. 3.52 Stepping over an obstacle of $25cm$ plus $3cm$ safety boundary zone: *ZMP* and waist position in walking direction (X) and horizontal and vertical foot positions.

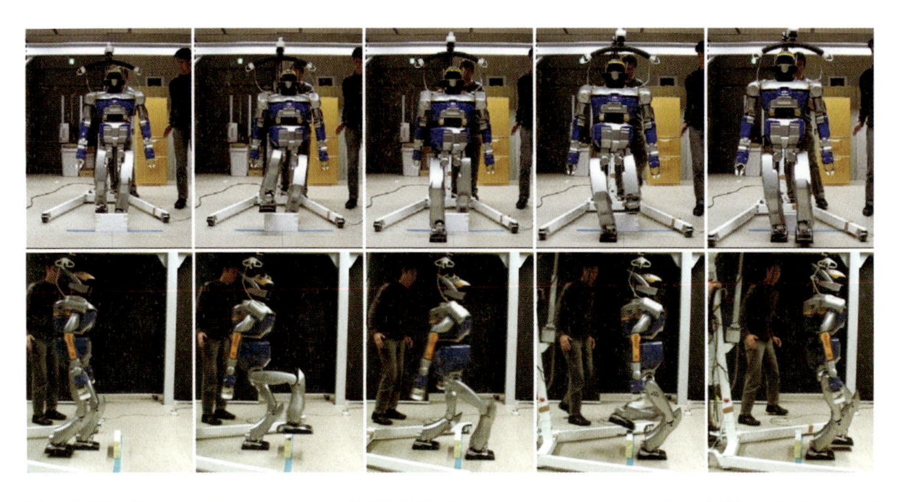

Fig. 3.53 Photograph sequence of HRP-2 stepping over an obstacle of $15cm$ height and $5cm$ width (18 height and 11 width including safety boundaries). The images are taken every $0.64s$.

of some specific motions. For stepping over an obstacle of $20cm$ this limitation was reached by the knee joint of the second swing leg stepping over the obstacle. For a video about this topic one is referred to **http://lucy.vub.ac.be/hrp2obstacle.wmv**.

The idea to change the hip height and to use the second stage of the preview controller to compensate for the generated errors is also used in (378). HRP-2, holding a bar of $2m$ in its right hand, is in front of a door opening. By using the embedded stereoscopic vision system the robot reconstructed the environment. KineoWorks^{TM} plans the motion and in order to pass through the door the robot had to descend, this is the sequence where the adapted preview controller was used.

3.8 Conclusion

The trajectory generator unit determines joint motion patterns to walk from one point to another while keeping the robot dynamically balanced. The Zero Moment Point (ZMP) is chosen as stability criterion. Two different approaches are presented.

"Objective Locomotion Parameters Inverted Pendulum Based Trajectory Generator" is a new trajectory generator based on the principles of inverted pendulum walking, modelling the robot dynamics as a single point mass. Important in the developed strategy is that the objective locomotion parameters (which are step length, intermediate foot lift and mean velocity) can be changed each step and that the trajectories are generated online. The motion of the hip during the single support phase is calculated in such a way that there is no ankle torque, meaning that the ZMP stays in the ankle joint. During the double support phase the accelerations are planned so that the next set of objective parameters is attained and that there is a smooth transition of the ZMP from rear to front ankle point. This approach is computationally very fast, suitable for realtime applications. Because this strategy does not include the complete multibody distributed masses the real ZMP differs from the desired ZMP. However, as long as the walking speed is moderate the ZMP stays in the foot contact region.

A second and improved version of the trajectory generator is based on the preview control method for the ZMP developed by Kajita (212) which has been adapted for use in the biped Lucy. The most important adaptation is that the robot uses the real step length instead of the desired step length at impact. The goal is to have the ZMP follow a predefined trajectory. This is not as straightforward as calculating the ZMP out of the joint trajectories. The main idea is to plan the motion of the COG, represented by the hip motion, in function of desired ZMP trajectories determined by the foothold sequences. The problem is regarded as a ZMP servo control implementation, trying to track the ZMP by controlling the horizontal jerk. Because often the hip has to move before the ZMP path changes, information about desired position of the ZMP in the future is needed, hence the use of a preview control method. The dynamics are simplified to a cart-table model, a cart that represents the global COG of the robot moving on a horizontally positioned pedestal table with negligible mass. Since the true robot is a multibody system the real and desired position of the ZMP will differ. In order to solve this issue, Kajita (212) proposed a re-feeding of the complete multibody calculated ZMP trajectory into the preview control by means of taking the error between the multibody calculated ZMP and the desired ZMP trajectory. This error is again presented as input to a second stage of preview control with the same cart-table model, resulting in deviations of the horizontal motion of the COG. By implementing this method the real ZMP tracks the imposed trajectory well, so a more stable walking motion is obtained. The cost is that this strategy is computationally more expensive. Another disadvantage of this method is that the trajectory is fixed for $2T_{prev} = 3.2s$ due to the use of the two stages of the preview controller. Depending on the walking speed and step length this means that some walking steps ahead cannot be changed anymore. This is not the case for the inverted pendulum based method. In the next chapter however it will be shown that at the beginning of a new step the real step length instead of the desired step length is taken and consequently in the coming $3.2s$ the trajectory is changed.

Finally, also a brief overview and results are given of the developed strategy for dynamically stepping over large obstacles by the humanoid robot HRP-2. Main difficulty is to cope with the large disturbances during the stepping over motion, as they threaten the dynamic stability. A combination of the generation of specific trajectories for the feet, waist and upper-body with the powerful and robots preview pattern generator has solved this problem. Included are strategies to prevent a collision of a part of the robot with the obstacle and to avoid an overstretch of the knee joint. HRP-2 can step over a $25cm$ height and $5cm$ width (with addition of the $3cm$ safety boundary at all sides) obstacle geometrically. However due to the joint torque and velocity limitations, $15cm$ (with addition of the $3cm$ boundary values) is currently the maximum height for HRP-2 experimentally. In view of this, one has to realize that an obstacle of $20cm$ height for a robot with stretched ankle to hip length of $60cm$ is comparable to an obstacle with a height of $30cm$ for a human. This is the highest obstacle a humanoid has currently ever stepped over without falling to the author's knowledge.

Chapter 4
Trajectory Tracking

The joint trajectory tracking controller has to control the torques generated in the joints so that the motion of the robot follows a specific trajectory as calculated by the trajectory generator.

For many robots this tracking is performed by servo controllers. For electrical motors such controllers are well-known and are commercially available. For example HRP-2 applies high gain PD position control, the joint trajectory is referenced every $5ms$ and the motor servo runs every $1ms$ (215). For the new HRP-3 humanoid robot a HP-RMT Processor is developed to do the control cycle 5 times faster (273). For KHR-3(HUBO), the joint motor controller receives every $10ms$ a reference value and the local PD tracking controller works at $1000Hz$ (317). A similar PD control makes it possible to ensure tracking of the trajectories of RABBIT (354). Main difficulty of joint motor controllers is to make them compact, lightweight and cheap, especially when the robot has many degree of freedoms. Also the cooling system has to be adequately designed.

The first walking experiments of Lucy were performed using an adapted feedback PI controller, implemented to set an angular position. The output of this low level controller generated a $\Delta\tilde{p}$ signal, which was added and subtracted in the antagonistic setup from a chosen mean pressure p_m (424). This results in the two desired pressure levels for both muscles in the antagonistic setup ($\tilde{p}_1 = p_m + \Delta\tilde{p}$ and $\tilde{p}_2 = p_m - \Delta\tilde{p}$). Subsequently a bang-bang pressure controller is used for controlling the actions of the on/off valves (see section 2.6.0.2) to set the desired pressures in the muscles. With this control strategy the robot was able to walk very slowly, about $20s$/step. The stability of the local PI joint controller was jeopardized when moving at higher frequency. So more advanced techniques to efficiently control the system are necessary to be able to attain higher walking speeds.

In particular the following difficulties are encountered when designing a controller for joints powered by artificial muscles:

- Non-linear force-contraction relation of the PPAM actuator
- Hysteresis in the force-contraction relation of the PPAM. Although this hysteresis effect is less pronounced in the PPAM than in other types of pneumatic artificial muscles, it still makes it difficult to estimate the actual force exerted by the actuator.

B. Vanderborght: Bipedal Walking by the Pneumatic Robot Lucy, STAR 63, pp. 143–175.
springerlink.com

- Imprecise knowledge of PPAM parameters.
- Non-linearity of the pressure regulating valves, choking effects.
- The coupling between actuator gauge pressures and link angles and angular velocities. This means that the system cannot be modelled as a cascade of a pneumatic system followed by a mechanical system.

Also electrical powered robots use nonlinear tracking control techniques to track given joint reference trajectories for a nonlinear system as a biped. A computed torque method is implemented in the robot Johnnie (264), the under-actuated robot Rabbit (108), the simulated 3-link (456) and 7-link robot (320). Tzafestaz et al. (405) compared a computed torque method with sliding mode control for a 5-link biped in simulation. Regarding robustness against parameter and modelling deviations, sliding mode control was found superior to a computed torque method at the cost of actuator control activity. Unfortunately, in this study actuator dynamics were not taken into account. Gorce and Guihard (140), on the other hand proposed a two level control method which combines a computed torque method with a dynamic control model of the pneumatic actuators in order to perform position and impedance control on the legs of the biped Bipman. Whatever control strategy is developed, it is obvious that the necessary calculations have to be finished within the control cycle.

When using high gain PID actions, the actuator force, determined by the position controller, can become excessively large due to a large difference between desired and real joint position. This can be caused by power failure to the actuators (valves don't open), external forces (for example when a foot is stuck) and so one. A proxy-based sliding mode control proposed by Kikuuwe and Fujimoto is a modified version of sliding mode control and is an extension of a PID control scheme to achieve safer overdamped recovery from large positional errors without sacrificing tracking accuracy during normal operation (240). This promising new control strategy has been implemented recently in a softarm actuated by PPAMs. Experiments show the good tracking performance together with a safe behaviour (410).

First the joint trajectory tracking controller is discussed consisting of an inverse dynamics control, a delta-p unit and a pressure bang-bang controller. The inverse dynamics control is different for the single and double support phase because the robot is overactuated during the double support phase. In the second half of this chapter the real walking experiments are discussed. First the global results of both implemented trajectory generators are compared, afterwards the local results of the method based on the preview control only are given.

4.1 Joint Trajectory Tracking Controller

The task of a joint trajectory tracking controller for Lucy is to apply pressures in the PPAMs ensuring the necessary torques such that the robot follows the trajectories as imposed by the trajectory generator. Due to the specific nature of the pneumatic actuation system, this tracking controller has several essential blocks which are depicted in figure 4.1 in order to cope with the highly nonlinear behaviour of

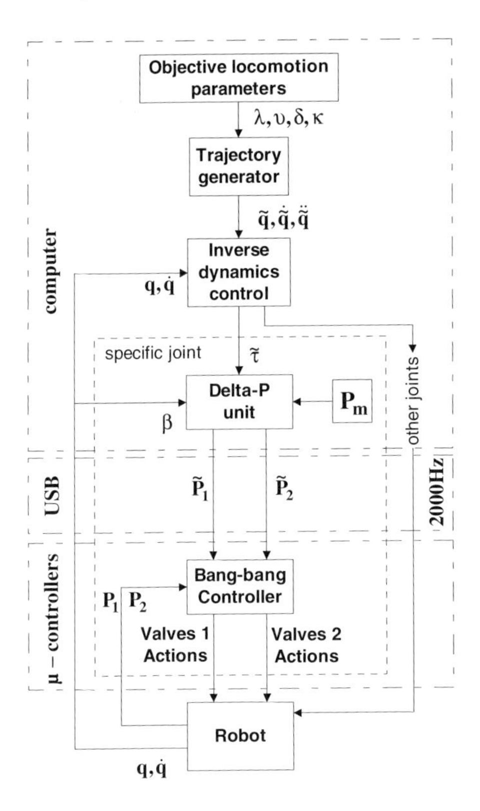

Fig. 4.1 Overview of the joint control architecture.

the complete system. The inverse dynamics unit determines the torque values required to track the imposed joint trajectories. These feedforward torque calculations are based on the robot dynamics for the single and double support phase, since the calculations demand a different approach for both phases.

For each joint a so called delta-p unit translates the required torques into pressure levels for the two muscles of the antagonistic set-up. Finally, a bang-bang controller determines the valve actions to set the pressures in the muscles. The trajectory generator, inverse dynamics and delta-p unit are implemented on a central PC, since these controllers require a substantial computational effort. The bang-bang pressure controller are locally implemented on micro-controller units (see chapter 2). In the next sections the different elements of the control structure are discussed in more detail.

4.1.1 Inverse Dynamics Control during Single Support

During the single support phase the robot's supporting foot is assumed to remain in full contact with the ground. This condition is guaranteed as long as the ZMP stays within the physical boundaries of the supporting foot and if the acceleration of the COG of the robot does not reach $-g$. Successful tracking of the generated joint trajectories should implicitly ensure the correct ZMP location. So during single support, the robot can be seen as a multi-link serial robot for which standard nonlinear tracking techniques of manipulator control are utilized. Here a computed torque method as described by (373) is proposed. This method, also called feedback linearization, linearizes the nonlinear input-output relation for the mechanical dynamic equations, describing the robot motion. The computed torque method determines the torque vector $\tilde{\tau}$. The calculation of these torques is performed by feeding forward the desired trajectory accelerations $\ddot{\tilde{\mathbf{q}}}$ and by feeding back measured positions $\mathbf{q}$ and velocities $\dot{\mathbf{q}}$ in order to cancel the nonlinear coriolis, centrifugal and gravitational terms in (2.19). A PID-feedback loop is added to improve control performance. This results in the following calculation:

$$\tilde{\tau} = \hat{C}(\mathbf{q},\dot{\mathbf{q}})\dot{\mathbf{q}} + \hat{G}(\mathbf{q}) + \hat{D}(\mathbf{q})\left[\ddot{\tilde{\mathbf{q}}} - K_p(\mathbf{q}-\tilde{\mathbf{q}}) - K_i\sum(\mathbf{q}-\tilde{\mathbf{q}}) - K_d(\dot{\mathbf{q}}-\dot{\tilde{\mathbf{q}}})\right] \quad (4.1)$$

The matrices $\hat{D}$, $\hat{C}$ and $\hat{G}$ contain estimated values of the inertial, coriolis, centrifugal and gravitational parameters. The feedback gain matrices K_p, K_i and K_d are manually tuned.

The torque vector τ contains the net torques acting on each link of the robot since the equations of motion are written in absolute coordinates (see section 2.9.1). The joint torques can then be calculated using equation 2.20.

$$\begin{aligned}
\tau_{A_a} &= \tau_6 \\
\tau_{K_a} &= \tau_5 + \tau_6 \\
\tau_{H_a} &= \tau_4 + \tau_5 + \tau_6 \\
\tau_{H_s} &= -\tau_3 - \tau_4 - \tau_5 - \tau_6 \\
\tau_{K_s} &= -\tau_2 - \tau_3 - \tau_4 - \tau_5 - \tau_6 \\
\tau_{A_s} &= -\tau_1 - \tau_2 - \tau_3 - \tau_4 - \tau_5 - \tau_6
\end{aligned} \quad (4.2)$$

4.1.2 Inverse Dynamics Control during Double Support

4.1.2.1 1st Approach: Using a Pseudo Inverse

Immediately after impact of the swing leg, three geometrical constraints are imposed on the motion of the system (see section 2.9.1). The number of DOF during double support is reduced to 3, but the same 6 Lagrange coordinates (2.18) are used. The equations of motion of single support are adapted with the three geometrical constraints as follows (201):

$$D(\mathbf{q})\ddot{\mathbf{q}} + C(\mathbf{q},\dot{\mathbf{q}})\dot{\mathbf{q}} + G(\mathbf{q}) = \tau + J^T(\mathbf{q})\Lambda \quad (4.3)$$

with $J(\mathbf{q})$ the Jacobian matrix, which is calculated by taking the derivative of the constraint equations with respect to the generalized Lagrange coordinates:

$$J(\mathbf{q}) = \begin{bmatrix} -l_1 sin(\theta_1) & -l_2 sin(\theta_2) & 0 & l_2 sin(\theta_4) & l_1 sin(\theta_5) & 0 \\ l_1 cos(\theta_1) & l_2 cos(\theta_2) & 0 & -l_2 cos(\theta_4) & -l_1 cos(\theta_5) & 0 \\ 0 & 0 & 0 & 0 & 0 & 1 \end{bmatrix} \quad (4.4)$$

and Λ the vector of Lagrange multipliers:

$$\Lambda = \left[\lambda_1\, \lambda_2\, \lambda_3\right]^T \quad (4.5)$$

Since each joint is actuated, the number of applied joint torques is 6. The number of DOF during double support is however reduced to 3, which makes the system overactuated during this phase. An infinite combination of torques can be applied to realize the tracking of a trajectory. In the following, one specific solution is selected. These calculations are based on an extended version of the method proposed by Shih and Gruver (371). The latter omitted the centrifugal and coriolis terms, which are taken into account in this work. Also, an adaptation of their pseudo-inverse calculation is proposed related to the specific goals of the trajectory generator.

The 6 Lagrange coordinates, depicted in figure 2.40, are divided into dependent and independent coordinates as follows:

$$\mathbf{q_1} = \left[\theta_1\, \theta_2\, \theta_3\right]^T \qquad \mathbf{q_2} = \left[\theta_4\, \theta_5\, \theta_6\right]^T \quad (4.6)$$

where $\mathbf{q_1}$ are the independent and $\mathbf{q_2}$ the dependent coordinates. The independent coordinates describe the absolute angle of the upper body and the orientation of the rear leg (stance leg of the previous phase), while the dependent coordinates describe the front leg and the front foot orientation. With these separate coordinates the constraints (2.21) can be rewritten in the following form:

$$Z(\mathbf{q}) = Z_1(\mathbf{q_1}) + Z_2(\mathbf{q_2}) = \mathbf{0} \quad (4.7)$$

with:

$$Z_1(\mathbf{q_1}) = \begin{bmatrix} l_1 cos(\theta_1) + l_2 cos(\theta_2) \\ l_1 sin(\theta_1) + l_2 sin(\theta_2) \\ 0 \end{bmatrix} \quad (4.8)$$

and

$$Z_2(\mathbf{q_2}) = \begin{bmatrix} -l_2 cos(\theta_4) - l_1 cos(\theta_5) - X_{A_F}^{td} \\ -l_2 sin(\theta_4) - l_1 sin(\theta_5) - Y_{A_F}^{td} \\ \theta_6 - C^{te} \end{bmatrix} \quad (4.9)$$

Analogously, the Jacobian matrix is also divided into two different parts J_1 and J_2:

$$J(\mathbf{q}) = \frac{\partial Z}{\partial \mathbf{q}} = (J_1\, J_2) \quad (4.10)$$

with

$$J_1(\mathbf{q_1}) = \frac{\partial Z_1}{\partial \mathbf{q_1}} = \begin{bmatrix} -l_1 sin(\theta_1) & -l_2 sin(\theta_2) & 0 \\ l_1 cos(\theta_1) & l_2 cos(\theta_2) & 0 \\ 0 & 0 & 0 \end{bmatrix} \quad (4.11)$$

and

$$J_2(\mathbf{q_2}) = \frac{\partial Z_2}{\partial \mathbf{q_2}} = \begin{bmatrix} l_2 sin(\theta_4) & l_1 sin(\theta_5) & 0 \\ -l_2 cos(\theta_4) & -l_1 cos(\theta_5) & 0 \\ 0 & 0 & 1 \end{bmatrix} \tag{4.12}$$

The constraint equation and the Jacobian matrix (4.10) are used to write the derivatives of the dependent coordinates as a function of the independent coordinates. Differentiating the constraint equation gives

$$\dot{Z}(\mathbf{q}) = \mathbf{0} \Leftrightarrow J_1(\mathbf{q_1})\dot{\mathbf{q}}_1 + J_2(\mathbf{q_2})\dot{\mathbf{q}}_2 = \mathbf{0} \tag{4.13}$$

The first derivatives of the dependent coordinates are then obtained:

$$\dot{\mathbf{q}}_2 = -J_2^{-1} J_1 \dot{\mathbf{q}}_1 \tag{4.14}$$

The Jacobian matrix J_2 is invertible when det $J_2 \neq 0$, or:

$$det(J_2) = l_1 sin(\theta_5) l_2 cos(\theta_4) - l_2 sin(\theta_4) l_1 cos(\theta_5) \neq 0 \tag{4.15}$$

If both lengths of upper and lower leg (l_1 and l_2) are identical, which is the case for the robot Lucy, then:

$$det(J_2) = l^2 sin(\theta_5 - \theta_4) \neq 0 \tag{4.16}$$

meaning that a fully stretched front leg corresponds to a singular configuration. For biped robots this situation can be avoided by walking with sufficiently bent knees (214).

Differentiating again (4.13) once more gives

$$\dot{J}_1\dot{\mathbf{q}}_1 + J_1\ddot{\mathbf{q}}_1 + \dot{J}_2\dot{\mathbf{q}}_2 + J_2\ddot{\mathbf{q}}_2 = 0 \tag{4.17}$$

The second derivatives of the dependent coordinates are then obtained:

$$\begin{aligned} \ddot{\mathbf{q}}_2 &= J_2^{-1}\left(-\dot{J}_1\dot{\mathbf{q}}_1 - J_1\ddot{\mathbf{q}}_1 - \dot{J}_2\dot{\mathbf{q}}_2\right) \\ &= \left(-J_2^{-1}\dot{J}_1 + J_2^{-1}\dot{J}_2 J_2^{-1} J_1\right)\dot{\mathbf{q}}_1 - J_2^{-1} J_1 \ddot{\mathbf{q}}_1 \end{aligned} \tag{4.18}$$

Additionally, the equations of motion (4.3) can be split as follows:

$$\begin{cases} D_{11}\ddot{\mathbf{q}}_1 + D_{12}\ddot{\mathbf{q}}_2 + C_{11}\dot{\mathbf{q}}_1 + C_{12}\dot{\mathbf{q}}_2 + G_1 = J_1^T\Lambda + \tau_1 \\ D_{21}\ddot{\mathbf{q}}_1 + D_{22}\ddot{\mathbf{q}}_2 + C_{21}\dot{\mathbf{q}}_1 + C_{22}\dot{\mathbf{q}}_2 + G_2 = J_2^T\Lambda + \tau_2 \end{cases} \tag{4.19}$$

where

$$D(\mathbf{q}) = \begin{bmatrix} D_{11} & D_{12} \\ D_{21} & D_{22} \end{bmatrix} \tag{4.20}$$

$$C(\mathbf{q},\dot{\mathbf{q}}) = \begin{bmatrix} C_{11} & C_{12} \\ C_{21} & C_{22} \end{bmatrix} \tag{4.21}$$

$$G(\mathbf{q}) = \begin{bmatrix} G_1 \\ G_2 \end{bmatrix} \tag{4.22}$$

and

$$\tau_1 = \left[\tau_1\ \tau_2\ \tau_3\right]^T \qquad \tau_2 = \left[\tau_4\ \tau_5\ \tau_6\right]^T \tag{4.23}$$

D, C and G are calculated by using estimated values of the inertial parameters. The equations of motion (4.19) are a set of 6 differential equations, containing 3 additional unknown variables of the Lagrange multiplier Λ. This set is transformed into three equations by eliminating the Lagrange multipliers in (4.19):

$$\begin{aligned} D_{11}\ddot{\mathbf{q}}_\mathbf{1} + D_{12}\ddot{\mathbf{q}}_\mathbf{2} - J_1^T(J_2^T)^{-1}D_{21}\ddot{\mathbf{q}}_\mathbf{1} - J_1^T(J_2^T)^{-1}D_{22}\ddot{\mathbf{q}}_\mathbf{2} \\ +C_{11}\dot{\mathbf{q}}_\mathbf{1} + C_{12}\dot{\mathbf{q}}_\mathbf{2} + G_1 - J_1^T\left(J_2^T\right)^{-1}(C_{21}\dot{\mathbf{q}}_\mathbf{1} + C_{22}\dot{\mathbf{q}}_\mathbf{2} + G_2) \\ = \tau_1 - J_1^T(J_2^T)^{-1}\tau_2 \end{aligned} \tag{4.24}$$

Next, the derivatives of the dependent coordinates are eliminated by substituting (4.14) and (4.18) in equation (4.24):

$$D'(\mathbf{q})\ddot{\mathbf{q}}_\mathbf{1} + C'(\mathbf{q},\dot{\mathbf{q}}_\mathbf{1})\dot{\mathbf{q}}_\mathbf{1} + G'(\mathbf{q}) = \tau_1 - J_1^T(J_2^T)^{-1}\tau_2 \tag{4.25}$$

with

$$D'(\mathbf{q}) = D_{11} - D_{12}J_2^{-1}J_1 - J_1^T(J_2^T)^{-1}D_{21} + J_1^T(J_2^T)^{-1}D_{22}J_2^{-1}J_1 \tag{4.26}$$

$$\begin{aligned} C'(\mathbf{q},\dot{\mathbf{q}}_\mathbf{1}) = & -D_{12}J_2^{-1}\dot{J}_1 + D_{12}J_2^{-1}\dot{J}_2J_2^{-1}J_1 \\ & - J_1^T(J_2^T)^{-1}\left(-D_{22}J_2^{-1}\dot{J}_1 + D_{22}J_2^{-1}\dot{J}_2J_2^{-1}J_1 + C_{21} - C_{22}J_2^{-1}J_1\right) \\ & + C_{11} - C_{12}J_2^{-1}J_1 \end{aligned} \tag{4.27}$$

$$G'(\mathbf{q}) = G_1 - J_1^T(J_2^T)^{-1}G_2 \tag{4.28}$$

In (4.25) $\mathbf{q}_\mathbf{1}$, $\dot{\mathbf{q}}_\mathbf{1}$ and $\ddot{\mathbf{q}}_\mathbf{1}$ are replaced by their desired values, $\tilde{\mathbf{q}}_\mathbf{1}$, $\dot{\tilde{\mathbf{q}}}_\mathbf{1}$ and $\ddot{\tilde{\mathbf{q}}}_\mathbf{1}$, computed by the trajectory generator. The three independent coordinates $\tilde{\mathbf{q}}_\mathbf{1}$ and their first and second derivatives are obtained from the trajectory generator. The dependent coordinates $\tilde{\mathbf{q}}_\mathbf{2}$ are analytically obtained from the geometrical constraint equations (2.21). Next, a feedforward torque $\tilde{\tau}_\mathbf{f}$, required to track these desired reference trajectories, is calculated.

The rhs of equation (4.25) can be rewritten as:

$$\tilde{\tau}_1 - J_1^T(J_2^T)^{-1}\tilde{\tau}_2 = \{W_1 - J_1^T(J_2^T)^{-1}W_2\}\tilde{\tau}_\mathbf{f} = W\tilde{\tau}_\mathbf{f} \tag{4.29}$$

with

$$W_1 = [I_{3x3}\, 0_{3x3}] \qquad W_2 = [0_{3x3}\, I_{3x3}] \tag{4.30}$$

and

$$\tilde{\tau}_\mathbf{f} = \left[\tilde{\tau}_1\ \tilde{\tau}_2\right]^T \tag{4.31}$$

Since W has dimensions 3×6 and the lhs of equation (4.25) is a three dimensional vector, the computed torque is calculated with a pseudo-inverse of matrix W:

$$\tilde{\tau}_{\mathbf{f}} = W^{+}\left[D'(\tilde{\mathbf{q}})\ddot{\tilde{\mathbf{q}}}_{\mathbf{1}} + C'\left(\tilde{\mathbf{q}},\dot{\tilde{\mathbf{q}}}_{\mathbf{1}}\right)\dot{\tilde{\mathbf{q}}}_{\mathbf{1}} + G'(\tilde{\mathbf{q}})\right] \tag{4.32}$$

In this case the rows of W are linearly independent, so WW^T is invertible. The pseudo-inverse can be calculated using the Moore-Penrose formula (346):

$$W \in \mathbb{R}^{3x6} \qquad W^{+} = W^T\left(WW^T\right)^{-1} \tag{4.33}$$

Expression (4.32) selects a certain solution (least square) that can be used to calculate the torque vector. Many trajectory generators demand zero or small ankle torques (to keep the ZMP in the ankle joint). This strategy can be expanded to have the same condition during double support. Moreover, small ankle torques allow these joints to be used by the ZMP observer as depicted in figure 3.1. So before applying a Moore-Penrose inverse in (4.32) an extra condition is added which expresses zero ankle torques during the double support phase. The front foot is taken into account in the equations of motion and this foot is forced to stay on the ground ($\theta_6 = C^{st}$). Consequently the calculated ankle torque of the front foot, represented by $\tilde{\tau}_{\mathbf{f}}(6)$, is already zero. Note that $\tilde{\tau}_{\mathbf{f}}$ represents net torques acting on each link. Thus, recalling (2.20), the ankle torque of the rear foot can be calculated by adding all the net torques. Demanding that the rear ankle torque has to be zero, is expressed thus by including the following condition:

$$\sum_{i=1}^{5} \tilde{\tau}_{\mathbf{f}}(i) = 0 \tag{4.34}$$

This results in the following expression:

$$\tilde{\tau}_{\mathbf{f}} = \begin{bmatrix} 1 & 1 & 1 & 1 & 1 & 0 \\ W_{11} & & & & & \\ & & \cdots & & & \\ & & & & & W_{36} \end{bmatrix}^{+} \begin{bmatrix} 0 \\ D'(\tilde{\mathbf{q}})\ddot{\tilde{\mathbf{q}}}_{\mathbf{1}} + C'\left(\tilde{\mathbf{q}},\dot{\tilde{\mathbf{q}}}_{\mathbf{1}}\right)\dot{\tilde{\mathbf{q}}}_{\mathbf{1}} + G'(\tilde{\mathbf{q}}) \end{bmatrix} \tag{4.35}$$

Finally, as was done for the computed torque during the single support phase, a PID-feedback loop is added to cope with modelling errors and to improve the tracking performance.

$$\tilde{\tau} = \tilde{\tau}_{\mathbf{f}} - K_p\left(\mathbf{q} - \tilde{\mathbf{q}}\right) - K_i\sum\left(\mathbf{q} - \tilde{\mathbf{q}}\right) - K_d\left(\dot{\mathbf{q}} - \dot{\tilde{\mathbf{q}}}\right) \tag{4.36}$$

The parameters of the gain matrices K_p, K_i and K_d of the feedback loop are tuned manually.

A disadvantage of this method are the discontinuities in desired torque at the transitions between single and double support phase. These discontinuities are shown in figure 4.2. Experiments on the real robot showed that due the slow dynamics of the valves, the discontinuities cause severe perturbations at the phase transitions, destabilizing the motion of the robot.

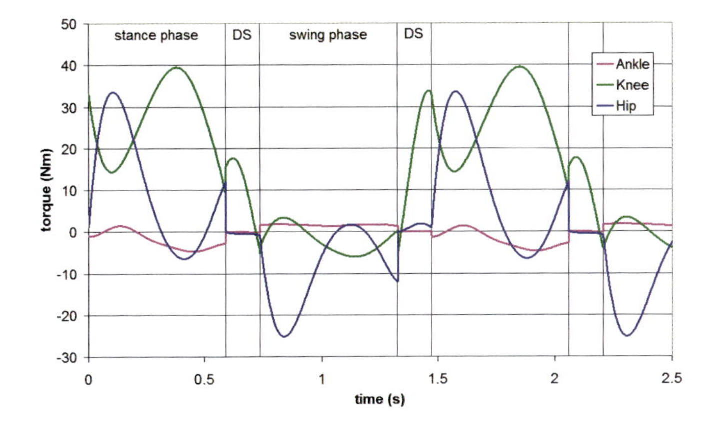

Fig. 4.2 Discontinuity in torque level at phase transitions.

4.1.2.2 2nd Approach: Linear Transition

An alternative approach to avoid the discontinuities at phase transitions, is to make a linear transition between the torques of the old and new single support phase, by calculating the applied torque as if the robot is in single support phase. The applied torque can be written into the following form:

$$\begin{aligned}\tilde{\tau} = &(1-s)\left(C(\mathbf{q},\dot{\mathbf{q}})\dot{\mathbf{q}} + G(\mathbf{q}) + D(\mathbf{q})\left[\ddot{\tilde{\mathbf{q}}} - K_p(\mathbf{q}-\tilde{\mathbf{q}})\right.\right.\\ &\left.\left. - K_i\sum(\mathbf{q}-\tilde{\mathbf{q}}) - K_d(\dot{\mathbf{q}}-\dot{\tilde{\mathbf{q}}})\right]\right)\\ &+ s\left(C'(\mathbf{q},\dot{\mathbf{q}})\dot{\mathbf{q}} + G'(\mathbf{q}) + D'(\mathbf{q})\left[\ddot{\tilde{\mathbf{q}}} - K'_p(\mathbf{q}-\tilde{\mathbf{q}}) - K'_i\sum(\mathbf{q}-\tilde{\mathbf{q}}) - K'_d(\dot{\mathbf{q}}-\dot{\tilde{\mathbf{q}}})\right]\right)\end{aligned} \tag{4.37}$$

with s going from $s = 0$ at impact until $s = 1$ at calculated lift off instant. C, G,...,K_d are computed as if the robot is in single support phase with the rear foot on the ground, C', G',...,K'_d are calculated as if the robot is in the next single support phase with the front foot on the ground.

The advantage of this strategy is that there are no torque discontinuities when switching between single and double support phase as can be seen in figures 4.28 - 4.30. The disadvantage is that the calculated torques are not dynamical correct, but the double support phase is rather short and a feedback loop is implemented. Simulated and experimental results show that this strategy works well for the motions considered.

4.1.3 Delta-p Unit

In the previous sections the net torque values for each link were calculated. These net torques can be transformed into the required joint torques with (4.2). On the other hand, the torques generated by each joint are obtained by the pressure differences in the antagonistic muscle system. Therefore the delta-p unit is used to transform the calculated torques into required pressure levels. For each muscle pair, such a controller is provided and dimensioned according to its specific torque characteristic.

The generated torque in an antagonistic muscle setup was already discussed in 2.5.1.2. For the sake of convenience, the formulation is repeated here. The generated torque is calculated with the kinematical model of the leverage and rod mechanism, combined with the estimated force function of the muscles (2.3) and the applied gauge pressures. This can be represented by the following calculation:

$$\begin{aligned} \tau &= p_1 l_{0_1}^2 r_1(\beta) f_1(\beta) - p_2 l_{0_2}^2 r_2(\beta) f_2(\beta) \\ &= p_1 t_1(\beta) - p_2 t_2(\beta) \end{aligned} \tag{4.38}$$

with τ the generated torque and β the locally defined relative joint angle. p_i is the applied gauge pressure in the respective muscle with initial unpressurized length l_{0_i} and $f_i(\beta)$ characterizes the force function of the respective muscle. The kinematical transformation from forces to torques are represented by $r_1(\beta)$ and $r_2(\beta)$ which results, together with the muscle force characteristics, in the torque functions $t_1(\beta)$ and $t_2(\beta)$. These nonlinear functions are determined by the choices made during the design phase and depend on the specific joint angle β.

Equation (4.38) is used to determine the two desired gauge pressure $\tilde{p}_1$ and $\tilde{p}_2$ for each muscle pair. These two pressures are generated starting from a mean pressure value p_m while adding and subtracting a $\Delta\tilde{p}$ value:

$$\tilde{p}_1 = p_m + \Delta\tilde{p} \tag{4.39a}$$

$$\tilde{p}_2 = p_m - \Delta\tilde{p} \tag{4.39b}$$

The mean value p_m normally influences the joint stiffness and can be controlled in order to influence the natural dynamics of the system. At this moment the controller of the complete biped does not yet incorporate the exploitation of natural dynamics as will be discussed in chapter 5 and consequently p_m is held constant. Combining the equations (4.39) with equation (4.38), allows one to calculate the $\Delta\tilde{p}$ value required to generate the torque originating from the inverse dynamics control module:

$$\Delta\tilde{p} = \frac{\tilde{\tau} + p_m\left[(\hat{t}_2(\beta) - \hat{t}_1(\beta)\right]}{\hat{t}_2(\beta) + \hat{t}_1(\beta)} \tag{4.40}$$

The delta-p unit is consequently actually a feed-forward calculation from torque level to pressure level using the kinematic model of the muscle actuation system.

4.1.4 Bang-bang Pressure Controller

For each joint the two desired pressure values $\tilde{p}_1$ and $\tilde{p}_2$ are sent to the respective local muscle pressure controller, which is responsible for tracking the required muscle pressure. In order to realize a lightweight rapid and accurate pressure control, fast switching on/off valves are used. For the inlet and the exhaust of a muscle respectively 2 and 4 valves are placed in a parallel configuration. The hardware of this valve system is described in section 2.6.0.2. The pressure controller itself is

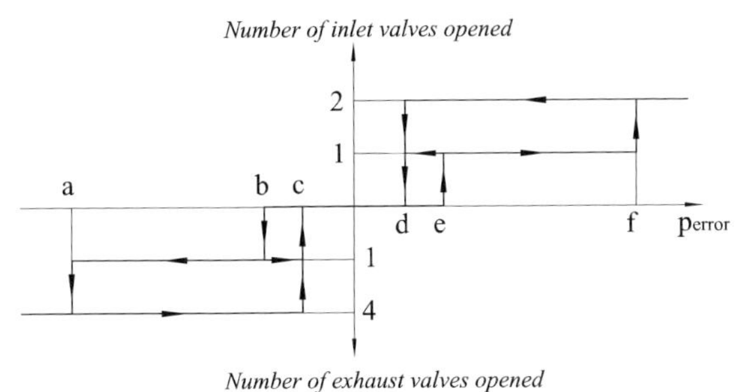

Fig. 4.3 Multilevel bang-bang pressure control scheme with dead zone.

Table 4.1 Currently applied reaction levels of the multilevel bang-bang pressure controlscheme

	$p_{error}(mbar)$	valve action
a	−60	open all exhaust valves
b	−25	open only one exhaust valve
c	−20	close all exhaust valves
d	20	close all inlet valves
e	25	open only one inlet valve
f	60	open all inlet valves

achieved by a multilevel bang-bang structure with various reaction levels depending on the pressure error. Figure 4.3 depicts the working principle of this control scheme and table 4.1 gives the currently applied reaction levels and the respective valve actions. The reaction levels were manually tuned. The pressure error is defined as $p_{error} = \tilde{p} - p$, with $\tilde{p}$ the desired pressure calculated by the delta-p unit and p the pressure measured inside the muscle. If this pressure error is small and stays within the boundaries b and e, no valve action is taken and the muscle volume stays closed. If p_{error} increases and reaches level e, one inlet valve is opened in order to make the pressure rise to the required level. If one opened inlet valve is not enough to track the required pressure and p_{error} becomes larger than f, a second inlet valve is opened. Whenever p_{error} drops again, the opened valves are closed only if the error drops below level d. This has been introduced since a considerable time delay exists to set pressures. The same approach is used for negative values of p_{error}, but beyond level a 4 exhaust valves are opened instead of 2. This asymmetrical situation is introduced since asymmetrical pneumatic conditions exist between exhaust and inlet as explained in section 2.6.0.2.

Figures 4.4-4.7 shows the tracking performances for 3 different pressure control set-ups. The experiment was performed on a pressure tank with a constant volume of $0.385l$, which is about the average volume for a muscle (see section 2.5). A desired sinusoidal trajectory with an amplitude of $1bar$ and a variable frequency was imposed. Figure 4.4 shows the average pressure error as a function of the imposed frequency. In figures 4.5-4.7 the real and desired absolute pressure in the volume are depicted together with the valve actions as calculated by the bang-bang controller. Closed valves are represented by a horizontal line depicted at $2bar$, $2.5bar$ or $3bar$ pressure level, while a small peak upwards represents one opened inlet valves, a small peak downwards one opened exhaust valves and the larger peaks represent two opened inlet or four opened outlet valves. In the set-up "1/1 valves no silencer" only one inlet and one outlet valve is used and no silencer is added at the exhaust. "2/4 valves with silencer" is the set-up used for Lucy containing 2 inlet and 4 outlet valves with a silencer and "2/4 valves no silencer" is without silencer. The best performances are off course with 2/4 valves and without silencer because a silencer lowers the exhaust airflow. However for Lucy a silencer is added because the immediate expansion to atmospheric conditions of the compressed air at the exhaust creates a lot of noise. Especially the exhaust valves have troubles tracking the decreasing pressure course, while the inlet valve can track the increasing pressure very well. The minimal error of figure 4.4 is at the lowest frequency and equals the error level where the bang-bang controller opens one valve. Compared to e.g. electrical motors the dynamics of this valve system is very slow and this might jeopardize the control of the robot. For more information about this topic one is referred to (415).

4.2 Walking Experiments

In this section the walking experiments of the biped Lucy are discussed. The two described trajectory generators of chapter 3 together with the joint trajectory tracking controller given in this section are compared. The global results are given for both trajectory generators, while the influence of the tracking controller is discussed only for the preview controller based generator because they are similar for the other one. For sake of simplicity the Objective Locomotion Parameters Based Inverted Pendulum Trajectory Generator is abbreviated *OLPIPM* and the method based on the Preview Controller of the Zero Moment Point *Preview Control*. It is important to mention that both trajectory generators calculate the trajectories for the different joints online.

4.2.1 Global Results

The experiments were performed using the set of objective locomotion parameters shown in table 4.2 for the *OLPIPM* based method and table 4.3 for the *PC* based method. In both tables the time instants are depicted when the objective locomotion parameters are changed. Intermediate foot lift κ was $0.04m$. Figures 4.8 and 4.9 show a sequence of photos of the biped Lucy, taken every $0.40s$ for both algorithms. For videos of the walking biped one is referred to **http://lucy.vub.ac.be/phdlucy.wmv**.

Figures 4.10 and 4.11 show the imposed and real step length for both methods. There is a substantial difference between the desired and the real step lengths because the tracking is not precise. For legged robots however, tracking precision is

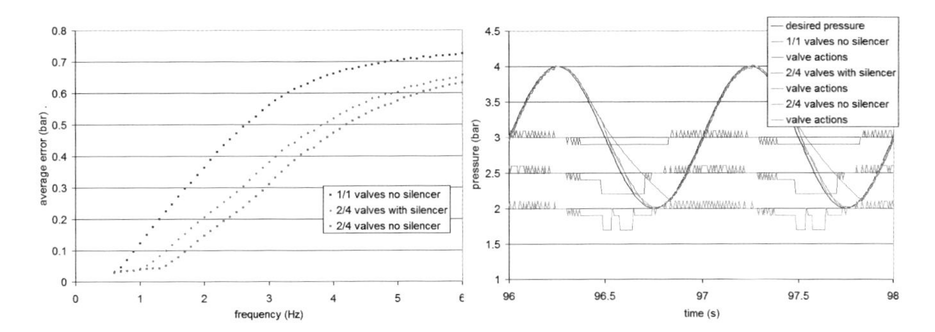

Fig. 4.4 Average pressure error as a function of imposed frequency for 3 different pressure control set-ups.

Fig. 4.5 Real and desired absolute pressure and valve actions for 3 different pressure control set-ups, frequency desired trajectory is $1Hz$.

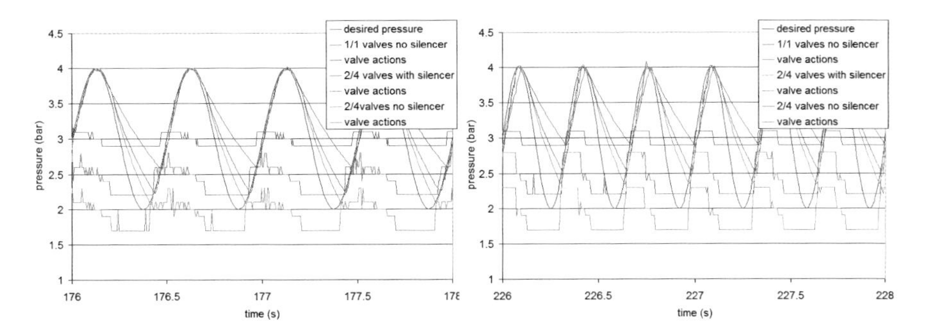

Fig. 4.6 Real and desired absolute pressure and valve actions for 3 different pressure control set-ups, frequency desired trajectory is $2Hz$.

Fig. 4.7 Real and desired absolute pressure and valve actions for 3 different pressure control set-ups, frequency desired trajectory is $3Hz$.

not as stringent as overall dynamic stability, which is achieved even while adapting the step length and forward speed of the robot. The maximum attained step length is $18cm$, then the angle between foot and lower leg exceeds the maximum joint angle which is $-30°$ for the ankle joint. This can also be found on graphs 4.23. Using a toe-joint, as discussed in section 4.3.1, makes it possible to increase the maximum step length.

Figures 4.12 and 4.13 depict the desired and real walking speed of the robot. Figure 4.12 additionally shows the speed of the treadmill and the X-position of the hip. The X-position of the hip was measured by a linear encoder placed on the horizontal rail of the guiding mechanism. The speed of the treadmill is controlled so the robot stays in the middle of the treadmill. If the position of the hip is behind the middle position, the speed of the treadmill is reduced and vice versa. The maximum possible speed for the *OLPIPM* based method is $0.11m/s$, while the *Preview Control*

Table 4.2 Set of objective locomotion parameters used for *OLPIPM* based method experiments (λ= desired step length, v= desired walking speed)

time (s)	λ (m)	v (m/s)
0	0.06	0.020
50	0.08	0.035
75	0.10	0.050
100	0.12	0.065
125	0.14	0.080
150	0.16	0.095
175	0.18	0.110
200	0.16	0.095
225	0.14	0.080
250	0.12	0.065
275	0.10	0.050
300	0.08	0.035
325	0.06	0.020

Table 4.3 Set of objective locomotion parameters used for *Preview Control* based method experiments (λ= desired step length, v desired walking speed)

time (s)	λ (m)	v (m/s)
0	0.08	0.025
75	0.12	0.050
100	0.14	0.075
125	0.16	0.100
150	0.18	0.125
175	0.18	0.150
200	0.18	0.125
225	0.16	0.100
250	0.14	0.075
275	0.12	0.050
300	0.08	0.025

based method attains $0.15m/s$. To the author's knowledge this is the fastest robot in the group of trajectory controlled pneumatic bipeds. The improvement of maximum speed is mainly due to the better control of the ZMP of the *Preview Control* based method compared to the *OLPIPM* based method as can be seen in figures 4.14 and 4.15. The maximum speed of the *Preview Control* based method is limited by the step length and the valves which cannot keep up with the desired pressure course anymore. At maximum speed the time spent in single and double support phase is $960ms$ and $240ms$ respectively.

Figures 4.14 and 4.15 are the desired and real multibody ZMP trajectory and the ZMP calculated out of the multibody model with desired trajectories. In the

Fig. 4.8 A sequence of photos of the walking biped Lucy. The images were taken every $0.40s$ (*OLPIPM*).

Fig. 4.9 A sequence of photos of the walking biped Lucy. The images were taken every $0.40s$ (*Preview Control*).

OLPIPM based method the desired ZMP is placed in the ankle joint during single support phases. For the *Preview Control* based method the ZMP is in the middle of the foot (at $6.5cm$ in front of the ankle joint), so the stability margin is the same as well forwards as backwards. It is clear that the *Preview Control* based method generates trajectories with a better ZMP curve than the *OLPIPM* based generator. This is logical because the *Preview Control* method includes the multibody model while this is not the case for the *OLPIPM* trajectory generator. The multibody ZMP curve of the *Preview Control* method has a spike at the beginning of each double support phase, when the ZMP is shifted from rear foot to the front foot. The reason is that the trajectory generator (both *OLPIPM* and *PC*) continues the calculations with the real step length instead of the desired step length. The real step length is calculated at impact out of the measured joint angles. If for example the real step length is different from the desired step length and the trajectory uses the desired step length instead of the real one, the tracking controller will create torques trying to achieve the desired step length. This is however impossible because both legs are on the ground and ground friction will prevent the step length from changing. When the foot lifts off at the end of the double support phase the torques will cause

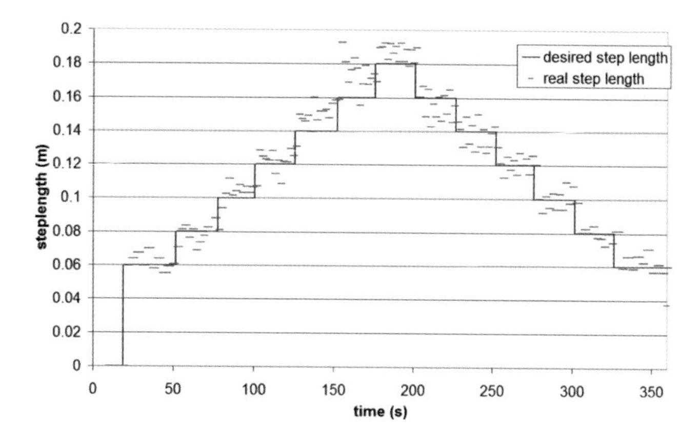

Fig. 4.10 Desired and real step length (*OLPIPM*).

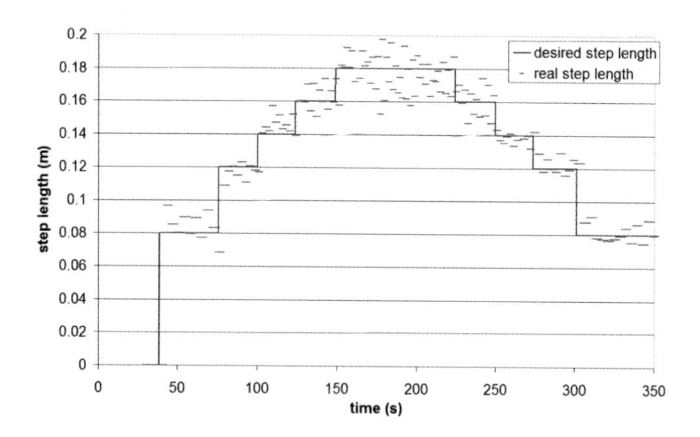

Fig. 4.11 Desired and real step length (*Preview Control*).

severe oscillations in the leg. This is unwanted so the strategy uses the real step length for its further trajectory generation of the double support phase and the next single support phase. For the *OLPIPM* based method the trajectory for position, velocity and acceleration is calculated using equations (3.12), (3.15) and (3.16). When a new step length is introduced suddenly a discontinuity in the trajectory will be observed which can for example be seen at joint angle level. However the ZMP is not influenced a lot and this effect can hardly be seen on figure 4.14. For the *Preview Control* based method the whole scheme of figure 3.42 has to be recalculated and the motion of the COG of the cart-table is treated as a servo tracking problem. Equation (3.75) implies a very fast response, making the jerk during a small time very high to track the desired ZMP again. This is felt on acceleration level causing the spike in figure 4.15 each time at the beginning of each double support phase. This leads to small discontinuities in joint angle as can be noticed in figures 4.22-4.24.

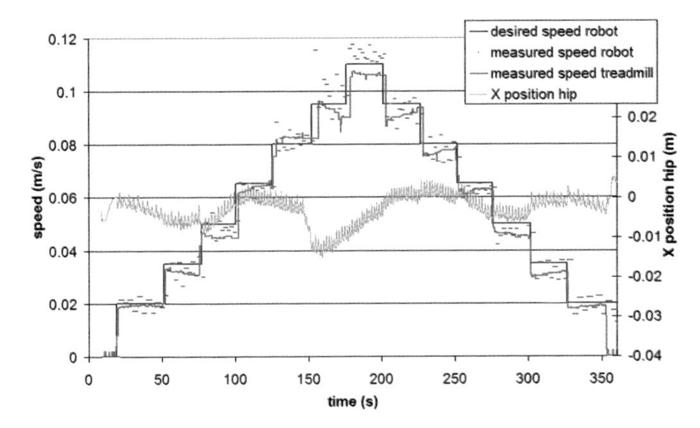

Fig. 4.12 Desired and real walking speed, real speed treadmill and X-position of hip on treadmill (*OLPIPM*).

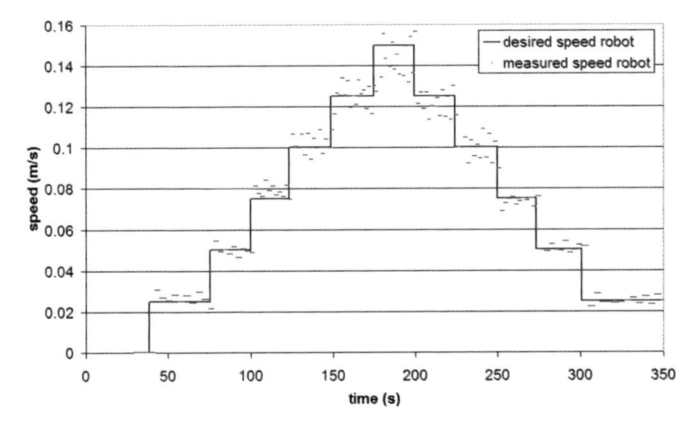

Fig. 4.13 Desired and real walking speed (*Preview Control*).

A discontinuity is for example noticeable at 214*s* in the desired ankle angle of the left foot. Because the real step length is now also the desired step length and due to the inertia of the robot this influence is filtered out in the real ZMP. At this moment the measurement of the ZMP is not yet used in the control loop, in section 4.3.3 stabilizers are discussed who exploit this sensory information to further stabilize the robot, especially for rough terrain and unexpected disturbances.

Figures 4.16 and 4.17 show the desired ZMP and the real and desired X path of the hip and calculated COG.

Figures 4.18 and 4.19 are the real and desired Y-position of the hip. The desired position of the hip for the *OLPIPM* based method is going up and down in order to achieve quasi-stretched walking during the single support phase. However, due to the small step length, these oscillations are small. For the *Preview Control* method the robot walks with constant hip height. Due to the compliance of the joints and

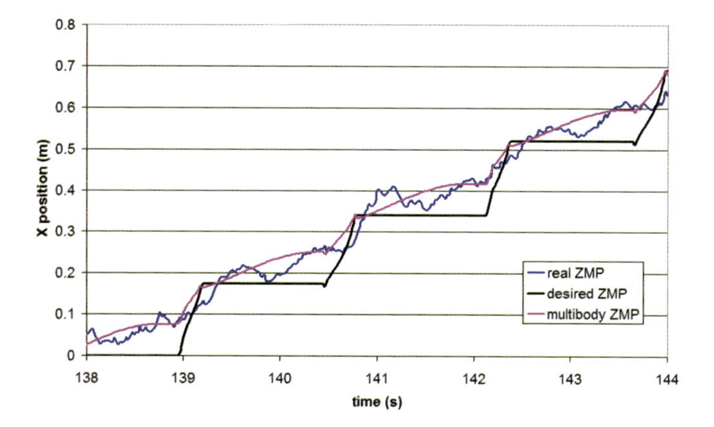

Fig. 4.14 Desired, real and multibody ZMP (*OLPIPM*).

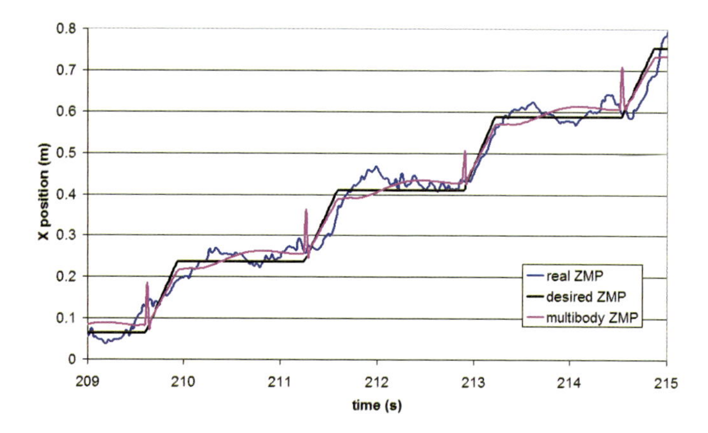

Fig. 4.15 Desired, real and multibody ZMP (*Preview Control*).

the non-perfect tracking the knees are more bent during single support phase and cause the reduced hip height. This effect can also be noticed on figure 4.23 were during stance phase the real knee angle is bigger than the desired one. But in fact the deviation between desired and real hip height is quite small, only $5mm$ on a total height of $0.85m$.

The desired objective parameters are attained as depicted in figures 4.20 and 4.21, showing the real and desired X and Y position of both feet for the *Preview Control* based method. In figure 4.20 the coordinate system is each time placed in the stance foot. These positions are obtained by the absolute position measured by the linear encoder on the rails and the measured joint angles. On 4.20 one can see that the real step length, measured at impact, is taken for the calculation of the trajectory. The impact times are shown by arrows. The deviation between desired and real foot and hip trajectory is due to tracking errors.

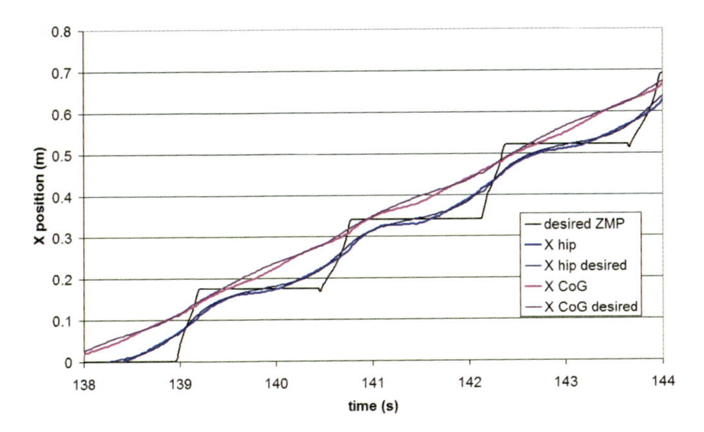

Fig. 4.16 Desired ZMP, real and desired COG and X-position hip (*OLPIPM*).

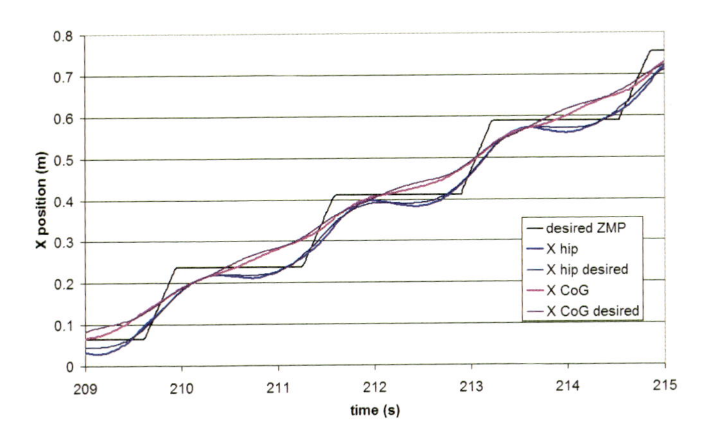

Fig. 4.17 Desired ZMP, real and desired COG and X-position hip (*Preview Control*).

4.2.2 *Local Results*

In this section the influence of the joint trajectory tracking controller on the results is discussed, this means the more local results. Only the results from the preview controlled method are discussed because the results of both methods are similar. The objective locomotion parameters are $\lambda = 0.18m$, $\kappa = 0.04m$ and $v = 0.125m/s$. So the results are taken at the boundary of the possibilities of the robot. This situation is shown because some interesting effects take place while at lower speeds the tracking etc will general not be worse.

Figures 4.22, 4.23 and 4.24 show the desired and real joint angle β_i of the hip ($i = 3$), knee ($i = 2$) and the ankle ($i = 1$) respectively of both legs. The definitions of the oriented relative joint angles are giving in figure 2.12 (counterclockwise positive). Vertical lines on all graphs show the phase transition instants. Due to the nature of the bang-bang pneumatic drive units and the imperfections introduced in the control

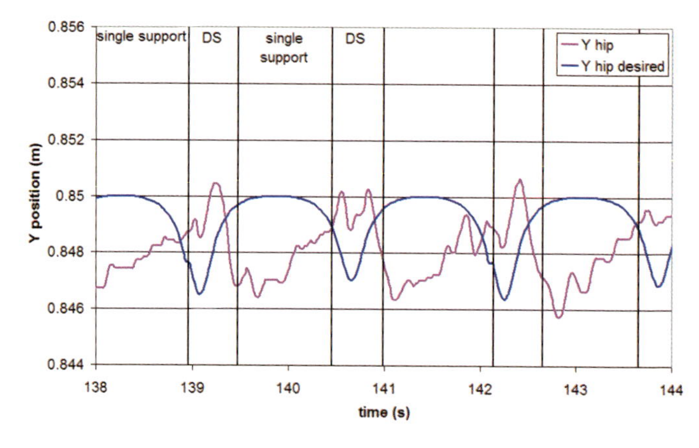

Fig. 4.18 Real and desired Y-position of hip (*OLPIPM*).

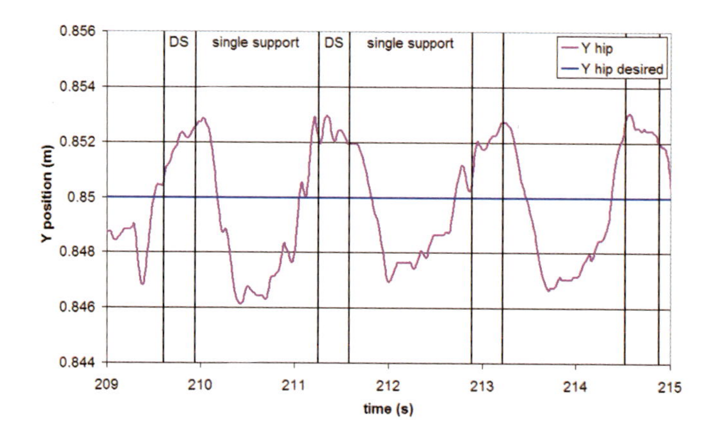

Fig. 4.19 Real and desired Y-position of hip (*Preview Control*).

loops, tracking errors can be observed. Also the phase transitions are responsible for tracking errors, since these introduce severe changes for the control signals. At the beginning of the double support phase also the discontinuity in the desired joint angle of the swing leg can be noticed caused by the trajectory generator who uses the real step length instead of the desired step length as explained some former paragraphs. For example at $214s$ the left leg was the swing leg and at the start of the double support phase the discontinuity in joint angle occurs. In figure 4.24 it can also be seen that the ankle joint angle reaches $-30°$ and puts a limit on the maximum step length.

Figures 4.25, 4.26 and 4.27 show the real and desired joint angle velocities of both legs. Especially the velocity during swing phase of the ankle joint oscillates severely. Reason is a combination of the bang-bang pressure control with the low inertia of the foot.

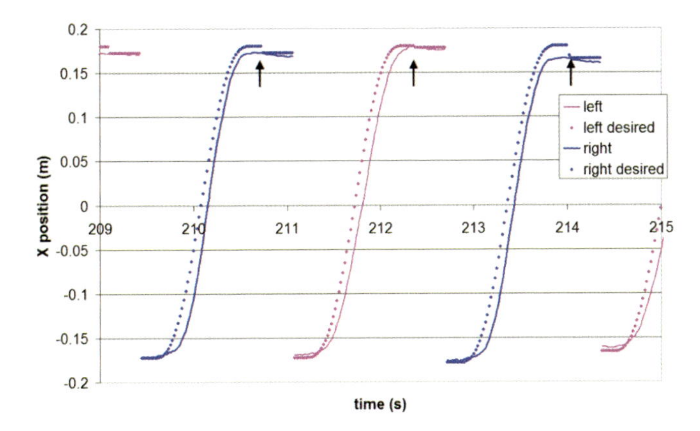

Fig. 4.20 X-position of foot (*Preview Control*).

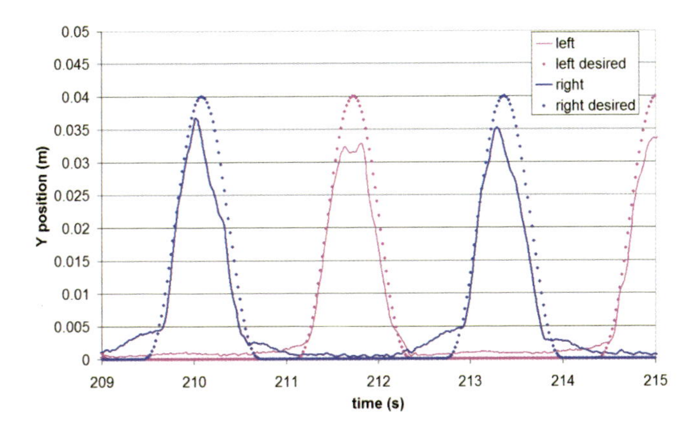

Fig. 4.21 Y-position of foot (*Preview Control*).

Figures 4.28, 4.29 and 4.30 visualize the torque calculated by the computed torque unit, which consists of a PID feedback part and a computed torque feed-forward part for the left hip, knee and ankle. For the next results only the data of the left leg are shown, since each leg takes all essential configurations over the time period shown. The definitions of the joint torques τ_H, τ_K and τ_A are giving in figure 2.41. The PID and computed torque component for the joint torques can be calculated using equation (4.2). The computed torque estimator is working well, but the robot parameters still have to be fine-tuned to lower the action of the PID controller, more information about this is given in section 4.3.2. The difference between stance leg and swing leg is clearly visible. For example the knee torque τ_K of figure 4.29 is positive (about $30Nm$) during stance phase to support the weight of the robot, while it is negative to support the weight of the leg during swing leg and of course the

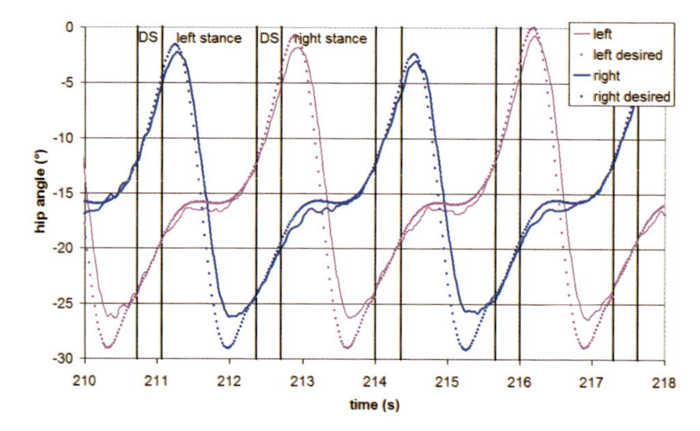

Fig. 4.22 Real and desired hip angle (*Preview Control*).

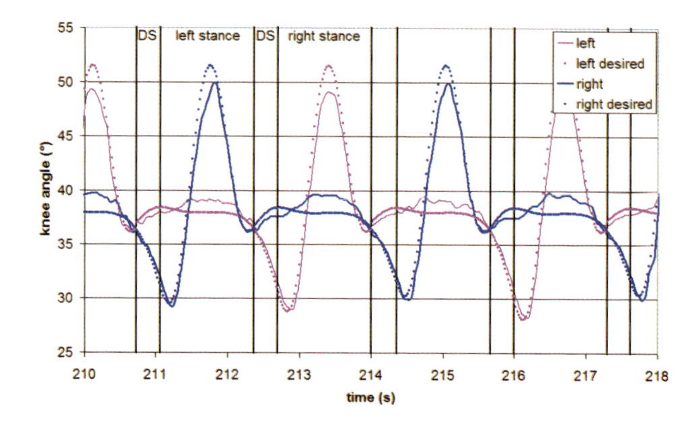

Fig. 4.23 Real and desired knee angle (*Preview Control*).

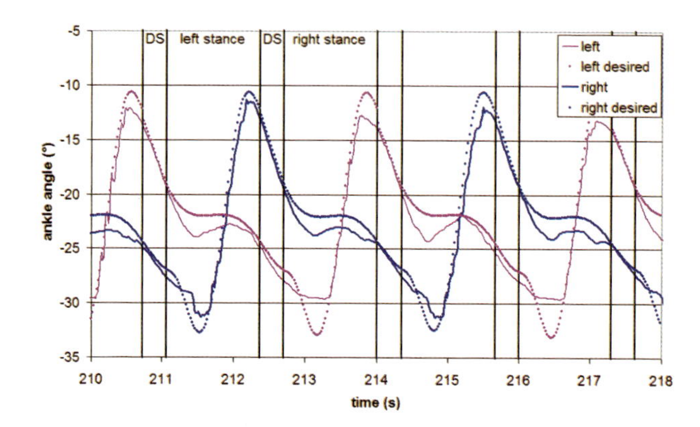

Fig. 4.24 Real and desired ankle angle (*Preview Control*).

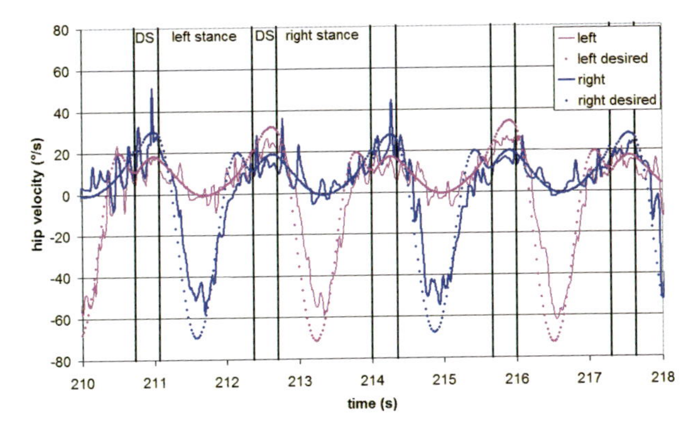

Fig. 4.25 Real and desired hip velocity (*Preview Control*).

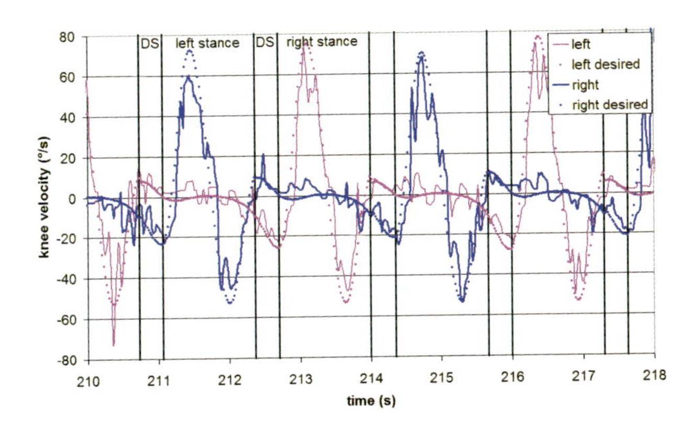

Fig. 4.26 Real and desired knee velocity (*Preview Control*).

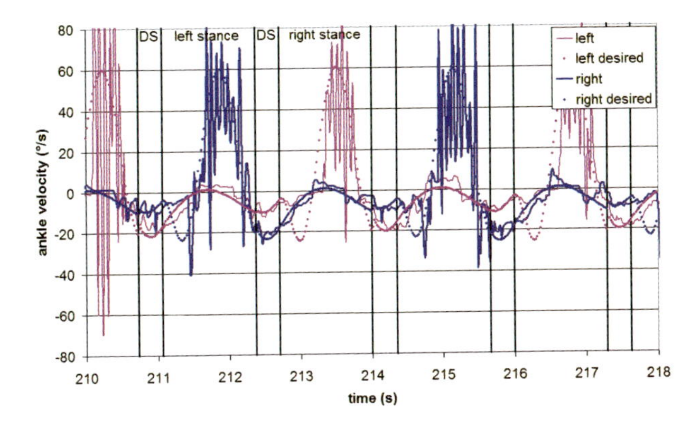

Fig. 4.27 Real and desired ankle velocity (*Preview Control*).

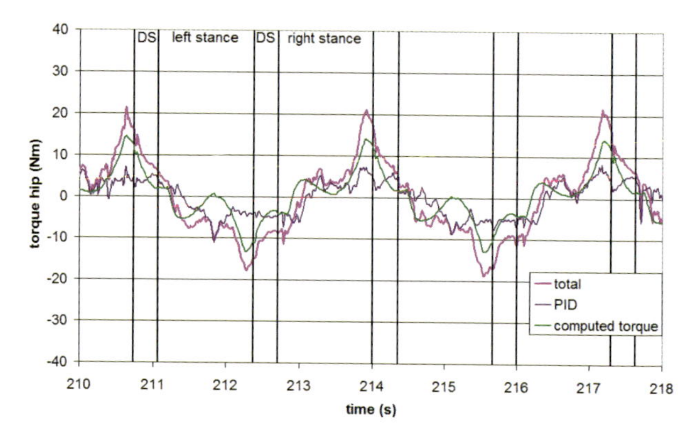

Fig. 4.28 Total hip torque τ_H, *PID* and computed torque component.

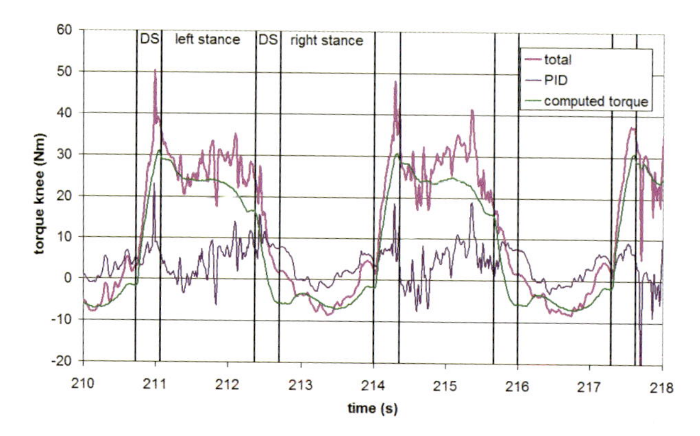

Fig. 4.29 Total knee torque τ_K, *PID* and computed torque component.

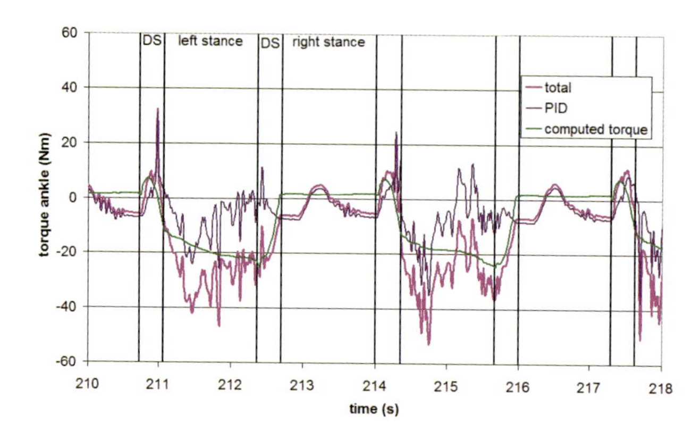

Fig. 4.30 Total ankle torque τ_A, P, I, D and computed torque component.

magnitude of the knee torque is lower in this phase. These torque values are sent to the delta-p unit.

Figures 4.31 - 4.36 depict desired and measured absolute pressure for the front and rear muscle of the different joints of the left leg. All these graphs additionally show the valve actions taken by the respective bang-bang pressure controller. Note that in these figures a muscle with closed valves is represented by a horizontal line depicted at the $2.5bar$ pressure level, while a small peak upwards represents one opened inlet valves, a small peak downwards one opened exhaust valves and the larger peaks represent two opened inlet or four opened outlet valves. The desired pressures are calculated by the delta-p unit. For this experiment the mean pressure p_m for all joints is taken at $3bar$, consequently the sum of the pressures in each pair of graphs, drawing the front and rear muscle pressures, is always $6bar$. It is observed that the bang-bang pressure controller is very adequate in tracking the desired pressure.

It is found that the ability to track the pressures limits the maximum obtainable speed of the robot. Especially the exhaust valves are causing this limitation, since the pressure gradient between muscle and atmosphere is low at some instants. For example at the start of the double support phase ($214s$), the back muscle of the knee joint (figure 4.34) cannot follow the pressure course although all the exhaust valves are open. Also the desired pressure inside both muscles of the ankle joint during the stance phase are not well tracked. This is partially solved by the PID part of the torque controller who will increase the pressure of the opposing muscle.

The ground reaction forces of the left foot are shown in figure 4.37. The green line is the sum of forces measured in the front and the rear. During single support on the left leg the total force is about $330N$, this is the total weight of the robot. During swing motion the total force is zero. During the double support phase the force gradually increases after impact and decreases before lift-off causing the ZMP to move from the old stance leg to the new stance leg. Out of these measured forces in the feet and the measured joint angles the position of the ZMP is calculated as shown in figures 4.14 and 4.15. On the same graph the state of the foot switches are depicted. They are used to determine the phase in which the robot is situated.

4.2.3 Adding Supplementary Mass

In figure 4.38 an extra mass of $6kg$ is attached at the hip of the robot at $t = 85s$ and released at $t = 91s$, this is repeated from $t = 101s$ till $t = 108s$. The robot was walking at $0.9m/s$ with a step-length of $13cm$. So the controller is able to handle severe disturbances. $6kg$ is 18% of the total weight of the robot. The parameters of the computed torque part are kept unchanged so this disturbance is mostly catched by the PID component of the inverse dynamics control unit which controls the pressures so the desired trajectories are tracked, while those trajectories remains unchanged. The role of a stabilizer as described in the next section should be to alter also the trajectories in order to guarantee stability. Figure 4.39 shows the robot with and without an extra weight of $6kg$ attached to the hip.

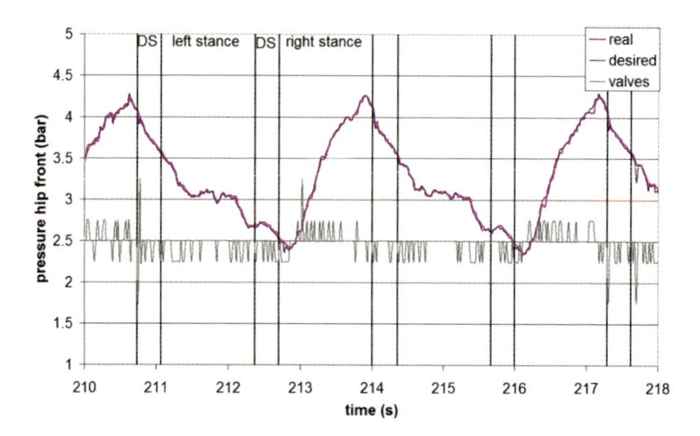

Fig. 4.31 Real and desired absolute pressure in front hip muscle, valve action.

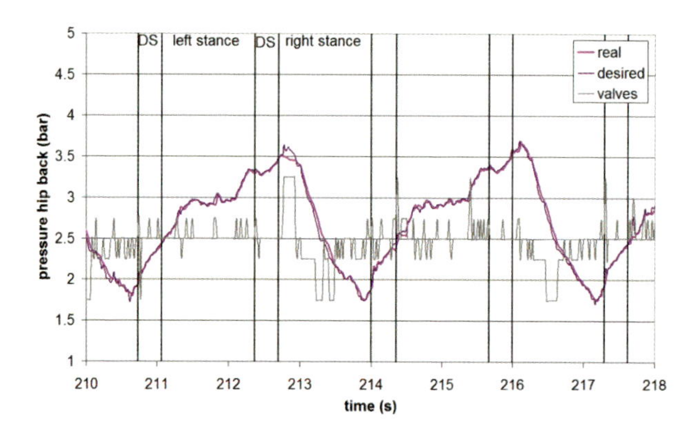

Fig. 4.32 Real and desired absolute pressure in back hip muscle, valve action.

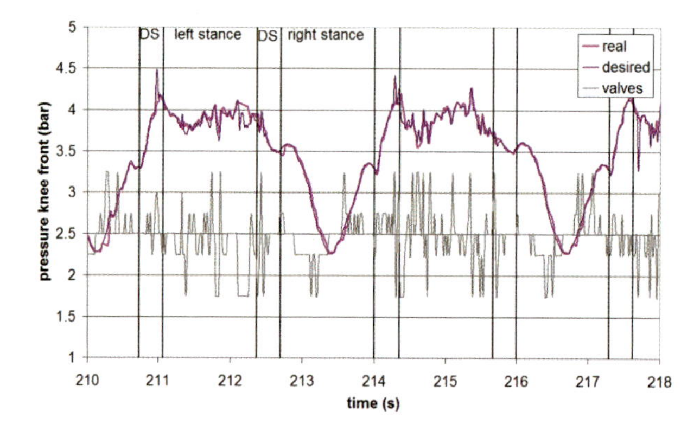

Fig. 4.33 Real and desired absolute pressure in front knee muscle, valve action.

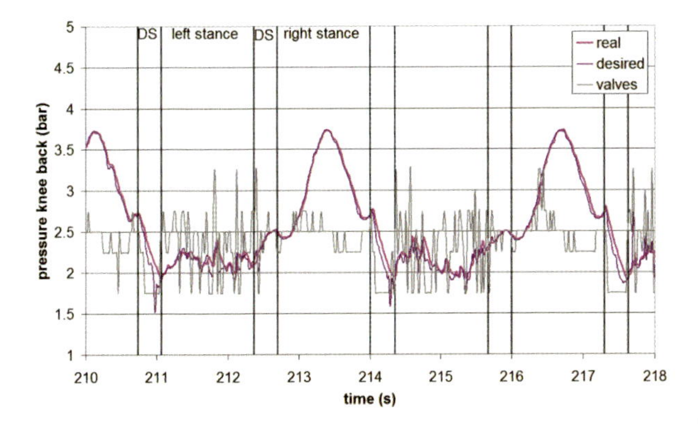

Fig. 4.34 Real and desired absolute pressure in back knee muscle, valve action.

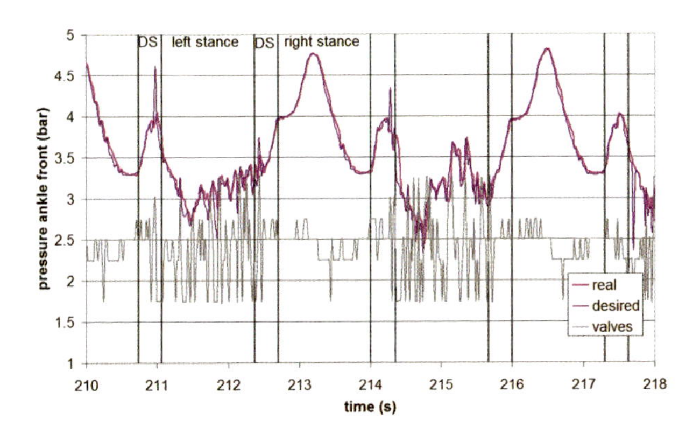

Fig. 4.35 Real and desired absolute pressure in front ankle muscle, valve action.

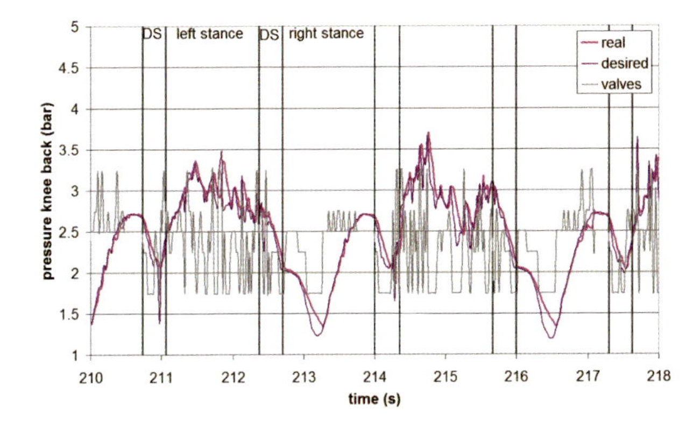

Fig. 4.36 Real and desired absolute pressure in back ankle muscle, valve action.

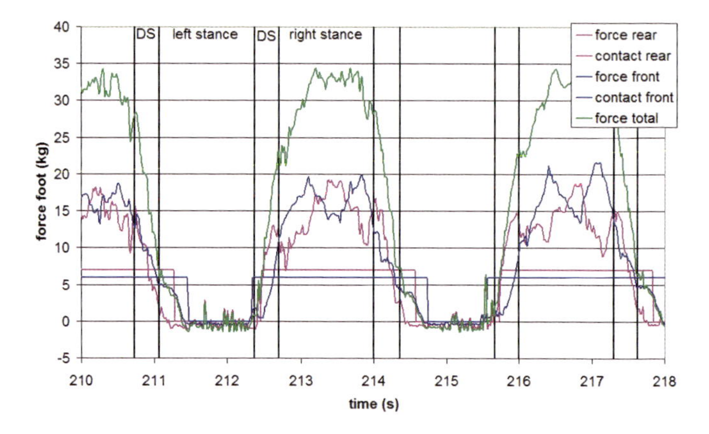

Fig. 4.37 Total, front and rear left foot force.

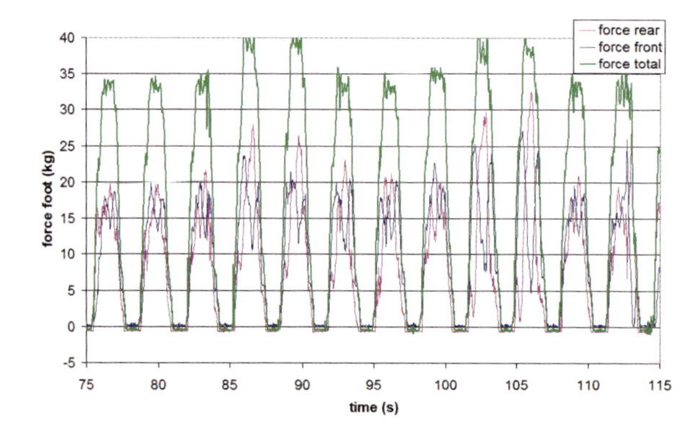

Fig. 4.38 Total, front and rear left foot force when adding extra mass.

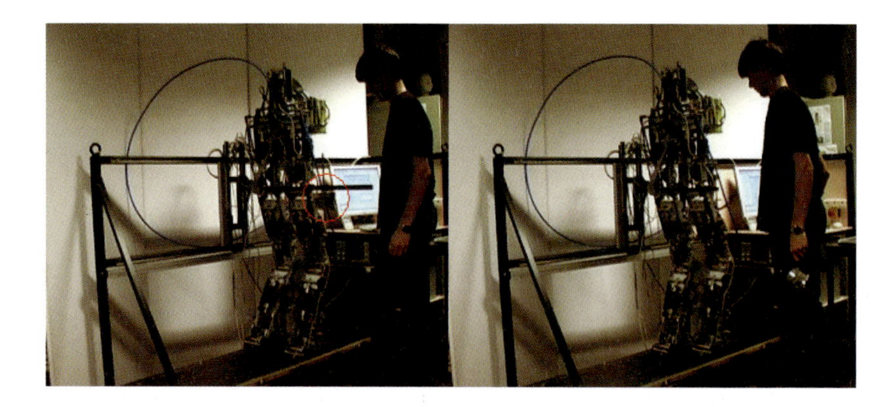

Fig. 4.39 Left: Robot Lucy with extra mass attached to the hip; right: no extra mass.

4.3 Improvements for Lucy

When building and performing experiments with the biped Lucy a lot of possibilities for improvement came up. In this section an overview of the most important items are presented and how other researchers tried to answer them.

4.3.1 Use of a Toe-Joint

The robot is walking with flat feet. The maximum step length of Lucy is $18cm$, for which the angle between foot and lower leg exceeds the limit. The next boundary is an overstretching of the knee joint. The rigid body structure of the foot makes it difficult to realize a heel lift-off to start the swing phase because small disturbances may lead to instabilities due to the line contact of the foot leading edge and the floor. Research performed by Ahn et al. (30), Nishiwaki et al. on humanoid robots H6 and H7 (300), Wang et al. (440) showed that a foot with actuated toe-joint results in a reduction of the maximum speed of knee joints and the walking speed could be increased as well as the maximum step-length. This extra link also contributes to more natural walking, which is similar to the human gait (393). For these reasons the robot Lola will also have actively driven toe joints (267). Studies are carried out also on passive toe-joints for HRP-2 (367) and WABIAN-2R (309) revealing that bigger steps and higher walking speeds are possible.

Passive walkers usually have arc-shaped feet rigidly mounted to the shank because this probably contributes to a positive effect on disturbance handling. As stated by Wisse (444) disadvantages of arc feet however are the non-human like nature. Moreover for 3D models friction torque against yaw (rotations around the vertical axis) is often insufficient for the arc foot walkers. At last it is not possible to stand still in an upright position with arc feet. An interesting thing is that the COP evolves from the heel to the ankle, while usually the COP stays in a fixed place for trajectory controlled robots. To be able to improve the walking performances of bipedal walking robots it is probably crucial to use the complete foot with a toe-joint.

4.3.2 Parameter Identification

Reference data for the dynamic parameters were obtained from conventional weighing and pendulum measurements. Disadvantage is that the robot has to be dismantled into its different links in order to be able to perform the experiments. Moreover this method is complex, time consuming and information about friction cannot be obtained. If the robot is physically changed afterwards, the experiments have to be redone to obtain the new parameters. The design of a computed torque is based on the robot model and its performance depends directly on the model accuracy (385). Errors made between the real robot and the model cause fairly high torques coming from the PID feedback part in the inverse dynamics control. More advanced parameter identification methods should be incorporated to better approach the reality. One can think of using CAD models to predict the mechanical parameters, but many parts of the robot cannot be modeled accurately, for example, the dynamics of the hosing and wiring, and the internal dynamics of the actuators (138). Additionally, there are many unknown forces acting within the robot, caused by friction, stiffness and damping of various elements. Therefore experimental methods are more suited. Parameter identification methods use the measurements of the motion and actuation data to extract the dynamic parameters. Off-line methods collect the input-output data prior to analysis in contrast to on-line identification. It is obvious that besides static also dynamic experiments have to be performed. MLE, Levenberg-Marquardt method, LSE, Kalman observers and pseudo-inverse are examples of well-known methods (385) (148). A problem of such parameter identification methods are that they require a lot of time to develope, while it is only a tool or implementation to improve the control and it is not new research.

4.3.3 Stabilizer

At this moment the measurement of the ZMP is not used in the control loop to increase the stability of the walking motion. HRP-1S can walk stable in simulation, but it falls in the experiment (163). Using a feedback stabilizer the robot is also able to walk in experiments because it is able to cope with eventual disturbances. For HRP-1S and HRP-2P a stabilizer is essential due to the soft spring-damper mechanism on its feet. The stabilizer used in HRP-1S consists of a Body Inclination Control, ZMP Damping Control and Foot Adjusting Control (459). The ZMP Damping Control accelerates the torso when the actual ZMP is forward of the desired ZMP. About the mathematics of this stabilizer little is known. It uses the measured ZMP, body inclination and joint angles coming from the robot together with the desired posture and ZMP to define the goal joint angles, tracked by the different links (312). The tracking of the desired horizontal motion of Johnnie is suspended whenever the ZMP approaches instability regions (327). Another interesting work is conducted by Mitobe et al. (286), where ZMP manipulation is used to control the angular momentum of a walking robot. The ZMP compensator developed by Okumura et al. (313) is based on altering the speed of walking instead of the walk pattern and is tested using the humanoid robot Morph3. The landing position of the foot is kept unchanged because this is normally determined according to exogenous environmental needs. The real time ZMP compensation control of the robot H5 consists of two units. The first one keeps the soles in contact with the ground and the second one is an inverted pendulum control is designed to maintain dynamic balance (295). Adding such a stabilizer is essential when the robot will leave the controlled environment of a laboratory because then one has less control about all kind of disturbances.

4.3.4 Reflexes and Emergency Stop Algorithm

4.3.4.1 Reflexes

A biped robot, certainly when working autonomously in the human environment, is at a permanent risk of loosing its balance. The concept of the ZMP is used to keep the robot dynamically balanced while walking and standing still. Larger perturbations, for example impacts against the robot, uneven or slippery terrain or a mistake of the robot (eg valves are not opened accordingly) can cause a fall down and at that moment the ZMP stability prediction turns sometimes out to be of little significance. In these situations a fast reaction or reflex has to be executed to prevent the robot from falling. Traditionally a reflex is defined as an involuntary movement which is triggered by a sensory stimulus (247). Höhn et al. (170) deals with a pattern recognition approach to detect and classify falls of bipedal robots according to intensity and direction. Reflex motions, which are initiated by the classified state, are intended to prevent the robot from falling. It turned out that the typical step execution time is about 400-500ms and that the time needed to detect and classify a fall should be shorter than 100ms. This study was extended in (171) were two different algorithms, Gaussian-Mixture-Models (GMM) and Hidden-Markov-Models (HMM), are presented that allow to distinguish exceptional situations from normal operations and these were verified on the biped BARt-UH.

In (182) a walk control consisting of a feedforward dynamic pattern and a feedback sensory reflex was proposed. The dynamic pattern is a rhythmic and periodic motion, considering the whole dynamics of the humanoid. The sensory reflex is a quick local feedback control to sensor input requiring no explicit modelling. The sensory reflex consists of a ZMP reflex, a landing phase reflex, and a body-posture reflex. These reflexive actions are organized online hierarchically to satisfy the dynamic stability constraint, to guarantee to land on the ground in time, and to keep a stable body posture for humanoid walking. This method was both verified by a dynamic simulator and an actual humanoid. Pratt et al. (336) defined the "capture region", the region on the ground where a humanoid must step to in order to come to a complete stop when the robot is pushed. To calculate this region the Linear Inverted Pendulum Model was extended to include a flywheel body. This rotational inertia enables the humanoid to control its centroidal angular momentum by lunging or "windmilling" with their arms. This enlarges the capture region significantly.

In case the robot cannot avoid a fall, the configuration of the robot has to adapt quickly to prevent as much damage as possible. Fujiwara et al. examined both the forward (132) as backward (133) falling motion. When a forward fall is detected the knees are bend at maximum angular velocity in order to make the potential energy of the robot smaller by converting it into kinetic energy. Afterwards the landing speed is braked by moving hip, waist and shoulder pitch joints and the feedback gains of the joints control are reduced to one-tenth of their original value to make the joints more compliant to impact at landing. For a backward falling motion, the neck, waist and arms are curled up into a landing posture and at a certain angle the legs of the robot are extended to decrease angular velocity. Additionally, the robot HRP-2P has impact absorbing materials mounted on the hip and knees because they are considered as the first impact points (133). Because it is not reasonable to use such a human-sized humanoid robot with full specifications for preliminary falling experiments Fujiwara et al. developed HRP-2FX (131). The robot is approximately one half the size of HRP-2P and has a simplified humanoid robot shape with 7 DOF and can emulate motions in the sagittal plane of a humanoid robot. Optimization techniques are used to minimize the landing impact of a falling motion with an inverted pendulum model to calculate backward falling motions and triple inverted pendulum to represent forward falling motion until landing on the knee. Also SDR-4X (255) of Sony is equipped with a Real-time Adaptive Falling over Motion Control which puts the robot into a secure pose. After falling over, the robot can make standing-up motion again.

The research of such falling down motions entails a high risk of seriously damaging the robot and has consequently not been performed on the robot. So this research has to be performed in simulation first. In case it is desired to make such a strategy for Lucy, then the developed simulator has to be extended to be able to calculate the impact forces at landing.

4.3.4.2 Emergency Stop Algorithm

There are many cases which force a walking robot to stop quickly without falling. Since an emergency occurs at an unpredictable moment and at any state of robot, the stopping motion must be generated in real-time. Morisawa et al. proposed an emergency stop algorithm that allows the robot to take a statically stable posture within one step for the humanoid robot HRP-2 (290). The signal to trigger the emergency stop motion was provided externally. An improved version is given in (226). Also Takana et al. (397) developed an emergent stop algorithm which can stop the robot immediately within one step.

4.4 Conclusion

The joint trajectory tracking controller controls the pressure in each muscle of the robot in order to track the different joint trajectories. This controller is multilayered and incorporates several feedforward structures in order to cope with the highly nonlinear behaviour of the complete system. The inverse dynamics unit calculates the required joint torques based on the robot dynamics. This dynamic model is different for the single and double support phase because during single support the robot has 6 DOF and during double support the number of DOF is reduced to 3 (which makes the system over-actuated). This block is based on the computed torque method consisting of a feedforward part and a PID feedback loop. For each joint a delta-p unit translates the calculated torques into desired pressure levels for the two muscles of the antagonistic set-up. This unit utilizes the nonlinear torque to angle relation. Finally, a local multilevel pressure bang-bang controller with dead zone commands the several on/off valves to set the required pressure in the respective muscles.

This chapter also contains the walking experiments performed on the robot Lucy. First both trajectory generators proposed in the previous chapter were compared. Both methods are able to change the objective locomotion parameters from step to step. The strategy based on the inverted pendulum method is computationally less but does not track the desired ZMP as well because the strategy doesn't include the complete multibody distributed masses. With this strategy the robot was able to walk up to $0.11m/s$. The other version of the trajectory generator is based on the preview control method for the ZMP developed by Kajita (212). The tracking of the ZMP is better and the maximum walking speed is $0.15m/s$. The maximum speed is limited by the maximum step length and the valves cannot keep up with the desired pressure evolution anymore. An elaborate discussion concerning the different characteristics of the walking system was presented, as well global results (e.g. step length, speed and ZMP) as local information (e.g. joint angles, torques and pressures). An indication of the robustness of the controller was shown by randomly adding and releasing a mass of $6kg$ (18% of robots weight) during walking.

Chapter 5
Compliance

In the introduction some reasons for using passive compliant actuators for bipedal locomotion were given. It seemed that by exploiting the natural dynamics of the system the energy consumption can be reduced. For passive walkers the mechanics are tuned so the motion is within the natural dynamics and the robot is able to walk down a slope without actuation. The only power source is gravity by means of the sloped surface to overcome friction and impact losses. Actuation in the joints can be provided as alternative to the slope, but in that case the actuation should be compliant in order to keep benefit of the exploitation of the natural dynamics. Disadvantage of this strategy is that such robots have difficulties or cannot start, change their speed and stop; contrary to a completely actuated robot as for example HRP-2. Such completely actuated robots which are joint-angle controlled, consume a lot of energy because they do not include the exploitation of natural dynamics (100). However, the advantage of these robots is that they can do many things. Probably the optimal will be somewhere in between those two approaches.

To have the same versatility as completely actuated robots the control strategy of the robot Lucy consists of the same parts as the actively controlled robots: a trajectory generator and a joint trajectory tracking controller. The advantage is that the robot can stand-up, start walking and this with different speeds and step lengths and finally come to a stop. The compliance is not yet adapted and a fixed compliance is chosen. The next step is to change the compliance in order to reduce the energy consumption by exploiting the natural dynamics. This research is the topic of the first part of this chapter.

Also for hopping and running robots compliance can be beneficial. Motion energy can be stored and released, which is impossible if stiff actuators or active (feedback) compliant actuators, were the compliance is introduced in the software, are used. Passive compliant actuators are also good to absorb impacts shocks, while severe impacts have to be avoided when for example harmonic drives are used. The first -very preliminary- experiments of the robot performing jumping motions are presented in the last part of this chapter.

In this chapter experimental results and simulations are mixed. In every caption is mentioned by "real" and "sim" whether the graphs are derived by doing real

B. Vanderborght: Bipedal Walking by the Pneumatic Robot Lucy, STAR 63, pp. 177–213.
springerlink.com

experiments or obtained by simulations respectively. When nothing is mentioned it is the same for real experiments as simulations.

5.1 Compliant Actuation for Exploitation of Natural Dynamics

This section is devoted to reduce energy consumption by exploiting the natural dynamics of the system. The idea of using compliance is a fairly new concept in robotics, for a long period the suggestion was "the stiffer the better". Now researchers are working on strategies to benefit from non-stiff actuation. In this section the basic idea of compliance control is studied on a reduced pendulum set-up actuated by an antagonistic pair of pleated pneumatic artificial muscles. The design is exactly the same as the limbs of the robot Lucy, only the connection parameters of the pull rod and lever mechanism are different (see 2.5.1.2).

Why study a pendulum motion for reducing the energy consumption? Walking is often likened to the motion of two coupled pendula, because the stance leg behaves like an inverted pendulum moving about the stance foot, and the swing leg like a regular pendulum swinging about the hip (252). When given an initial push to a simple gravity pendulum, the pendulum will swing back and forth under the influence of gravity and this at a certain frequency. The swing amplitude will gradually decrease due to the friction losses. When the pendulum is powered by an electrical motor a controller is able to follow the same trajectory as the freely swinging pendulum. The advantage of the first approach is that, when friction is neglected, the pendulum motion consumes no energy. It is however obvious that the second approach can follow whatever desired trajectory, of course within the limits of the controller and generally requires more energy. When the pendulum is equipped by a torsion spring the pendulum will oscillate at a higher frequency, but still only at a fixed frequency. If the pendulum is powered by an actuator with adaptable passive compliance different resonant frequencies can be selected. This is the main idea to have a minimal energy consumption with different desired trajectories while applying torques to deviate the trajectory from the unforced swing motion.

First sinusoidal functions as imposed trajectory are studied on the real pendulum. These motions are studied first because they ressemble the natural unforced motion of a pendulum. The results show that for a certain frequency an optimal compliance can be found for which the airmass consumption is minimum. At this optimal compliance the number of valve actions is strongly reduced. A mathematical formulation is proposed to calculate the optimal compliance dependent on the physical properties of the pendulum and the frequency of the imposed trajectory. The idea of the formulation is to fit the controllable actuator compliance to the "natural" compliance of the desired trajectory, and combine that with trajectory tracking control. This means that the torque of the joint is calculated so a desired trajectory is tracked, the compliance is calculated to reduce the energy consumption. For a sinusoidal function the "natural" compliance of the desired trajectory is a constant over the trajectory. For more complex trajectories this is not the case anymore and will be studied which strategy is the best.

While doing the experiments with the antagonistic setup of muscles some questions arose whether an antagonistic setup or maybe other designs of compliant actuators are more suitable for reduced energy consumption. To study this, different designs of compliant actuators are compared with each other in section 5.1.6. Comparing different designs is of course very dangerous because the results will strongly depend on the design choices. Therefore it was decided to do the experiments in simulation and to evaluate only ideal situations. Although no proof is provided some important remarks can be made definitely triggering further research.

5.1.1 Pendulum Powered by PPAMs

The complete pendulum set-up is shown in figure 5.1. The design is exactly the same as the limbs of the robot Lucy, only the connection parameters of the pull rod and lever mechanism are different (see 2.5.1.2). The physical properties of the pendulum are:

- Length of the link: $l = 0.45m$
- Coefficient denoting COG from rotation point: $\alpha = 0.77$
- Mass: $m = 6.81kg$
- Inertia in COG: $I = 0.1105kgm^2$

The sensors are an Agilent HEDM6540 incremental encoder for reading the joint position and two pressure sensors (Honeywell CPC100AFC), mounted inside each muscle. The controller is implemented on a PC and 2 data acquisition cards of National Instruments are used. The NI PCI-6602 Counter/Timer with 8 up/down, 32-bit counter/timers is used to measure the joint angles and a NI PCI-6220 with 16 analog inputs and 24 digital I/O are used to control the valves, measure the pressures and joint velocity. The control loop is performed at $400Hz$. Both PC cards are unable to measure the speed out of the encoder signal, so a PIC16F876A micro-controller, working at $2MHz$ is used to measure the time between the pulses and detect the sign of the speed. The velocity signal is sent as an analog signal ($-10V \rightarrow 10V$) to the data acquisition card.

The control architecture is similar as the one used for the biped Lucy (see section 4.1). The formulation of the delta-p unit, calculating the pressures in one joint, is slightly adapted:

$$\tilde{p}_1 = \frac{\tilde{p}_s}{t_1(\tilde{\theta})} + \Delta\tilde{p} \tag{5.1a}$$

$$\tilde{p}_2 = \frac{\tilde{p}_s}{t_2(\tilde{\theta})} - \Delta\tilde{p} \tag{5.1b}$$

with now $\tilde{p}_s$ a parameter that is used to influence the sum of pressures and consequently the joint stiffness instead of p_m. $\tilde{\theta}$ is the desired joint angle. The calculation of the required $\Delta\tilde{p}$, corresponding to the required joint torque $\tilde{T}$, for the new formulation becomes:

$$\Delta\tilde{p} = \frac{\tilde{T}}{t_1(\tilde{\theta}) + t_2(\tilde{\theta})} \tag{5.2}$$

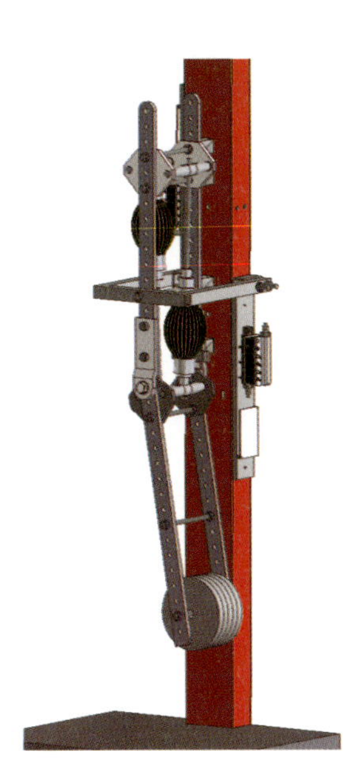
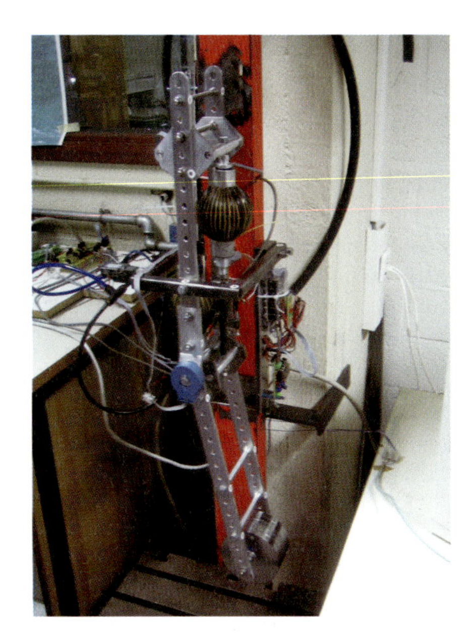

Fig. 5.1 CAD drawing and photograph of the physical pendulum.

Equation (5.2) shows, that with the new formalism the stiffness parameter $\tilde{p}_s$ does not affect the generated torque, consequently joint stiffness can be set without changing joint position, while in equation (4.40) the mean pressure value p_m is present.

When the equation (2.15) is substituted in equation (2.12), while using the required pressures (equation (5.1)) for substituting p_i, then $\tilde{p}_s$ is derived as a function of the desired stiffness K.

$$\tilde{p}_s = \frac{K - g_1 \Delta\tilde{p} - g_2}{g_3} \tag{5.3}$$

with

$$g_1(\tilde{\theta}, \dot{\tilde{\theta}}) = (-\frac{nt_1}{V_1}\frac{dV_1}{d\tilde{\theta}} - \frac{nt_2}{V_2}\frac{dV_2}{d\tilde{\theta}} + \frac{dt_1}{d\tilde{\theta}} + \frac{dt_2}{d\tilde{\theta}})$$

$$g_2(\tilde{\theta}, \dot{\tilde{\theta}}) = P_{atm}(-\frac{nt_1}{V_1}\frac{dV_1}{d\tilde{\theta}} + \frac{nt_2}{V_2}\frac{dV_2}{d\tilde{\theta}})$$

$$g_3(\tilde{\theta}, \dot{\tilde{\theta}}) = (-\frac{n}{V_1}\frac{dV_1}{d\tilde{\theta}} + \frac{n}{V_2}\frac{dV_2}{d\tilde{\theta}} + \frac{1}{t_1}\frac{dt_1}{d\tilde{\theta}} - \frac{1}{t_2}\frac{dt_2}{d\tilde{\theta}})$$

Each time the controller calculates new pressures, an adaptation of $\tilde{p}_s$ should be made in order to control the compliance. The control of the compliance is consequently a feedforward calculation.

The airmass consumption is measured by a compressed air meter SD6000 of IFM Electronics.

5.1.2 Experimental Results: Airmass Consumption

In a first real experiment the desired trajectory is a sine wave at a certain frequency and amplitude. The experiments have been repeated for different stiffness settings. Figure 5.2 and 5.3 show the total measured average airmass consumption over 5 swing motions as a function of the stiffness and frequency for sinusoidal trajectories with an amplitude of 5° and 10°. The frequency ranges from $1.5Hz$ to $2.5Hz$, in steps of $0.1Hz$, the stiffness goes from $50Nm/rad$ to $150Nm/rad$ in steps of $5Nm/rad$. The stiffness is limited because of the minimum and maximum pressure inside the muscles. At higher and lower stiffness settings the necessary torques cannot be generated anymore and the good tracking performances is threatened. This can be seen in equation (5.1). $\Delta\tilde{p}$ has a certain range to attain the desired torque, consequently $\tilde{p}_s$ is limited. It is clear there exists an optimal stiffness value and it is logical that for increasing frequencies the stiffness has to increase as well. When the two amplitudes are compared, the optimal stiffness stays nearly the same as expected for a pendulum. The airmass consumption is however higher for larger amplitudes. The actual passive trajectory of the pendulum deviates from a pure sine-wave. This deviation increases for larger amplitudes analogues for the friction losses. Consequently the airmass consumption will be higher.

Another important factor influencing the airmass consumption is the dead volume in the muscle and tubing. This volume has to be pressurized and depressurized without contributing to the output force. So the tubes should be taken as short as possible, thus the valve system should be placed close to the muscle. Another improvement is to add material in the muscles to reduce the dead volume. Davis et al.

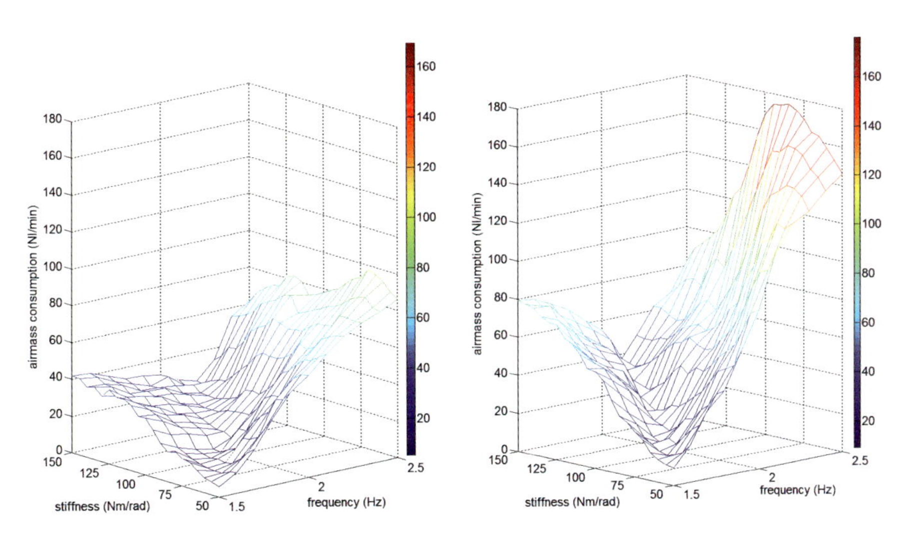

Fig. 5.2 Total airmass consumption vs stiffness and frequency for pendulum powered by PPAMs (sine of 5°) (real).

Fig. 5.3 Total airmass consumption vs stiffness and frequency for pendulum powered by PPAMs (sine of 10°) (real).

(109; 110) experimented with different filler materials which gave a higher bandwidth and reduced air consumption.

5.1.3 Experimental Results: Valve Action

The effect of choosing the optimal stiffness (so with the lowest airmass consumption) can also be witnessed in figure 5.4. These figures show the valve actions taken by the bang-bang pressure controller and the real and desired pressure course for a desired trajectory of $1.8Hz$ and $A = 5°$. The stiffness on the left was $50Nm/rad$, the middle $85Nm/rad$ and the right $150Nm/rad$. Note that in these figures closed valves are represented by a horizontal line depicted at respectively 1.7, 2.4 and 3.6 *bar* pressure level, while a small peak upwards represents one opened inlet valve, a small peak downwards one opened exhaust valve. The number of valve actions is significantly lower when the stiffness setting is at the optimal value which is the situation on the middle of figure 5.4. In the optimal case the desired pressure course is nearly the natural pressure already present in the muscle. Only at certain instants a little energy input has to be provided to the system to overcome the friction losses and to adapt to the deviation of the natural trajectory. In the cases where the compliance setting is not optimal (right and left of figure 5.4), significantly more valve actions are required. The imposed trajectory differs a lot from the natural movement of the pendulum causing a lot of valve switching and consequently energy dissipation. The actual motion, however, is the same because it is controlled by the joint trajectory tracking controller which can be seen in figure 5.5. At $t = 10s$ the controller is stopped by closing all the valves. The pendulum, for the situation in the middle of figure 5.5 with $K = 85Nm/rad$ (which is the optimal compliance for $1.8Hz$), will keep swinging with almost the same frequency as the imposed trajectory, after closing the valves. A higher stiffness compared to the optimal one means a stiffer joint, consequently the frequency increases; a lower stiffness makes the joint more compliant and thus the frequency decreases. One can see that when $K = 150Nm/rad$ (figure 5.5, right), the pendulum starts oscillating after $t = 10s$ with a frequency of about $2.1Hz$, and when $K = 50Nm/rad$ (figure 5.5, left) the frequency is about $1.5Hz$. These are the natural swing motion frequencies. The unforced amplitude of course decreases due to friction. The higher the pressures inside the muscles, the higher the damping. Figure 5.4 shows the real pressure inside the muscles after the valves are closed.

5.1.4 How to Choose the Optimal Compliance?

The previous experiments showed that each time an optimal compliance could be found for which the airmass consumption was minimal. This optimal compliance is dependent on the imposed trajectory and the physical properties of the pendulum. The idea is to fit the actuator compliance to the natural compliance of the desired trajectory. The natural stiffness of the desired trajectory $K_{trajectory}$, the inverse of the compliance, is calculated as the derivative of the torque $\tilde{T}$ necessary to track the

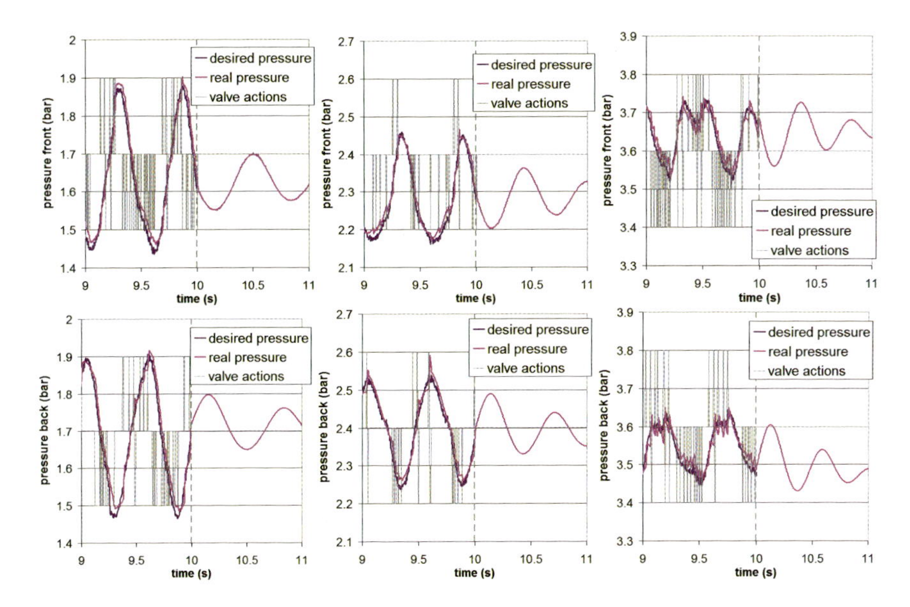

Fig. 5.4 Valve action and detail of the pressure courses in front and back muscle for optimal (middle: $85Nm/rad$) and non optimal (left: $50Nm/rad$, right: $150Nm/rad$) stiffness setting (real)

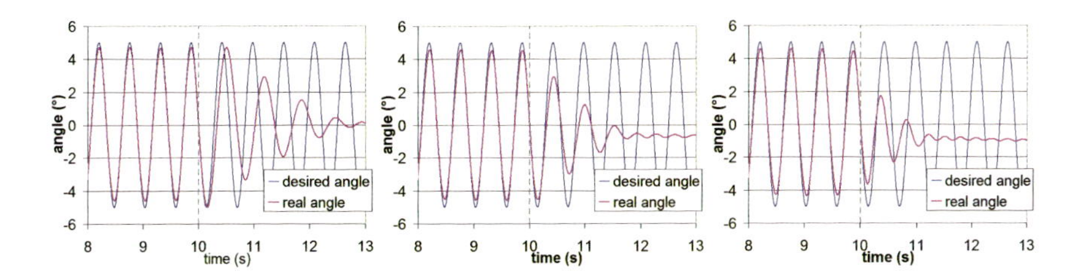

Fig. 5.5 Effect of closing all valves at $t = 10s$ when left: $K = 50Nm/rad$, middle: $K = 85Nm/rad$ and right: $K = 150Nm/rad$ (real).

desired trajectory with respect to the joint angle $\tilde{\theta}$. The torque $\tilde{T}$ is given by the inverse dynamics:

$$K_{trajectory} = \frac{d\tilde{T}}{d\tilde{\theta}} = \frac{d}{d\tilde{\theta}}\left(\hat{D}(\tilde{\theta})\ddot{\tilde{\theta}} + \hat{C}(\tilde{\theta},\dot{\tilde{\theta}})\dot{\tilde{\theta}} + \hat{G}(\tilde{\theta})\right) \tag{5.4}$$

where $\hat{D}$ is the inertia matrix, $\hat{C}$ is the centrifugal and coriolis term and $\hat{G}$ is the gravity term containing estimated values. $\tilde{\theta}$ is the desired trajectory. This stiffness $K_{trajectory}$ is substituted in equation (5.3) as a value for K. So the overview of the joint control architecture (figure 4.1) can be expanded with a compliance controller as shown in figure 5.6. This is a major improvement over strategies where an arbi-

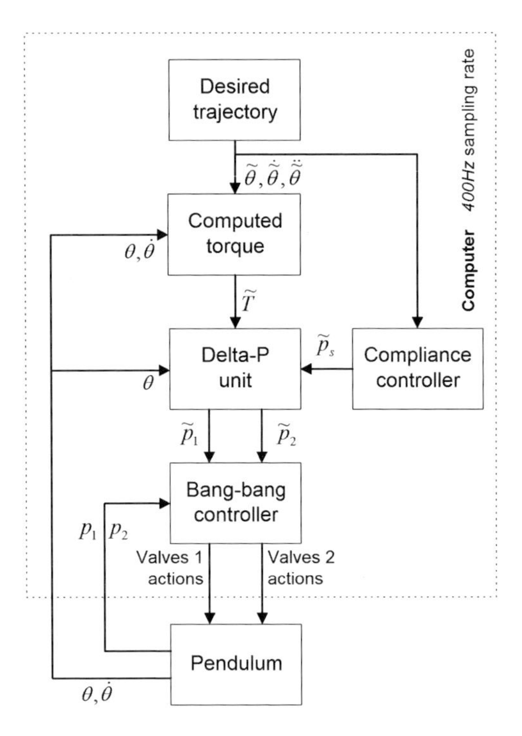

Fig. 5.6 Overview of the control architecture with compliance controller.

trary compliance value is taken as is the case with most of the robots powered by pneumatic muscles (360; 157).

For a pendulum the optimal stiffness $K_{trajectory}$ becomes:

$$
\begin{aligned}
K_{trajectory} &= \frac{d\tilde{T}}{d\tilde{\theta}} \\
&= \frac{d}{d\tilde{\theta}}\left(d_{11}\ddot{\tilde{\theta}} + g_1 sin(\tilde{\theta})\right) \\
&= d_{11}\frac{\dddot{\tilde{\theta}}}{\dot{\tilde{\theta}}} + g_1 cos(\tilde{\theta})
\end{aligned}
\tag{5.5}
$$

with $d_{11} = m\alpha^2 l^2 + I = 0.92kgm^2$ and $g_1 = gm\alpha l = 23.45Nm$ for this pendulum.

For a sinusoidal trajectory $\tilde{\theta} = Asin(\omega t)$, the optimal stiffness becomes:

$$
\begin{aligned}
K_{trajectory} &= d_{11}\frac{-A\omega^3 cos(\omega t)}{A\omega cos(\omega t)} + g_1 cos(Asin(\omega t)) \\
&= -d_{11}\omega^2 + g_1 cos(Asin(\omega t)) \\
&\approx -d_{11}\omega^2 + g_1
\end{aligned}
\tag{5.6}
$$

Table 5.1 Experimental and calculated optimal values of K_{opt}

Sine wave frequency (Hz)	$K_{trajectory}^{calculated}$ (Nm/rad)	K_{opt}^{exp} (Nm/rad) amplitude = 5°	K_{opt}^{exp} (Nm/rad) amplitude = 10°
1.5	48	50	50
1.6	59	65	60
1.7	70	80	80
1.8	82	85	85
1.9	95	105	90
2.0	108	125	120
2.1	122	145	125
2.2	137	150	150
2.3	153	150	150
2.4	169		
2.5	186		

with A the amplitude of the motion, $\omega = 2\pi f$ the angular frequency and f the frequency. The approximation of equation (5.6) is valid if $\tilde{\theta}$ is small. So the optimal stiffness approximates a constant value dependent on the physical properties of the pendulum and the frequency of the imposed motion in case $\tilde{\theta}$ is small.

Table 5.1 gives the experimentally determined stiffness K_{opt}^{exp} and the natural calculated stiffness of the desired trajectory $K_{trajectory}$ for different frequencies. One can conclude that the calculated stiffness gives a good approximation of the stiffness that is needed in order to reduce airmass consumption. At frequencies above $2.2Hz$ the optimal stiffness is outside the range the muscles can cover. So the stiffness of the trajectory $K_{trajectory}$ can be considered as the optimal compliance, so $K_{opt} = K_{trajectory}$.

5.1.5 Non-natural Trajectories

In the previous section it was found that the optimal compliance K_{opt} for a sinusoidal function is the derivative of the torque $\tilde{T}$ with respect to the joint angle $\tilde{\theta}$ and that this approximates a constant value. This value can be visualized in a torque-angle graph which for a sine function is a straight line under a certain angle. The slope represents the stiffness. This is shown for different frequencies in figure 5.7. At low frequencies (e.g. $0.5Hz$) the slope of the this curve is positive or the actuator has to generate a positive force for positive angles to decelerate the pendulum. Reason is that the desired motion is slower than the natural motion. Adapting the stiffness is of no issue because stiffness can only increase the natural frequency. When the frequency increases the slope of the swing period will become negative and stiffness adaptation can be used to exploit the natural dynamics.

For more complex trajectories than a sine function, equation (5.5) for $K_{trajectory}$ will not give a constant anymore but will change over time, also the torque-angle relation will not be a straight line anymore. Moreover at some instants $\frac{d\tilde{T}}{d\tilde{\theta}}$ will be infinite, meaning a stiff desired connection. This is of course impossible because

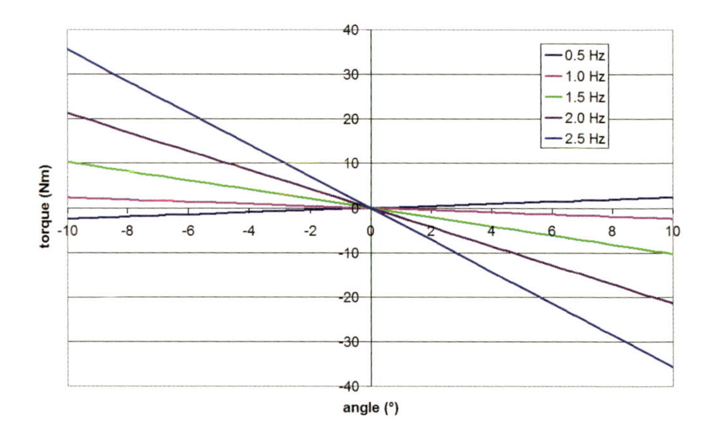

Fig. 5.7 Torque-angle relation for sinusoidal trajectory for different frequencies.

the maximum pressure inside the muscles has to be limited. This problem arises for example when the desired trajectory is a combination of two sinusoidal functions, $\tilde{\theta} = Acos(\omega t) + A/2cos(\omega/2t)$. A is the amplitude and $\omega = 2\pi f$ the angular frequency. The natural stiffness $K_{trajectory}$ of this trajectory is:

$$K_{trajectory} = -d_{11}\omega^2 \frac{\frac{1}{32} + cos(\omega t/2)}{\frac{1}{8} + cos(\omega t/2)} + g_{11}cos(\tilde{\theta}) \tag{5.7}$$

At $\theta = -10.3°$ the stiffness $K_{trajectory}$ becomes infinite, when the term $\frac{1}{8} + cos(\omega t/2)$ equals zero. This can also be visualized in figure 5.8, showing the torque-angle relation for different frequencies ranging from $0.5Hz$ till $4.5Hz$. At $\theta = -10.3°$ the slope of the tangent at this point is infinite. So tracking $K_{trajectory}$ is impossible. A possible strategy can be to choose 2 stiffnesses and switch between them. So between $\theta = 5°$ and $\theta = -10.3°$ stiffness K_1 is taken and between $\theta = 15°$ and $\theta = -10.3°$ stiffness K_2. A possibility is that the stiffness of the two red dotted lines shown in figure 5.8 comes out as optimal. In figure 5.9 the airmass consumption for such a trajectory with $A = 10°$ and $f = 3.5Hz$ is given while all the possible combinations are measured for K_1 and K_2 between $50Nm/rad$ and $250Nm/rad$. This is a simulation with extra valves because the real valves cannot follow the desired pressure courses, especially for switching between the desired stiffness values. The valley of minimal airmass consumption is clearly at the values were $K_1 = K_2$. This means that it is not interesting to change the compliance for a certain trajectory. Main reason is that changing the compliance costs energy to increase and decrease the mean pressure $\tilde{p}_s$ and this without delivering torque at the joint. Therefore it will probably be better to select a fixed compliance for a certain repetitive motion and when this motion changes the compliance has to be adapted.

The previous experiment showed that a fix stiffness setting is preferred above a changing stiffness. The logical next question is which constant stiffness should be selected as optimal? The same combination of two sinusoidal functions

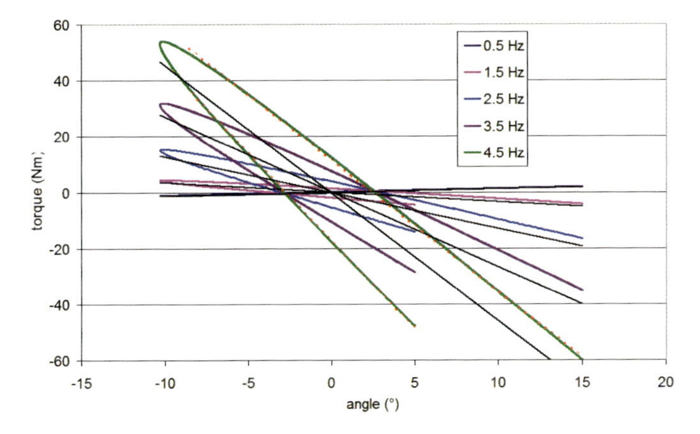

Fig. 5.8 Torque-angle relation for sum of two sinusoidal trajectories for different frequencies with first order linear regression lines.

($\tilde{\theta} = Acos(\omega t) + A/2cos(\omega/2t)$) is taken with $A = 10°$ and the frequency is increased in steps of $0.25Hz$ from $0.5Hz$ till $5.25Hz$. Figure 5.10 shows the airmass consumption versus frequency and stiffness in simulation. A clear minimum is observed for each frequency and the optimal stiffness is plotted in figure 5.12 versus stiffness. Under $50Nm/rad$ and above $200Nm/rad$ the measured optimal stiffness is meaningless because they are at the borders of the stiffness range. Using figure 5.8 the value of the average stiffness of each frequency is plotted in the same figure. This average stiffness is calculated by taken a first order linear regression of the torque-angle curve (shown by the straight lines), then the slope is the value for the average stiffness. When the experimentally optimal stiffness and calculated average stiffness are compared (figure 5.12), then one can concluded that this is a good approximation for the optimal stiffness. A part of this experiment is performed on the real pendulum as shown in figure 5.11. The maximum stiffness is $150Nm/rad$ and maximum frequency is $2.6Hz$. One can notice that the valley of minimum airmass consumption starts also at $2Hz$ as is the case in simulation.

In the following experiments the trajectory for the hip, calculated by the inverted pendulum based trajectory generator (see section 3.4), is imposed as trajectory on the pendulum. To be able to do this motion at higher speeds too, the mass at the end of the pendulum was reduced by $3.2kg$. The new physical parameters are $\alpha = 0.66$, $m = 3.58kg$ and $I = 0.08kgm^2$. Figure 5.13 shows the computed torque-angle relation for different walking speeds. A remark that has to be made is that the trajectories are generated for walking, but the torques are only for swinging in the air for this specific pendulum. While the actual torques during walking will be different because walking consists of stance phases, double support phases, swing phases and impacts. The labels "swing", "stance" and "DS" are put there only to be able to imagine the motion. "swing" is consequently a fast forward motion, while the "stance" period is a slow backward motion. The airmass consumption is measured for different walking speeds going from $v = 0.1m/s$ till $v = 0.6m/s$ in steps of

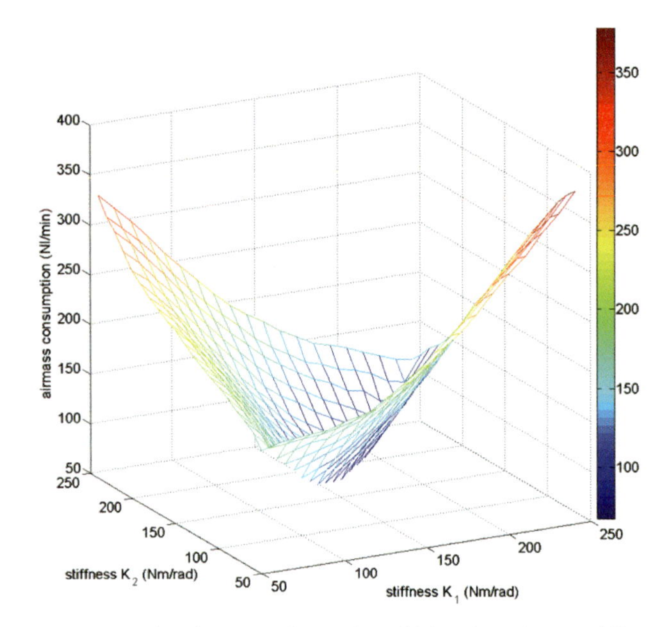

Fig. 5.9 Power consumption for sum of two sinusoidal trajectories vs stiffness K_1 and K_2 (sim).

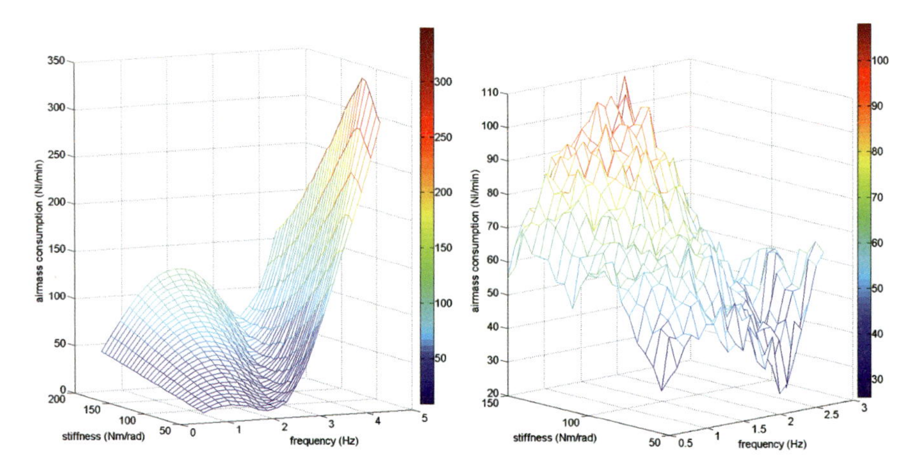

Fig. 5.10 Airmass consumption for sum of two sinusoidal trajectories vs frequency and stiffness (sim).

Fig. 5.11 Airmass consumption for sum of two sinusoidal trajectories vs frequency and stiffness (real).

$0.02m/s$ with a constant step length of $\lambda = 0.2m$. The stiffness range goes from $50Nm/rad$ till $150Nm/rad$ in steps of $5Nm/rad$. A valley of minimal airmass consumption can be found which starts from a speed of $0.4m/s$. The minima are not so pronounced anymore, for more complex trajectories, such as a hip trajectory

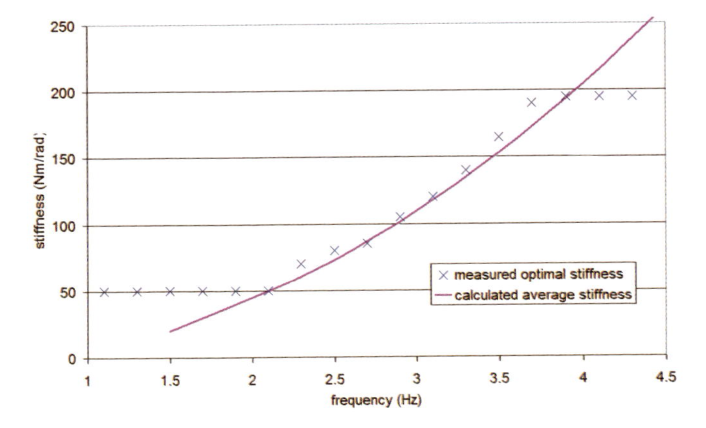

Fig. 5.12 Measured and calculated optimal stiffness in function of frequency (sim).

obtained by the trajectory generator, the airmass consumption will increase, because the imposed trajectory differs more from the natural movement of the pendulum. For each walking speed above $0.4m/s$ the airmass consumption at the lowest point is about 30% lower than the maximum airmass consumption at that walking speed. The optimal stiffness for the different walking speeds is plotted in figure 5.14. The minimal stiffness was $50Nm/rad$, this explains the flat line between $0.1m/s$ and $0.4m/s$. The position of this curve can be explained with a similar strategy as the previous case. A first order linear regression has been performed on each curve of the torque-angle relation of figure 5.13 and is shown by a dashed line. The slope of the linear regression line is a stiffness value which is plotted in figure 5.15. One can see that both curves have a similar course. At low walking speeds the motion is slower than the natural motion (average slope is positive) and using compliance is not possible. The strategy can neither be used when the slope is negative but is lower than the lowest possible stiffness, which is here $50Nm/rad$. So this average stiffness strategy to calculate the optimal stiffness seems to be interesting to apply, but further research is certainly necessary. A big advantage of this strategy is that it is also applicable for other designs of passive compliant actuators, for which an overview is given in section 1.4.2. The complete strategy calculates a torque T so a desired trajectory is tracked and a stiffness K so the energy consumption is minimized. These are provided as input to the compliant actuator. Using equations 5.1a and 5.1b together with equations 5.2 and 5.3 the pressures inside the muscles can be calculated if an antagonistic setup of two muscles is used as compliant actuator. In section 5.1.7 the equations are given to calculate the motor positions out of T and K when an antagonistic setup of two series elastic actuators, the AMASC and MACCEPA actuator are used.

When the torque-angle relation of the hip joint is plotted for the real walking robot Lucy (by eliminating the time out of figures 4.22 and 4.28) figure 5.16 is obtained. A strategy as proposed for the free swinging pendulum can be used. Doing this for the torque-angle relation of the hip of the walking robot a stiffness of

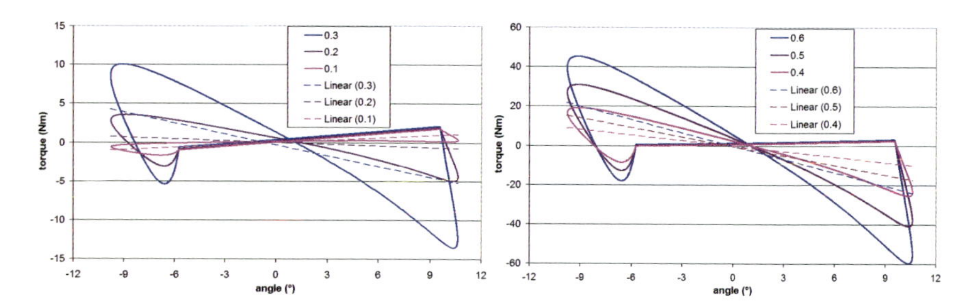

Fig. 5.13 Torque-angle relation for inverted pendulum hip trajectory for different speeds with first order linear regression lines.

$42Nm/rad$ is obtained. This is lower than the minimal stiffness. So the maximum walking speed of the robot is still too slow to have the optimal compliance in the possible stiffness range. So probably it makes no sense to implement this strategy in the real biped, due to the limited walking speeds which can be obtained currently.

Experiments concerning compliance adaptation were also performed on the walking robot WL-14. A reduction of 25% of energy consumption during the swing phase was observed compared to the case when the stiffness was not varied actively (450). This was measured when walking at $1.28s/step$ and $0.15m/step$. Strategies on how the optimal stiffness was chosen were not discussed. Mao et al. have repeated the experiment proposed in this section by using Festo muscles (271). The trajectory was based on the method developed by Huang et al. (183). In simulation it was shown that a reduced energy consumption could be obtained by selecting an optimal stiffness. The eigenfrequency of the trajectory, obtained by frequency analysis, is used to derive the optimal stiffness. The stiffness was defined in a different way.

5.1.5.1 Comparison with an Ideal Stiff Actuator

In the previous section a strategy was discussed to change the compliance in order to reduce control efforts and energy consumption by fitting the natural compliance on the desired compliance. So the natural dynamics are adapted as a function of the imposed trajectories. The effectiveness of the proposed control strategy has been shown through experimental results on a pendulum setup. But these are only relative values and are not compared with for example a traditional electrical drive. To do this research, the same experiments with sinusoidal trajectories are repeated as in section 5.1.2. Simulations are performed only whereby ideal situations are compared. It is supposed there are no estimation errors on the inertial parameters of the pendulum and no errors on the estimated force function of the pneumatic muscle. A pendulum with the same physical properties as the real pendulum is taken.

Measuring the energy consumption of a stiff actuator is not so difficult. It is supposed that the servo-motor is ideal. In this case this means that the motors can reach in one sample period the exact desired position. The servo-motors are supposed to be non-backdrivable. When applying a constant force at a fixed position no

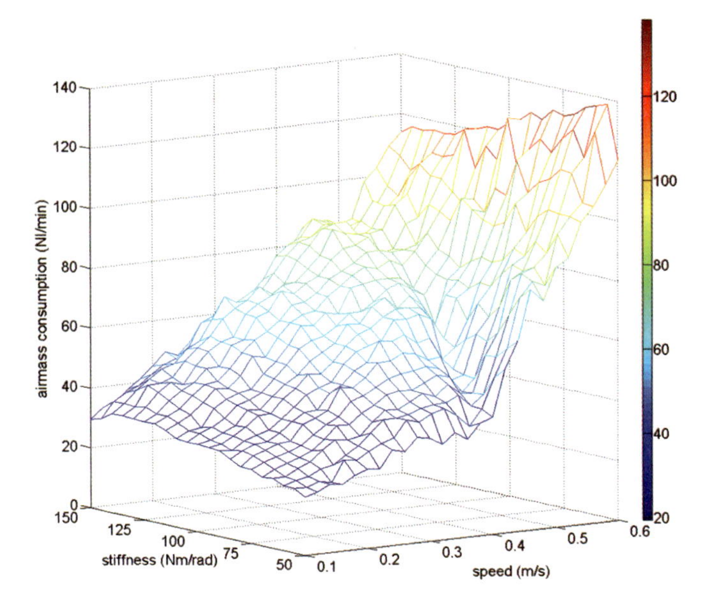

Fig. 5.14 Airmass consumption vs walking speed and stiffness for hip trajectory (real).

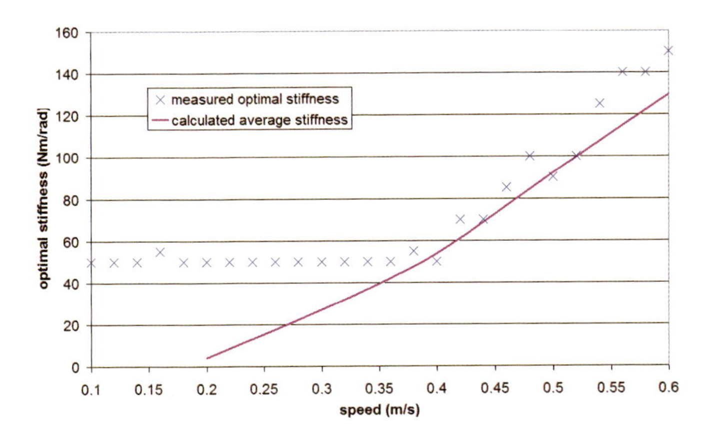

Fig. 5.15 Optimal compliance in function of walking speed (real).

energy is consumed. No energy can be recovered in the servo-motor. So the following formulation is used to calculate the energy consumption in case the pendulum is powered by an ideal stiff actuator.

$$E = \int |T\dot{\theta}|dt \tag{5.8}$$

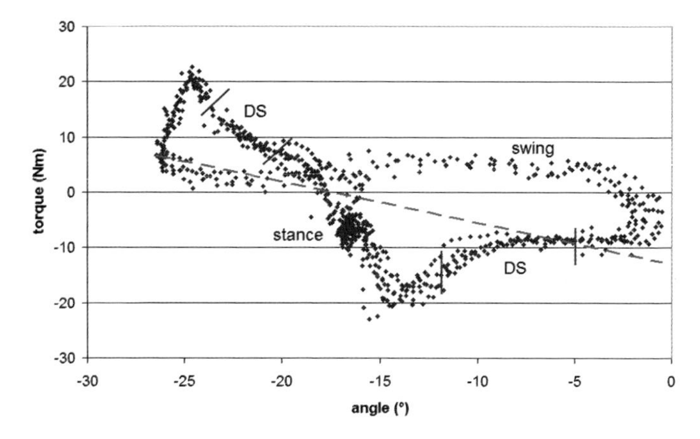

Fig. 5.16 Torque-angle relation of hip joint for real walking with first order linear regression line (real).

The calculation of the energy consumption of compressed air is not so obvious, since energy consumption depends on how air was pressurized by the power source. In (418) a possible way is described by taking the exergy associated with the pneumatic air mass flow as an estimation for the energy consumption. Exergy is the maximum amount of energy, with respect to the surrounding environment, which can be transformed into useful work. Because ideal conditions are presumed in the next experiments, the power of the airmass flow is calculated out of the airmass consumption with $W_{air} = rT_{air}^{sup}\dot{m}_{air}^{in}$ with the dry gas air constant $r = 287Jkg^{-1}K^{-1}$ and the temperature of the supplied air $T_{air}^{sup} = 293K$. Only the ingoing air is taken because the outgoing air is not recuperated and blown off.

Figure 5.17 shows a simulation of the power consumption versus frequency and the stiffness of an antagonistic setup of two pleated pneumatic artificial muscles. In the region of high frequencies and low stiffness settings a reduced power consumption can be observed. During fast motions the torques have to be high so $\Delta\tilde{p}$ is big and the joint is compliant so $\tilde{p}_s$ is small, causing the desired pressure to be below atmospheric pressure, which of course cannot be attained. In this region the controller has difficulties tracking the desired trajectory. There is no problem at high stiffness settings because the inlet pressure was taken high enough.

Figure 5.18 shows the power consumption in case an ideal stiff actuator is used. So no motion energy can be stored. This graph doesn't show the stiffness-axis because for a stiff actuator the stiffness is infinite. This curve is proportional to the third order with respect to the frequency. This is easy to find. $|T\dot{\theta}| \approx I\ddot{\theta}\dot{\theta}$ and $\theta = Asin(\omega t)$. So $|T\dot{\theta}| \sim \omega^3$.

When the minimal for each frequency of figure 5.17 is compared with the power consumption of an ideal stiff actuator of figure 5.18, then the pneumatic muscles consume much more power! This opens the discussion whether an antagonistic setup of two pleated pneumatic artificial muscles is good for reduced energy

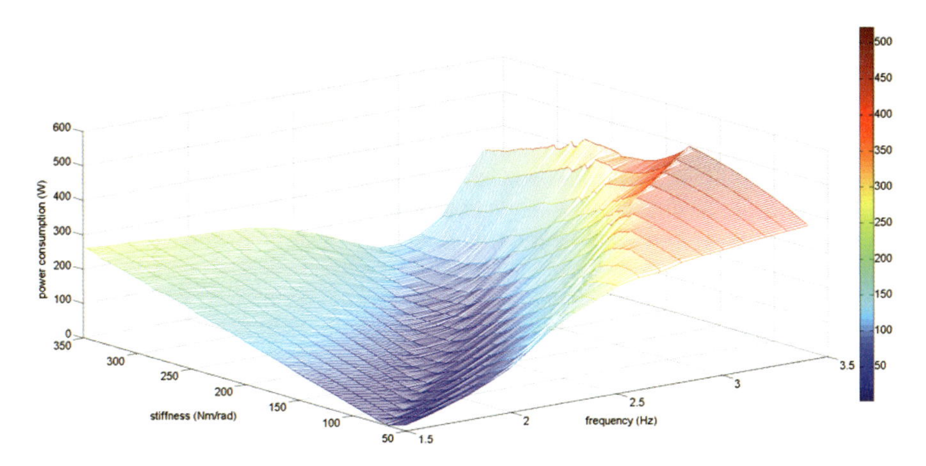

Fig. 5.17 Power consumption vs frequency and stiffness of design 1a (sim).

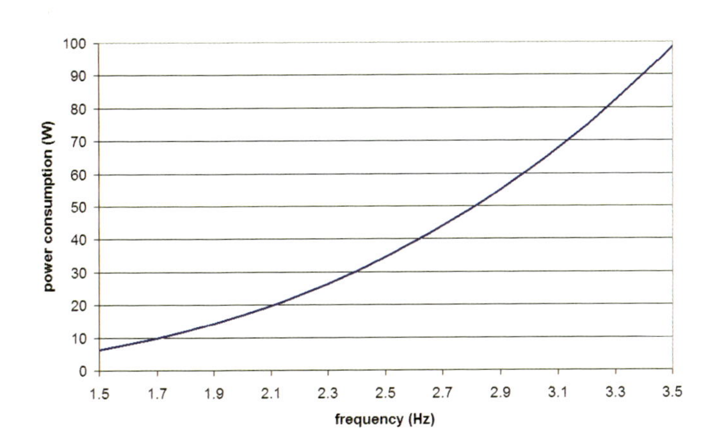

Fig. 5.18 Power consumption vs frequency of ideal stiff actuator in the joint (sinusoidal trajectory with 10° amplitude) (sim).

consumption by exploiting the natural dynamics of the system or if maybe other compliant actuators are more suitable for this purpose.

The remaining of this section is devoted to a comparison on energy consumption between different designs of compliant actuators while a sine wave is imposed. First a presentation is given about the different compliant actuators that will be used. The power consumption is measured in function of different frequencies and stiffness settings.

5.1.6 Presentation of the Different Compliant Actuators

Van Ham et al. (413) categorized passive compliant actuators in 4 groups, depending on their working principle:

1. Equilibrium-controlled stiffness. Design is based on adjusting the equilibrium position of springs. This concept is simple to control but constantly requires energy to regulate its actuator position.
2. Antagonistic setup of two non-linear springs. The idea is that 2 actuators with a non-linear force-displacement characteristic, are coupled antagonistically, which means that they work against each other. By controlling both actuators the compliance and equilibrium position of this antagonistic setup can be set.
3. Mechanically Controlled Stiffness. By varying the position of the attachment points of a compliant element to the structure of a joint, the compliance and torque of this joint can be changed.
4. Structural Controlled Stiffness. By varying the dimensional properties of a (leaf) spring, like length or moment of inertia, the stiffness constant of a spring can be adapted.

Figure 5.19 depicts mechanisms belonging to the first two categories and it is shown that there is a link between them. The explanation will start from an antagonistic setup of two pneumatic muscles and will go step by step towards the MACCEPA design (414). Design 1a and 8 are developed by the Robotics & Multibody Mechanics Research Group.

5.1.6.1 Design 1a and 1b (PPAM and PAM)

Figure 5.19-1b is the configuration which is mostly used by other groups working with pneumatic muscles and rotational joints (79; 438; 324), while the setup of figure 5.19-1a shows the way how PPAMs are used for the robot applications developed at the R&MM research group (see 2.5.1) and previously discussed in this work. The use of lever arms instead of a pulley introduces a non-linearity in the joint. This can be exploited by changing the angle and length of the lever arms to compensate the strong non-linearity of the force-angle characteristic of the muscles in order to flatten the torque-angle characteristics of the joint as shown in section 2.5.1. For the following discussion however it is assumed in the first designs (1a till 5) that both lever arms are of equal length and are placed on 180° relatively to each other.

5.1.6.2 Design 2a and 2b (VSA and design by Migliore)

The pneumatic muscles can be replaced by a stiff electrical motor and a non-linear spring. The non-linearity of the spring is necessary to be able to vary the compliance of the joint as explained in (413). The design of figure 5.19-2b represents the VSA (variable stiffness actuator) by Tonietti (55), and also the design of "biological inspired joint stiffness control" by Migliore (283).

5.1.6.3 Design 3a and 3b (AMASC)

In this design the endpoints of the two springs are mechanically coupled either by a lever arm or by a pulley. This important step decouples the setting of the equilibrium

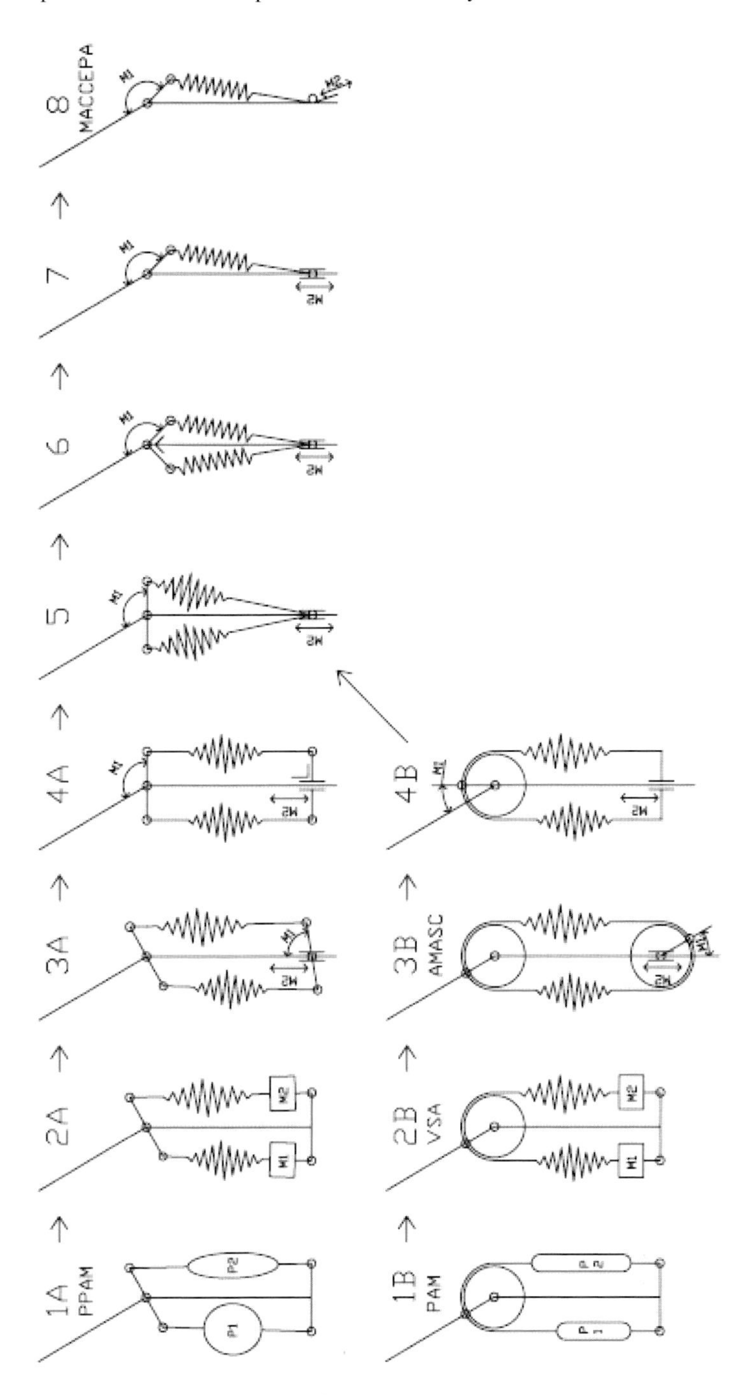

Fig. 5.19 Link between different compliant actuators.

position (this is the position where no torque is generated) and the compliance. This means that one motor is used to set the compliance by changing the pretension of both non-linear springs (servo-motor $M2$) and one to set the equilibrium position ($M1$). Figure 5.19-3b represents a simplified design of the AMASC actuator (186).

5.1.6.4 Design 4a and 4b

In both cases in figure 5.19-3a and b the motor to set the equilibrium position $M1$ is not on the joint but on the other side of the non-linear springs. It is also possible to move this motor to the joint, which e.g. would make the design of the AMASC considerably less complex. In this case the equilibrium position of both lever arms is horizontal, and the motor for the equilibrium position of the actual joint sets the relative position of the arm of the joint with respect to the lever arms. This design has been proposed by English and Russell (120).

5.1.6.5 Design 5

By joining both attachment points of the non-linear springs near motor drive M_2 a slightly compacter design can be obtained. Of course this introduces again some minor (depending on the dimensions) non-linearities.

5.1.6.6 Design 6

Until the previous step, non-linear springs and lever arms of equal length under an angle of 180° were used in order to be able to vary the compliance, so the adaptable compliance was implemented in the springs. If the angle of the lever arms is changed, a non-linearity is introduced at this level, and linear springs can be used instead. For example, when we set the angle between both lever arms at 90°, and the joint is pulled out of the equilibrium position (e.g. 45°), one spring generates no torque. The torque generated by the other one can be changed by changing the pretension of the springs, and thus the compliance can be altered.

5.1.6.7 Design 7

When the angle between both lever arms is made smaller, until they converge, they can be replaced by one lever arm and one spring. This simplifies the design significantly.

5.1.6.8 Design 8 (MACCEPA)

A last step is to fixate the point where the spring is guided. This can be done by guiding a cable, which is attached to the spring, around a fixed point to a pretension mechanism. This is the MACCEPA actuator (Mechanically Adjustable Compliance and Controllable Equilibrium Position Actuator) (414).

5.1.6.9 Summary

So the most important changes in design are:

1. Replacement of pneumatics by electrical power ($1 \rightarrow 2$)
2. Mechanical coupling of the endpoints of the two springs ($2 \rightarrow 3$)
3. Replacement of non-linearity in the springs by non-linearity in lever mechanism ($5 \rightarrow 6$)

5.1.7 Equations of Force and Compliance

In this section only designs 2a, 3a, 4a and 8 will be compared with an ideal stiff actuator because they have been built for real. The energy consumption of the others (except the one powered with artificial muscles) is in line with the proposed logical evolution. The generated torque and compliance of the compliant actuators under study and the way they will be controlled in the simulation are described.

It is supposed that the servo-motors used in the different setups are ideal and have the same characteristics as the ideal stiff actuator placed in the joint itself as described in section 5.1.5.1. The energy consumed by a linear motor is consequently:

$$E = \int |F\dot{x}|dt \tag{5.9}$$

and for a rotational actuator:

$$E = \int |T\dot{\theta}|dt \tag{5.10}$$

5.1.7.1 Equations for Design 2a (Ideal Antagonistic Setup of Two SEA)

The first design is the antagonistic setup of two series elastic actuators (SEA) with non-linear springs as shown in figure 5.20. For the non-linear springs quadratic springs are used. This gives a simple and straightforward equation to calculate the positions of the motors. This has also an advantage on the energy consumption as will be explained in section 5.1.8. Several methods have been proposed to produce non-linear elasticity, e.g. (283).

The generated non-linear forces by the springs are (see figure 5.20 for parameters used):

$$F_1 = k(x_1 - Asin(\theta))^2 \tag{5.11}$$

$$F_2 = k(x_2 + Asin(\theta))^2 \tag{5.12}$$

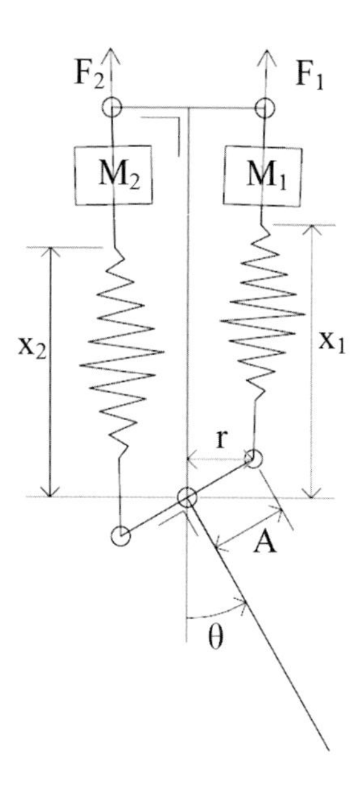

Fig. 5.20 Scheme of an antagonistic setup of two SEA (design 2a).

This approximation is valid if $x_1, x_2 >> A$. k is the spring constant. Joint torque:

$$\begin{aligned} T &= r(F_1 - F_2) \\ &= kAcos(\theta)\big((x_1 - Asin(\theta))^2 - (x_2 + Asin(\theta))^2\big) \\ &= kAcos(\theta)\big(x_1^2 - x_2^2 - 2(x_1 + x_2)Asin(\theta)\big) \end{aligned} \tag{5.13}$$

Joint compliance:

$$\begin{aligned} K = \frac{dT}{d\theta} &= \frac{dr}{d\theta}F_1 + r\frac{dF_1}{d\theta} - \frac{dr}{d\theta}F_2 - r\frac{dF_2}{d\theta} \\ &= -kAsin(\theta)(x_1 - Asin(\theta))^2 \\ &\quad - 2kAcos(\theta)(x_1 - Asin(\theta))Acos(\theta) \\ &\quad + kAsin(\theta)(x_2 + Asin(\theta))^2 \\ &\quad - 2kAcos(\theta)(x_2 + Asin(\theta))Acos(\theta) \\ &= -kAsin(\theta)\big(x_1^2 - x_2^2 + 2(x_1 + x_2)Asin(\theta)\big) - 2kA^2cos^2(\theta)(x_1 + x_2) \\ &= -tan(\theta)T - 2kA^2cos^2(\theta)(x_1 + x_2) \end{aligned} \tag{5.14}$$

Combining equations (5.13) and (5.14) it is possible to calculate the sum and difference of both motor positions.

$$S_{x_1x_2} = x_1 + x_2 = -\frac{K + tan(\theta)T}{2kA^2cos^2(\theta)} \tag{5.15a}$$

$$D_{x_1x_2} = x_1 - x_2 = \frac{\frac{T}{kAcos(\theta)} + 2(x_1 + x_2)Asin(\theta)}{x_1 + x_2} \tag{5.15b}$$

The motors have to be controlled in the following way in order to fulfill the desired torque and stiffness.

$$x_1 = \frac{S_{x_1x_2} + D_{x_1x_2}}{2} \tag{5.16}$$
$$x_2 = \frac{S_{x_1x_2} - D_{x_1x_2}}{2} \tag{5.17}$$

Power consumption:

$$W_{M1} = |F_1\dot{x}_1| \tag{5.18}$$
$$W_{M1} = |F_2\dot{x}_2| \tag{5.19}$$

Energy consumption:

$$E_{M1} = \int |F_1\dot{x}_1|dt \tag{5.20}$$
$$E_{M2} = \int |F_2\dot{x}_2|dt \tag{5.21}$$

5.1.7.2 Equations for Design 3a

The generated forces are (see figure 5.21 for parameters):

$$F_1 = k(x_1 - Asin(\theta))^2 \tag{5.22}$$
$$F_2 = k(x_2 + Asin(\theta))^2 \tag{5.23}$$

with $x_1 = P + Asin(\alpha)$ and $x_2 = P - Asin(\alpha)$. α and P are imposed by servo-motors.

Joint torque:

$$T = kAcos(\theta)\left(x_1^2 - x_2^2 - 2(x_1 + x_2)Asin(\theta)\right) \tag{5.24}$$

Joint compliance:

$$K = -tan(\theta)T - 2kA^2cos^2(\theta)(x_1 + x_2) \tag{5.25}$$

The position of the motors is calculated with:

$$x_1 + x_2 = -\frac{K + tan(\theta)T}{2kA^2cos^2(\theta)} \tag{5.26}$$
$$x_1 - x_2 = \frac{\frac{T}{kAcos(\theta)} - 2(x_1 + x_2)Asin(\theta)}{x_1 + x_2} \tag{5.27}$$

$$P = \frac{(x_1 + x_2)}{2} \tag{5.28}$$
$$\alpha = asin(\frac{x_1 - x_2}{2A}) \tag{5.29}$$

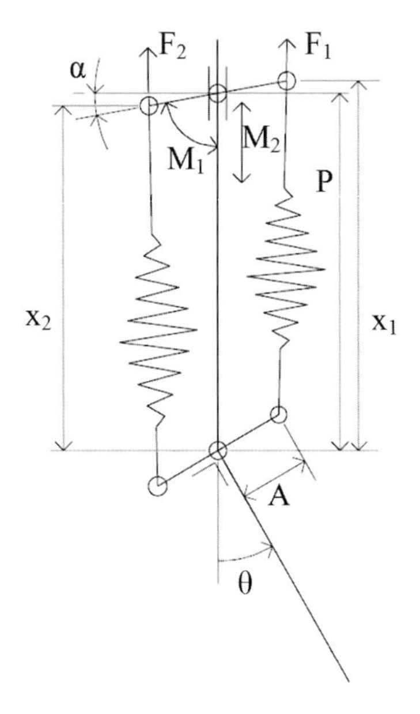

Fig. 5.21 Scheme of design 3a.

Energy consumption:

$$E_{M1} = \int |T\dot{\alpha}|dt \tag{5.30}$$

$$E_{M2} = \int |(F_1 + F_2)\dot{P}|dt \tag{5.31}$$

5.1.7.3 Equations for Design 4a

The generated forces are (see figure 5.22 for parameters):

$$F_1 = k(P - Asin(\alpha))^2 \tag{5.32}$$

$$F_2 = k(P + Asin(\alpha))^2 \tag{5.33}$$

with $\alpha = \theta - \beta + \pi/2$. β and P are imposed by servo-motors. Joint torque:

$$\begin{aligned} T &= r(F_1 - F_2) \\ &= kAcos(\alpha)\big((P - Asin(\alpha))^2 - (P + Asin(\alpha))^2\big) \\ &= -4kA^2cos(\alpha)sin(\alpha)P \\ &= -2kA^2sin(2\alpha)P \end{aligned} \tag{5.34}$$

Joint compliance:

$$K = -4kA^2cos(2\alpha)P \tag{5.35}$$

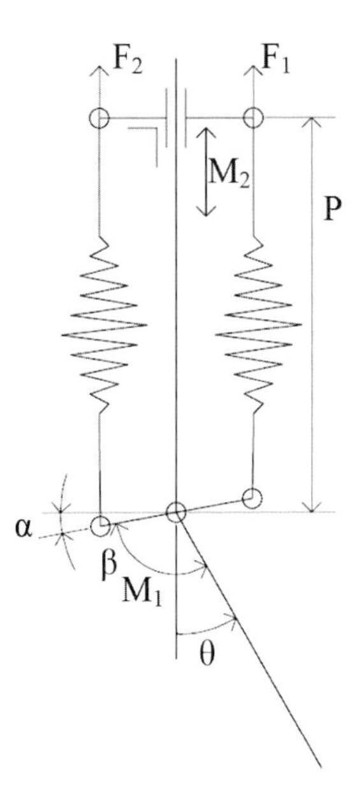

Fig. 5.22 Scheme of design 4a.

The position of the motor to control the stiffness K is calculated by:

$$P = -\frac{K}{4kA^2 cos(2\alpha)} \tag{5.36}$$

The angle β to control the equilibrium position is:

$$\beta = \theta - \frac{1}{2} atan(\frac{2T}{K}) + \pi/2 \tag{5.37}$$

Energy consumption:

$$E_{M1} = \int |T\dot{\beta}| dt \tag{5.38}$$

$$E_{M2} = \int |(F_1 + F_2)\dot{P}| dt \tag{5.39}$$

5.1.7.4 Equations for Design 8 (MACCEPA)

Figure 5.23 shows the scheme of design 8, the MACCEPA actuator. B is the lever arm, which sets the equilibrium position. At the end of the lever arm a linear spring is attached which is connected to a cable. C is the distance between the joint rotation point and a fixed point on the pendulum. The cable between the spring and the pretension mechanism is guided around this point. P is the extension of the spring caused by pre-tensioning and equals the total extension of the spring when $\alpha = 0$. α

is the angle between lever arm and the pendulum. φ is the angle between the vertical and the lever arm and is also the equilibrium position. The extension of the spring, equal to $\sqrt{B^2+C^2-2BCcos(\alpha)}-|B-C|+P$, has two independent causes: the variation of α, and the setting of the pre-tensioning P.

In (412) the torque T generated by a MACCEPA actuator is calculated as:

$$T = kBCsin(\alpha)\big(1+\frac{P-|B-C|}{\sqrt{B^2+C^2-2BCcos(\alpha)}}\big) \tag{5.40}$$

The stiffness K is:

$$\begin{aligned} K =&kBCcos(\alpha)\big(1+\frac{P-|B-C|}{\sqrt{B^2+C^2-2BCcos(\alpha)}}\big) \\ &-kB^2C^2sin^2(\alpha)\big(\frac{P-|B-C|}{\sqrt{(B^2+C^2-2BCcos(\alpha))^3}}\big) \end{aligned} \tag{5.41}$$

In the same work is shown that a linearization is justified for angles smaller than $\pm 45°$ if $B/C>5$, due to the quasi linear torque characteristic. The torque can be written:

$$T = \alpha\frac{kBC}{|C-B|}P \tag{5.42}$$

The constants B, C and k, which are fixed during the design, can be combined into a single constant μ:

$$\mu = \frac{kBC}{|C-B|} \tag{5.43}$$

The linearized torque now becomes:

$$T = \alpha\mu P \tag{5.44}$$

which is the torque-angle relation of a torsion spring with spring constant $K=\mu P$.

The pretension P has to be calculated out of the stiffness setting:

$$P = \frac{K}{\mu} \tag{5.45}$$

The angle between the vertical and lever arm (equilibrium position) φ is:

$$\varphi = \frac{T}{P\mu}+\theta \tag{5.46}$$

Energy consumption:

$$E_{M1} = \int |T\dot{\varphi}|dt \tag{5.47}$$

$$E_{M2} = \int |F\dot{P}|dt \tag{5.48}$$

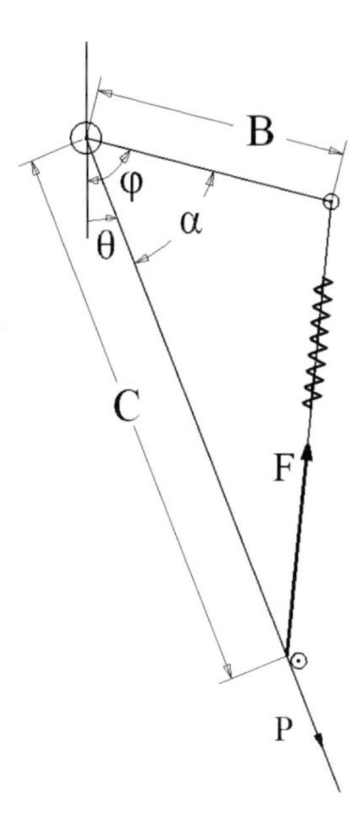

Fig. 5.23 Scheme of the MACCEPA (design 8).

with

$$F = k(\sqrt{B^2 + C^2 - 2BCcos(\alpha)} - |B - C| + P) \tag{5.49}$$

5.1.8 Simulation Experiments

All the following experiments were performed in simulation. Common to all the studied designs is the fact that when the position of the joint is moved out of the equilibrium position and subsequently released again, the pendulum will perform a swing motion without consuming energy. The imposed trajectory for these experiments is a sine wave with an amplitude of $10°$. The necessary torque to track this imposed trajectory is calculated using the computed torque method. In a first experiment the stiffness K is held constant. The energy consumption is measured over 5 swing motions and divided by the elapsed time span to have the average power consumption. These experiments were repeated for different stiffness settings going from $50Nm/rad$ till $350Nm/rad$ and frequencies between $1.5Hz$ and $3.5Hz$ and visualized in mesh plots (figures 5.25-5.29). Figure 5.25 depicts the power consumption of an antagonistic setup of two SEA (design 2a). One can see that the power consumption is very high with values up to $8000W$, especially when a high stiffness was set and that there is no minimum in energy consumption. Compared to a stiff actuator (figure 5.18) it is consequently better -on energy level- to use a stiff actuator instead of design 2a. The reason is that, in order to obtain the high stiffness, the springs have to generate high forces and in order to maintain a fix compliance setting the motors $M1$ and $M2$ have to move continuously while generating these high forces. The compliance is dependent of the joint angle as can be seen in equation (5.14). The relation between energy consumption and stiffness is a third power function. When the desired stiffness is increased with a factor 2, the sum of the motor positions $x_1 + x_2$ is also 2 times bigger as is the difference $x_1 - x_2$, which can be seen in equations (5.15a) and (5.15b). This causes the forces generated in the springs to be increased by a factor 4, due to the quadratic springs, and the motor speeds to be increased by a factor 2. Because the power consumption is a multiplication of spring forces and motor speeds the total power consumption is 8 times bigger.

A first improvement for the energy consumption is to allow the stiffness to change over the angle range instead of asking for an accurate tracking of a fix stiffness. It is chosen to take the stiffness at $\theta = 0°$ for the complete angle range. This means $cos(\theta) = 1$ and $sin(\theta) = 0$ in equation 5.14, which results in:

$$K = 2kA^2(x_1 + x_2) \tag{5.50}$$

The improved results can be found in figure 5.26. A clear minimum for each frequency can be found. In the further simulation experiments a similar angle-independent stiffness strategy is taken. The position of this valley of all the following graphs (5.26-5.29) are the same and can be found using the strategy proposed in section 5.1.4. Compared to a stiff actuator the energy consumption at the optimal stiffness is much better. So in this case using a compliant actuator is better than a stiff actuator. Using equation (5.50) has the effect that the real stiffness (calculated with equation (5.14)) deviates from the desired stiffness as can be seen in figure 5.24. The stiffness formulation of equation (5.50) corresponds to the actual stiffness related to design 2b, using a pulley with radius A instead of a lever mechanism. English and Russell (120) demonstrated theoretically that for this design only quadratic springs can decouple the joint stiffness from joint deflexion. This is an important reason to choose quadratic springs for a joint actuated by two antagonistic SEA.

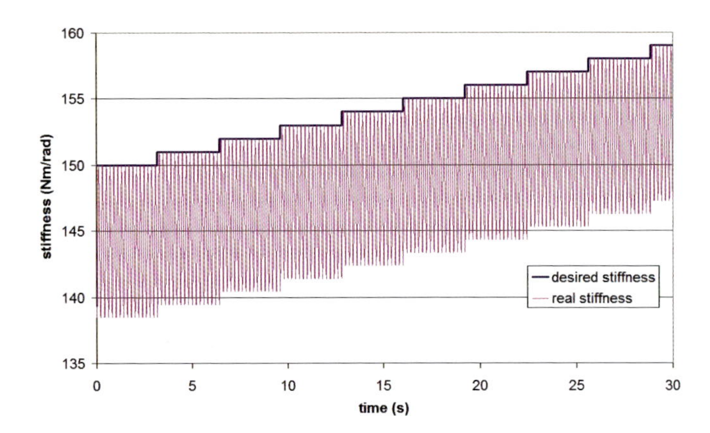

Fig. 5.24 Real and desired stiffness over time of design 2a (sim).

Although the new control method, with the average stiffness, consumes less energy at the optimal compliance, the energy consumption increases very fast at non-optimal values and goes much above the level needed when stiff actuators are used. The main reason for the high energy consumption is that both springs are not mechanically coupled. To produce a torque without affecting stiffness, both motors ($M1$ and $M2$) must move synchronic in opposite direction. This movement costs energy which cannot be recovered in the other motor. When the endpoints are mechanically coupled to each other energy only has to be supplied for the difference as is the case for design 3a and further. Moreover in design 2a, to generate a certain torque T, the forces generated by the two motors are dependent of the stiffness, while for the other designs (3a, 4a and 8) only the motor controlling the stiffness ($M2$) has to generate a force dependent on the stiffness. The power consumption for this motor is zero because during the motion the stiffness is unchanged ($\dot{P} = 0$). The other motor controlling the equilibrium position ($M1$) has only to generate the necessary torque to follow the desired motion which is independent of the stiffness.

Another disadvantage of design 2a is that the stiffness of the springs k has an influence on the power consumption, while this is not the case for the other designs. If the stiffness of the springs k in equation (5.50) is 10 times smaller, with the same desired joint stiffness K then $x_1 + x_2$ is 10 times bigger. This value remains unchanged during the motion due to the fixed desired compliance setting. The difference $x_1 - x_2$ (equation (5.15b)) remains unchanged and is independent of the spring stiffness. This means that the velocity of the servo-motors are the same. The forces in the springs on the other hand are also 10 times bigger, because it is a multiplication of k (10 times smaller) and the quadratic power of x (10 times bigger). This means that the total power consumption is 10 times bigger. So stiff springs should be preferred above compliant springs. The stiffness cannot be increased infinitely because then the motors have to do motions which are impossible.

The levels of power consumption for designs 3a and 4a are comparable as can be seen in figures 5.27 and 5.28. So it does not matter on energy level where motor $M1$ is positioned, or directly at the joint or at the end of the springs. In case the stiffness setting is more compliant than the optimal stiffness setting the power consumption increases fast. The forces in the springs are low and motor $M1$ has to make large motions to generate the required torque. Design 4 has as advantage that the joint position is not limited, while for design 3 the joint range is limited by the maximum stretch or compression of the springs. The construction of design 4 is also considerable less complex compared to design 3. Because for these designs the equilibrium position and compliance can be set more or less independently, each by a dedicated servo-motor, both motors can be dimensioned appropriately. This is not possible with design 2. Visser et al. (433) developed a metric for comparing different designs of variable stiffness actuators on the energy efficiency of the actuators with results in accordance with this result.

With stiffness settings lower than the optimal stiffness the power consumption of design 8 (figure 5.29) is better than design 3a and 4a. For the region of stiffness settings higher than the optimal stiffness the power consumption is similar. The

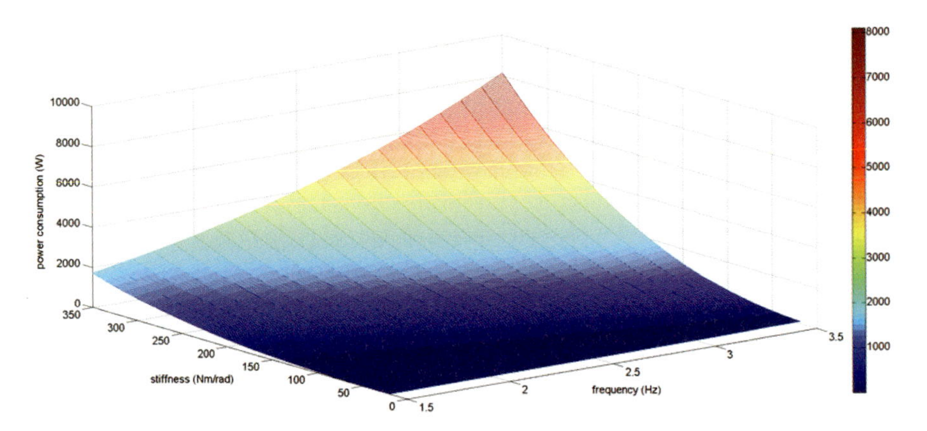

Fig. 5.25 Power consumption vs frequency and stiffness of design 2a (sinusoidal trajectory with 10° amplitude) (sim).

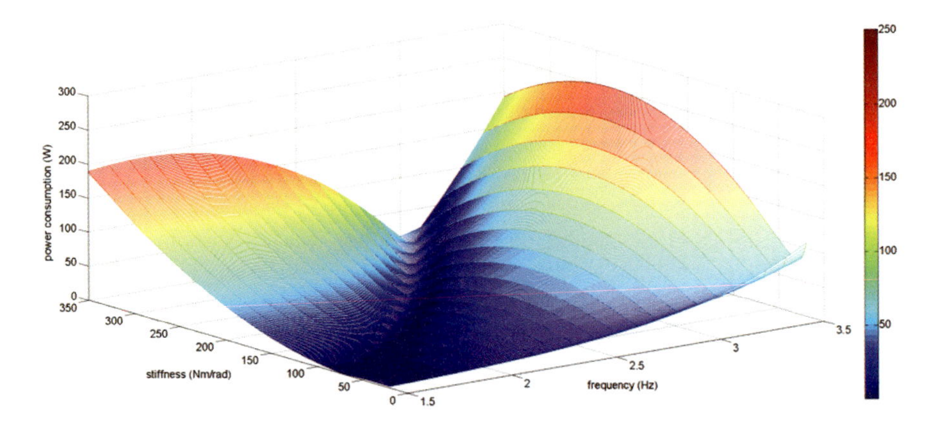

Fig. 5.26 Power consumption vs frequency and stiffness of design 2a, angle-independent stiffness (sinusoidal trajectory with 10° amplitude) (sim).

MACCEPA actuator has as advantage that the construction is much more straightforward than the other designs, especially the realization of quadratic springs is difficult.

Comparing the results of the designs equipped with springs with the setup powered by artificial muscles is not so easy, although the physical appearance is very similar with design 2. Measuring the energy consumption when using a spring and motor is very straightforward as shown in section 5.1.7. This is not the case when using artificial muscles. Measuring the energy in the same way is impossible because the attachment point of the muscle is rigidly connected to the structure and there the energy consumption is zero. For artificial muscles the spring and motor are combined in one element, the pressurized air is both responsible for the compliance and the force generation. But most remarks of design 2 can be repeated for a setup

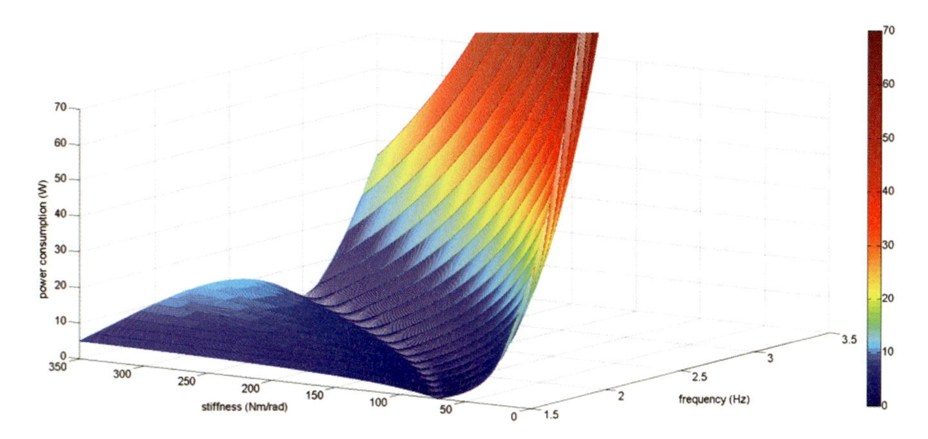

Fig. 5.27 Power consumption vs frequency and stiffness of design 3a (sinusoidal trajectory with 10° amplitude) (sim).

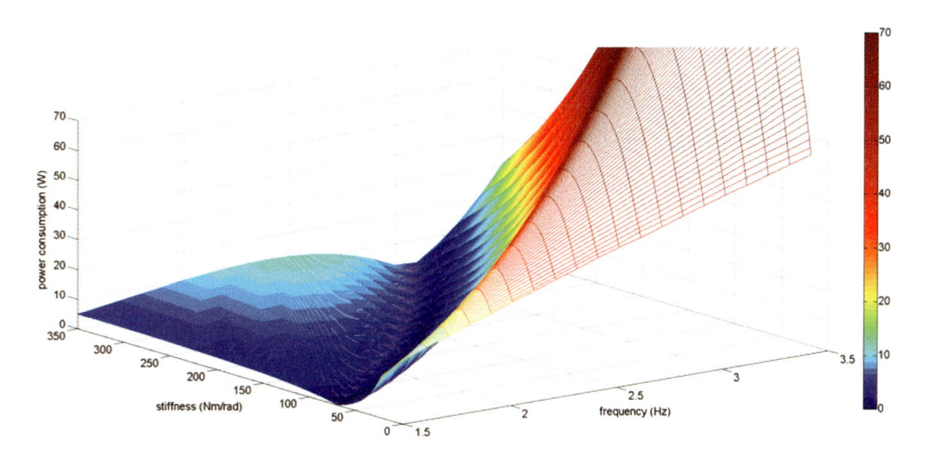

Fig. 5.28 Power consumption vs frequency and stiffness of design 4a (sinusoidal trajectory with 10° amplitude) (sim).

of artificial muscles. Both artificial muscles are not connected to each other. One can think of putting an extra tube with valve between them but this will only increase the complexity of the system. Both muscles have to be pressurized or depressurized to change the compliance and/or the torque. Also dimensioning one actuator for the compliance and the other for the torque is not possible. The maximum range of motion is limited by the maximum contraction and rest length of the muscle.

5.1.9 Conclusion

Actuators with adaptable compliance are gaining interest in the field of robotics. Few research, however, has been carried out on how to control the compliance.

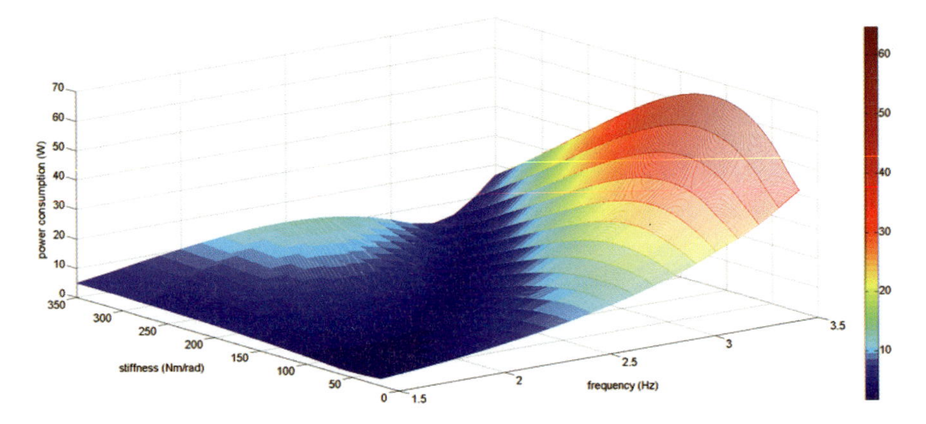

Fig. 5.29 Power consumption vs frequency and stiffness of design 8 (sinusoidal trajectory with 10° amplitude) (sim).

In this chapter a study is performed considering adapting the natural dynamics by compliance control. It was shown that for sine trajectories an optimal stiffness can be found with minimal energy consumption. A strategy was proposed to find this optimal compliance. The idea behind the mathematical formulation is to fit the controllable stiffness of the actuator to the natural stiffness of the desired trajectory. The natural stiffness of the desired trajectory is calculated as the derivative of the torque necessary to track the desired trajectory with respect to the joint angle. A first more complex trajectory which has been studied was a sum of two sinusoidal functions. Experiments showed that changing the stiffness during the trajectory costs a lot of energy so this should be avoided and a strategy with a fixed stiffness for a certain trajectory is preferable. A good stiffness seemed to be the slope of the first order linear regression line of the torque-angle curve. This strategy also worked when a hip trajectory, as calculated by the trajectory generator, was used as imposed trajectory. The minima were not so distinct anymore since for more complex trajectories the energy consumption increases because it differs more from the unforced motion.

The strategy is not implemented in the real biped Lucy. The current maximum walking speeds are too slow to use the proposed average stiffness strategy to be able to exploit the natural dynamics. Another important remark is the fact that the trajectory generator for the biped Lucy does not include the possibilities of compliance adaptation. At this moment the strategy is to generate first the trajectories and perform afterwards the compliance control. So besides choosing an optimal compliance also energy optimal trajectories have to be generated for which also the compliant characteristics of the actuator are incorporated.

Another problem of using pneumatic muscles in an antagonistic setup is the limited stiffness range because of the minimum and maximum pressure inside the muscles. By choosing other attachment points of the muscles the stiffness range can be changed, but not at no cost. A possibility to increase the maximum stiffness is to make the lever arm bigger, but then the movable joint angle range is reduced. The dimensioning of the actuation system is a complex mechanical design problem with multiple non-linear objectives and non-linear constraints. An exhaustive search approach is proposed by Beyl et al. (53) to solve this multi-objective optimization problem. Also not included in this discussion is the power consumption to produce compressed air.

At the end of this section a comparison was made between different compliant actuators equipped with a spring. It can be concluded that a precise tracking of the desired stiffness consumes a lot of energy and that a stiffness setting of for example the middle of the trajectory is more interesting. An antagonistic setup of two SEA each actuated by a servo-motor consumes more energy than designs were both energy buffers are coupled. The MACCEPA actuator was the best regarding energy consumption.

5.2 Compliant Actuation for Jumping

As presented in the introduction (chapter 1), the function of a human leg in hopping or running is comparable to a spring. Motion energy is stored during the landing phase and released during the push-off phase. However, stiff actuators and active compliant actuators cannot store energy. The forces generated by the impacts have to be limited in order not to damage the gearboxes. On the other hand compliant actuators are able to store motion energy and absorb impact shocks. The compliance also influences peak ground reaction force, ground contact time, center of mass displacement and stride frequency (122). Pneumatic artificial muscles can generate high torques fast, making it possible for a robot to jump as will be shown.

The experiments in this section show the robot Lucy is to jump. All the figures shown in this section are real experiments. One has to consider that these experiments are very preliminary. The COG of the robot is accelerated vertically to a certain take-off speed needed to attain a desired jumping height. Out of the motion of the COG it is straightforward to calculate the trajectories for the knee and ankle joint. The orientation of the upper body is kept fixed. In fact the upper body should also be controlled using an equation to control the angular momentum at take-off. But the range of motion of the joints is too limited. At take-off the feet are lifted and at the highest position the control of the robot is stopped and all the valves are closed. This to examine the passive effects of the pneumatic muscles.

5.2.1 *Equations*

The jumping motion is divided in a jumping phase, air phase and landing phase. During the first phase the COG is accelerated vertically from a position with bent knees to a nearly stretched position. Because the motion range of the joints is limited the start position of the COG is $X_{COG}^{0} = 0.023m$ and $Y_{COG}^{0} = 0.563m$, the take-off (TO) position is $Y_{COG}^{TO} = 0.617m$. So within $Y_{COG}^{TO} - Y_{COG}^{0} = 0.054m$ the COG has to reach its necessary take-off speed $\dot{X}_{COG}^{TO} = \sqrt{2g(Y_{COG}^{max} - Y_{COG}^{TO})}$, with Y_{COG}^{max} the desired maximum jumping height. When the start velocity and acceleration is taken zero, because at that moment the robot is standing still it is straightforward to calculate the trajectory of the COG:

$$
\begin{aligned}
\ddot{X}_{COG}(t) &= \frac{2\dot{X}_{COG}^{TO}}{t_{TO}^2}t \\
\dot{X}_{COG}(t) &= \frac{\dot{X}_{COG}^{TO}}{t_{TO}^2}t^2 \\
X_{COG}(t) &= X_{COG}^0 + \frac{\dot{X}_{COG}^{TO}}{3t_{TO}^2}t^3
\end{aligned}
\tag{5.51}
$$

This trajectory of the COG is the same as the vertical jumping motion of HRP-2 (356). t_{TO} is the time spent in the jumping phase and is calculated out of the distance the robot can travel in the jumping phase.

$$
t_{TO} = \frac{3(Y_{COG}^{TO} - Y_{COG}^0)}{\dot{X}_{COG}^{TO}}
\tag{5.52}
$$

5.2.2 *Jumping Experiments*

For a jumping height of $5cm$ the necessary take-off speed becomes $\dot{X}_{COG}^{TO} = 0.98m/s$ as can be seen in figure 5.31. The desired and real position of the hip and COG are shown in figure 5.30. The desired trajectories stops at the moment of take-off because they remain unchanged. At take-off the knees are retracted. This can be seen in figure 5.32, showing the knee angle which is extended during the jumping phase and bent again during the first half of the air phase. The definition of the angles is shown in figure 2.12, an angle of $0°$ for the knee means completely stretched. This causes the feet to be lifted higher ($9cm$ as shown in figure 5.33) than the jumping height. As can be seen in figure 5.34-5.35, showing the desired and real pressure inside the front and back muscle of the knee joint, the pressure has to change very fast in a short time period during the jumping phase. The pressure of the back knee muscle has difficulties tracking the desired pressure course, but there are possibilities to increase the jumping height because not all the inlet valves of the front muscle are open during the whole time. At maximum jumping height all the valves are closed to see the impact effects of the passive system falling down.

Figure 5.38 shows screenshots of the feet from jumping phase until impact, taken every $0.08s$. They have to be compared with the foot forces measured by the loadcells in the sole, figures 5.36-5.37. One can see that due to the large generated torque in the ankle at take-off the biggest reaction forces are generated in the tip and consequently the heel is lifted first. The desired absolute angle of the feet is always set horizontal, but due to large torques the foot rotates down at take-off. At touch-down, first the tips touch the ground which can be noticed by the increase in front force at $t = 84.3s$, then the heel creates a large impact of about $110kg$. Performing such experiments using harmonic drives is not advisable.

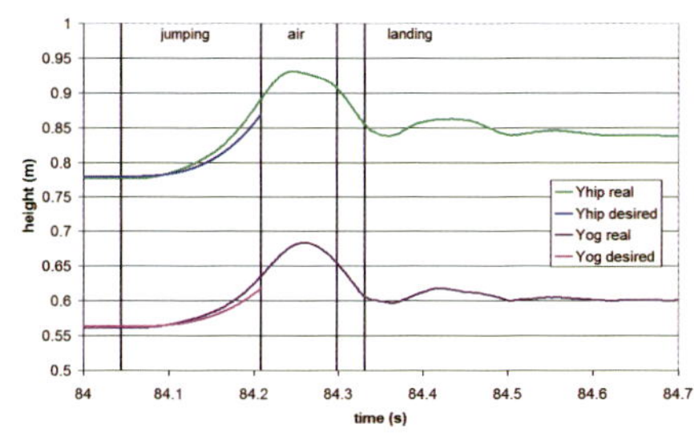

Fig. 5.30 Vertical position of hip joint and COG.

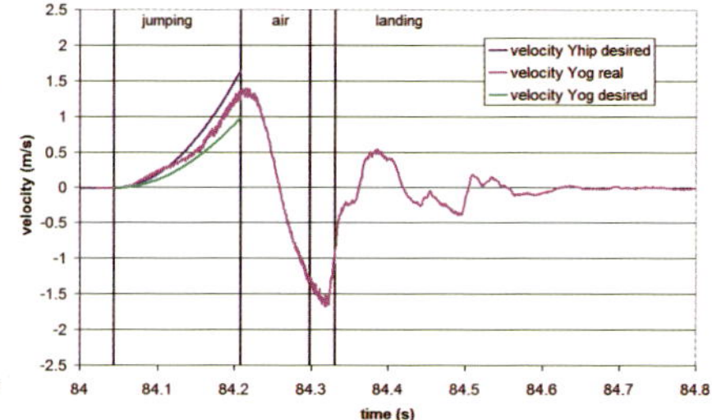

Fig. 5.31 Vertical velocity of hip joint and COG.

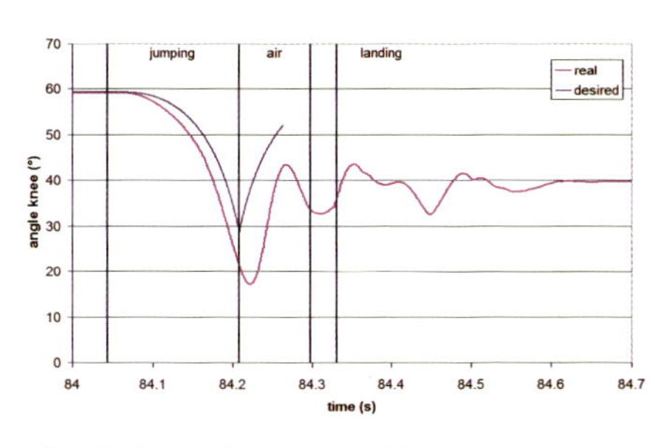

Fig. 5.32 Desired and real knee angle.

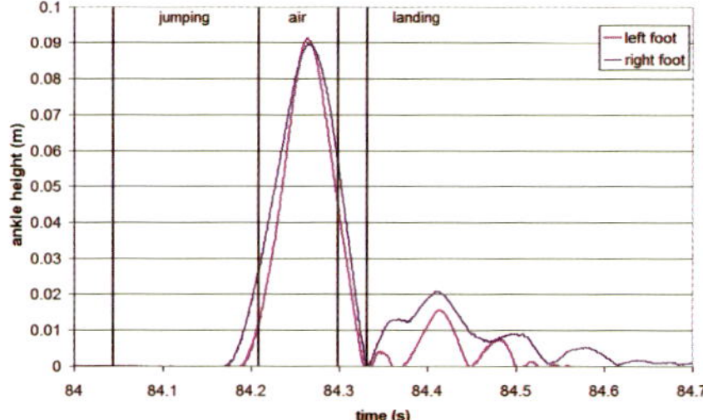

Fig. 5.33 Foot height left and right foot.

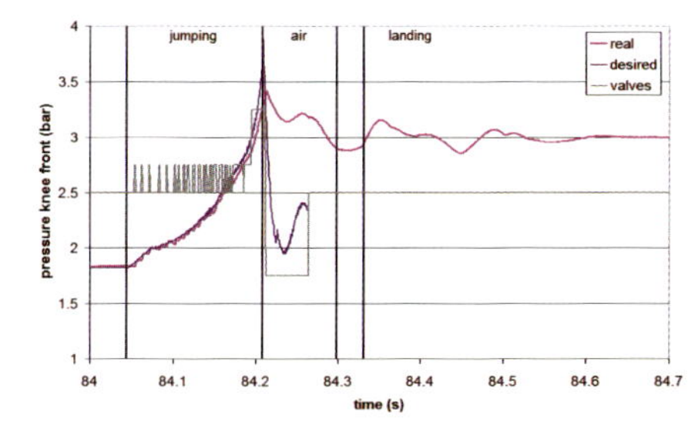

Fig. 5.34 Real and desired pressure in front knee muscle, valve action.

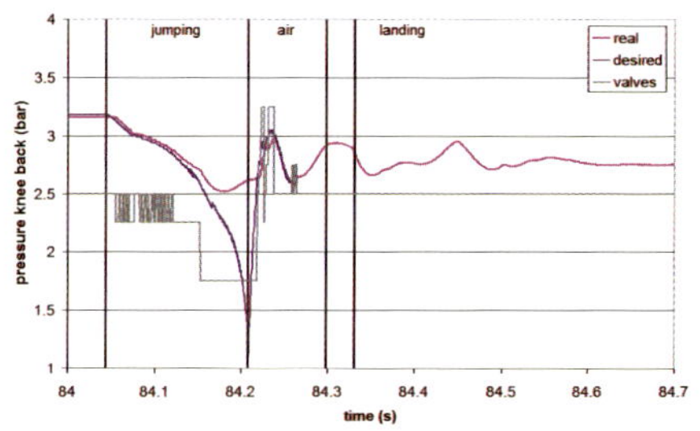

Fig. 5.35 Real and desired pressure in back knee muscle, valve action.

5.2.3 *Future Work*

It is clear that these experiments are preliminary and further research should be done. First a trajectory generator has to be developed to be able to hop while stability is maintained. Next a strategy has to be developed for the period from impact to the next lift-off. During this period the compliance will store motion energy and release it during the second phase. The strategy has to set an optimal compliance depending on hop-height/frequency, add energy to compensate for the friction and impact losses and one has to include control to maintain stability.

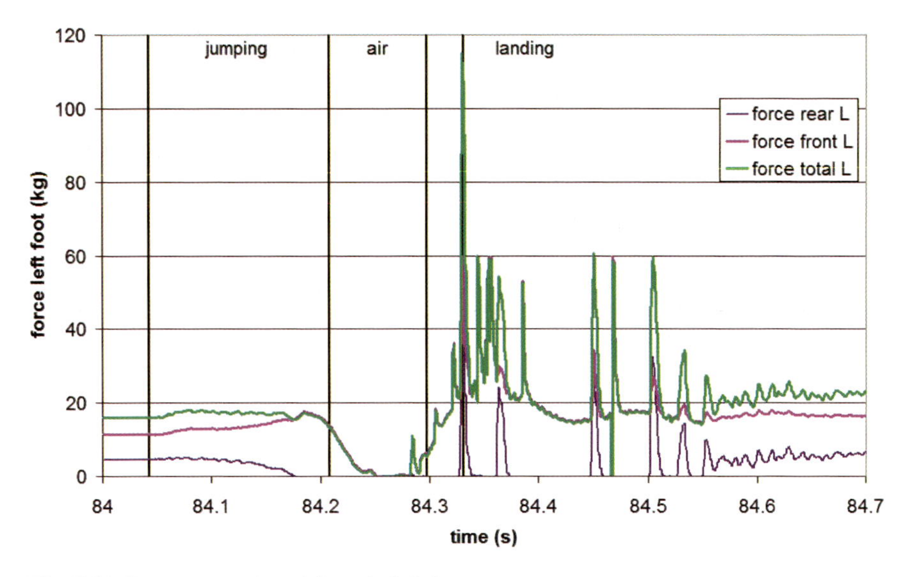

Fig. 5.36 Front, rear and total force in left foot.

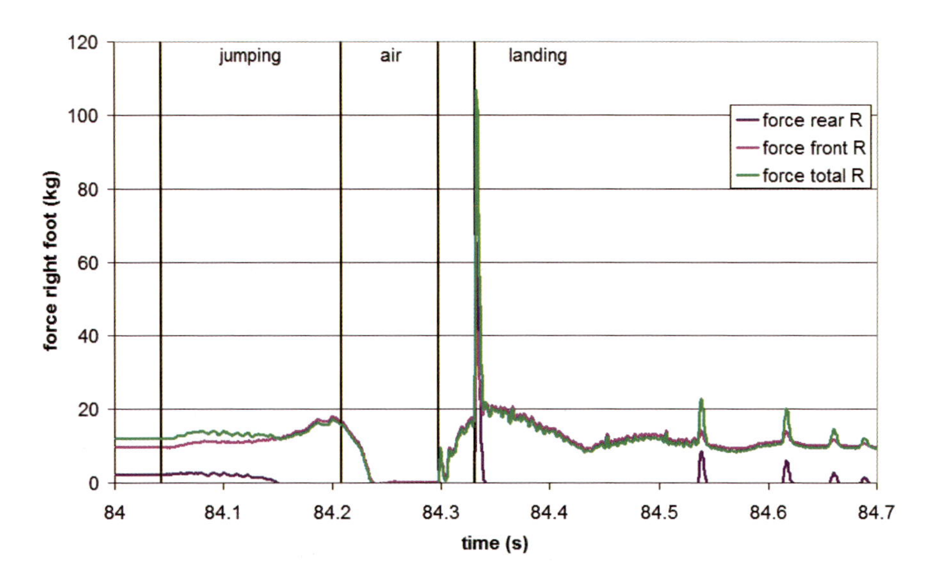

Fig. 5.37 Front, rear and total force in right foot.

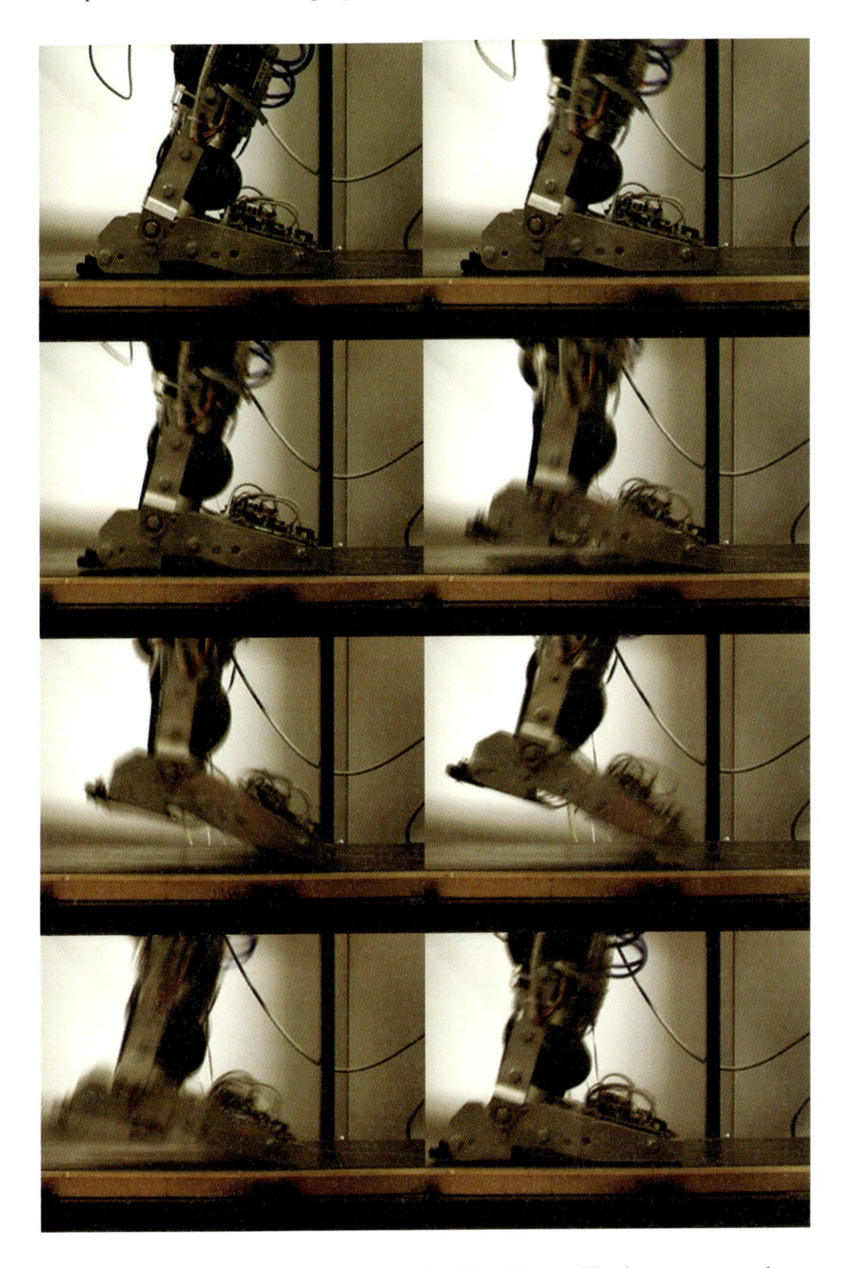

Fig. 5.38 A sequence of photos of the jumping biped Lucy. The images were taken every $0.08s$.

Chapter 6
General Conclusions and Future Work

6.1 General Conclusions

This book reports on the development and control of the bipedal walking robot Lucy. The main purpose of the biped is to evaluate the implementation of compliant actuators and to develop control strategies for bipedal locomotion. Compliant actuators are currently studied and applied in walking systems for research in the field of low-energy consumption walking. Often the approach is to start from passive walkers while adding control to be able to walk on level ground, to be more robust against disturbances and so on. Disadvantage of this group is that the number of walking motions is limited. The approach intended for Lucy is to start from dynamic stable trajectories which are tracked by a tracking controller. Afterwards a compliance controller should select an appropriate stiffness so the motion of the natural dynamics correspond as much as possible to the reference trajectories. Advantage is that the robot will be capable of starting, stopping and walking at different walking speeds and step lengths.

An interesting compliant actuator for this purpose is an antagonistic setup of two pleated pneumatic artificial muscles. Pneumatic artificial muscles have some characteristics which can be beneficial towards actuation of legged locomotion. These actuators have a high power to weight ratio and they can be coupled directly to the structure without complex gearing mechanism. Due to the compressibility of air, a joint actuated with pneumatic drives shows a compliant behavior, which can be employed to reduce shock effects at touch-down of a leg. Moreover, in a joint setup with two muscles positioned antagonistically, the joint compliance can be adapted while controlling the position. This joint compliance adaptation can be used to influence the natural dynamics of the system for reduced energy consumption.

Chapter 2 started with a description of the pleated pneumatic artificial muscles and the antagonistic muscle setup. This antagonistic setup is used in a modular unit which forms a link of the biped. Such a modular unit has two valve islands consisting of on/off valves. The opening and closing times of these valves were reduced by an electronic valve speed-up circuitry. Every modular unit is controlled by a 16-bit micro-controller which measures the pressures in both muscles, the angular position

B. Vanderborght: Bipedal Walking by the Pneumatic Robot Lucy, STAR 63, pp. 215–221.
springerlink.com

and angular velocity and controls the on/off valves. Six modular units are linked to each other and together with the feet they form a complete robot. So the robot consists of a an upper-body and two articulated legs with one dimensional joints. To prevent the sagittal robot from falling sidewards, a guiding mechanism consisting of a horizontal and a vertical rail is chosen. These rails are mounted on a frame which also incorporates a treadmill so that the robot is able to walk longer distances. The robot is controlled by a central PC which communicates at $2000Hz$ with the micro-controllers using the USB 2.0-protocol. An interface program with GUI is written allowing the user to control the robot. Besides the real biped also a hybrid simulator is developed in which the mechanics of the robot and the thermodynamical processes in the muscles are combined in one set of differential equations. The simulator is used to debug control programs and evaluate them before implementation in the real biped.

The current control architecture of the biped Lucy consists of two components: a trajectory generator and a joint trajectory tracking unit.

The trajectory generator, described in chapter 3, calculates trajectories for the different joints so that the robot can walk from a certain position to another while keeping the Zero Moment Point (ZMP) in the stability region, thus ensuring dynamic balancing of the robot. A first developed strategy is based on the inverted pendulum model, which represents the complete mass of the robot as a single point mass. For each step the objective locomotion parameters (step length, intermediate foot lift and speed) can be chosen. The motion of the hip during the single support phase is calculated in such a way that there is no ankle torque, meaning that the ZMP stays in the ankle joint. During the double support phase the accelerations are planned so that the next set of objective parameters is attained and that there is a smooth transition of the ZMP from rear to front ankle point. This strategy does not include the complete distributed masses of a real robot and consequently the real and desired ZMP will differ. When walking at moderate speeds the ZMP stays in the support area so the robot maintains its stability. For higher walking speeds however a stabilizer should be implemented or a trajectory generator has to be implemented which includes the complete multibody mass distribution.

The second version of the trajectory generator is based on the preview control method for the ZMP developed by Kajita (212), which has been successfully used in the humanoid robot HRP-2. The goal is to have the ZMP follow a predefined trajectory. This is not as straightforward as calculating the ZMP out of the joint trajectories. The main idea is to plan the motion of the COG in function of desired ZMP trajectories determined by the foothold sequence. The problem is regarded as a ZMP servo control implementation, trying to track the ZMP by controlling the horizontal jerk. Because the hip has to move before the ZMP path changes, information about desired position of the ZMP in the future is needed, hence the use of a preview control method. The dynamics are simplified to a cart-table model, a cart that represents the global COG of the robot moving on a horizontally positioned pedestal table with negligible mass. Since the true robot is a multibody system the real and desired position of the ZMP will differ. In order to solve this issue, Kajita (212) proposed a re-feeding of the complete multibody calculated ZMP trajectory into a second stage of the preview control with the same cart-table model by means of taking the error between the multibody calculated ZMP and the desired ZMP trajectory. This results in deviations of the horizontal motion of the COG. By implementing this method it is observed that the real ZMP tracks the imposed trajectory well, so a more stable walking motion is obtained. It is important to have enough mass in the upper body of the robot so the COG is positioned near or above the hip, in order to minimize the influence of the swinging leg on the dynamics. Normally the trajectory cannot be changed anymore in the time period up to twice the preview period ahead. To be able to implement the strategy for Lucy, the whole trajectory is recalculated at impact to correct for the real step length instead of the desired one. Reason is that the real and desired step length can differ.

This preview control approach was also used in a strategy to let the humanoid robot HRP-2 dynamically walk over large obstacles. Trajectories for the feet, waist and upper-body were generated that fulfilled the following requirements. All the trajectory had to be collision-free. Obstacles require large steps which threaten the dynamic stability of the robot and the knee of HRP-2 cannot be put in an overstretched position. Impacts at touch down of the foot have to be minimized in order not to damage the harmonic drives for example. The method of the preview control presumes the robot walks with a constant hip height, but to walk over large obstacles the hip has to be lowered. The second stage of the preview controller is robust enough to tackle this besides the disturbances coming from the swing leg. Experiments showed the robot is capable of negotiating an obstacle of $15cm$ (plus $3cm$ safety boundary zone) in height and $5cm$ (plus 2x$3cm$ safety boundary zone) in width. For higher obstacles the speed limits of the knee joints are reached. This is the highest obstacle a humanoid has currently ever stepped over to the author's knowledge.

The task of the joint trajectory tracking controller (chapter 4) is to apply appropriate muscle pressures to let the robot follow the joint trajectories as required by a trajectory generator. Due to the specific nature of the pneumatic actuation system and the highly non-linear character of the system, this tracking controller has several essential units. The inverse dynamics unit calculates the required joint torques based on the robot dynamics. This dynamic model is different for the single and double support phase because during single support the robot has 6 DOF and during double support the number of DOF is reduced to 3 (which makes the system over-actuated). This unit is based on the computed torque method consisting of a feedforward part and a PID feedback loop. Subsequently for each joint a delta-p unit translates the calculated torques into desired pressure levels for the two muscles of the antagonistic set-up. Finally, a bang-bang controller with dead zone determines the necessary valve signals that control the actions of a set of on/off valves to set the correct pressures in the muscles. The complete control structure, trajectory generator and joint trajectory tracking controller, was experimentally validated on the robot. First a comparison is made between the method based on the inverted pendulum model and the preview control method. The former requires less computations but the real ZMP differs more from the desired one in comparison with the preview control method. The use of the latter method results in a higher attainable walking speed of $0.15m/s$ while the first method can attain $0.11m/s$. This is quite fast for an actively controlled pneumatic biped. This speed limitation has two causes. The robot always has to walk with flat feet due to the lack of a toe-joint so the maximum step length is $18cm$. Another reason is due to the ability of the valves to follow the desired pressure course. Especially the exhaust valves cannot follow the imposed values above a speed of $0.15m/s$. An indication of the robustness of the controller was shown by randomly adding and releasing a mass of $6kg$ (18% of robots weight) during walking.

Chapter 5 concerns the adaptability of the compliance of the actuators. A strategy is developed to combine active trajectory control with the exploitation of the natural dynamics in order to reduce energy consumption. This study was not performed on the biped Lucy, but on a single pendulum structure powered by pleated pneumatic artificial muscles. First sinusoidal trajectories were studied. By changing the stiffness an optimal constant stiffness could be found for which the airmass consumption was minimal. A mathematical formulation was derived to calculate this optimal value which is dependent of the physical properties of the pendulum and the frequency of the imposed motion. The idea behind the mathematical formulation is to fit the actuator compliance to the natural compliance of the desired trajectory. The natural stiffness of the desired trajectory is calculated as the derivative of the torque necessary to track the desired trajectory with respect to the joint angle. For trajectories more complex than a sinusoidal function, the optimal stiffness as calculated with the previous strategy is not a constant anymore. It was shown however that changing the compliance costs a lot of energy and a fixed compliance strategy should be preferred. Both for a trajectory consisting of a sum of two sinusoidal functions and a hip trajectory calculated by the trajectory generator the average stiffness seems to be a good approximation of the optimal stiffness. The average stiffness is defined here as the slope of the first order linear regression line fitted to the torque-angle curve. This strategy could not be implemented in the real biped because the walking speed of the robot is too slow to benefit from a compliance adaptation. In the last part of the section about compliant actuation for exploitation of natural dynamics different designs of compliant actuators with a spring element are compared. For all the designs holds that the energy consumption for a sinusoidal trajectory is much less than in the case stiff actuators. If a compliance is chosen away from the optimal then the energy consumption for an antagonistic setup were the endpoints of the springs are not coupled mechanically increases very fast and becomes much higher than in the case where stiff actuators are used. The conclusion is that an antagonistic setup is less appealing regarding energy consumption. The reason is that in order to change the torque or compliance both actuators have to work, while energy from one motor cannot be recovered in the other motor. This is not the case when the equilibrium position and compliance can be set independently.

In the same chapter, preliminary jumping motions were analyzed, thereby showing the capability of muscles to absorb impact shocks. Further research in this field was made impossible by the limited joint range. Strategies for exploiting the capability of the muscles to store and release motion energy therefore were not developed.

The goal of this work was to give an answer to the following questions:

Can pneumatic artificial muscles be used for dynamic balanced bipedal locomotion in a trajectory controlled manner? The robot Lucy has been built and a control architecture has been developed to dynamically stabilize the pneumatic biped. It is currently the fastest and one of the most advanced robots in the field of trajectory controlled pneumatic bipeds.

How to control the adaptable compliance of a joint powered by passive compliant actuators? A compliance controller has been developed to control the compliance of the actuators to reduce the energy consumption. Because the current walking speed of the robot is too slow to benefit from the exploitation of the natural dynamics, it is not yet implemented in the biped.

Please go to **http://lucy.vub.ac.be/phdlucy.wmv** to watch a video of the biped Lucy. The video starts with a brief history of some milestones in the construction of the robot. Afterwards videos of the experiments described in this work are shown. **http://lucy.vub.ac.be** is the site of the biped Lucy containing publications, press coverage and other information.

6.2 Future Work

The knowledge gathered during this project is currently extended towards other applications developed at the Robotics & Multibody Mechanics Research Group. Two new projects are situated in the emerging field of medical rehabilitation robotics. The ALTACRO project (54), which stands for "Automated Locomotion Training using an Actuated Compliant Robotic Orthosis", focuses on the design, construction and testing of a step rehabilitation robot for patients suffering from gait disorders. Introducing compliance in the actuation opens up the possibility for new therapies, for example in combination with Functional Electrical Stimulation (FES). Another project is the design of an intelligent transtibial prosthesis actuated by the pneumatic actuators (429). The use of these pneumatic actuators allows both the incorporation of adaptable compliance in the prosthesis by regulating the internal air pressure and the generation of the required plantar flexion torque for obtaining a normal gait pattern.

The more fundamental research towards the use of compliant actuators for bipedal locomotion is certainly not finished. The optimal choice between the trajectory-controlled robots on one side and the group of robots derived from the passive walkers on the other side is not yet found. This will be a robot, like a human, which can execute all the desired motions in combination with the exploitation of the natural dynamics to reduce energy consumption. A lot of research activity is currently observed to extend the capabilities of powered passive walkers. However there is little research going on in the path followed by the author: combining trajectory tracking with compliance adaptation to exploit natural dynamics. In chapter 5 a strategy is proposed to adapt the compliance in function of the desired trajectory. This is not yet tested on a complete biped. The trajectories of chapter 3 have been built to ensure dynamic stability, but no considerations about energy consumption have been made. Thus, a new trajectory generator is needed which combines the possibilities of compliance adaptation and trajectory generation.

Few robots are yet able to walk on surfaces were legged robots do have a real advantage over wheeled robots: rough terrain and uneven structures. When the actually most advanced humanoid robot Asimo gives a show, the technical requirements are that the floor surface has irregularities of at most $2mm$ and the horizontal deviation is at most 1°. No slippery or springy floors are allowed. For HRP-2 a stabilizer has been developed that can cope with slightly uneven terrain. The surface may have gaps smaller than $20mm$ and slopes $< 5\%$. Terrain maps can be built using stereo vision. However, methods for 3D reconstruction of surfaces of a real environment are computationally very expensive. This is a disadvantage because the reconstruction has to be performed in real-time and together with other processes such as motion planning, trajectory generation and stabilization. Given a height map of the terrain and a discrete set of possible footstep motions, planners are developed to generate a sequence of footstep locations to reach a given goal state. Typically errors of such stereo vision data in height are, for the humanoid robot HRP-2, of $20mm$ when the robot stands still (462). One can conclude that for a range of sizes of obstacles on one side the robot cannot see them and on the other side the robot is unable to walk over rough terrain without initially knowledge of the structure of the surface. Compliance can be the key to solve this gap. A control method using stiff actuators will reach the desired position whatever the external forces are and will reject any disturbances. In case of an unknown structure the desired position will be reached, this does not mean that a good and firm contact between sole and ground is established. Here flexibility is required. The author believes that a strategy based on active or passive compliant actuators can give good results. The latter has the advantage that it is able to absorb impact chocks occurring at higher walking speeds, but the control is probably more difficult.

Compliant actuators are very crucial for hopping and running robots to store and release motion energy. Humans run energy efficiently by storing motion energy mainly in the Achilles tendon and the stored energy is released during the next hop. A strategy from which robots with stiff actuators cannot benefit. At impact all the energy is lost. Moreover such stiff structure creates big impact shocks, possibly damaging electronics and mechanics. Consequently strategies have to be developed to reduce impact effects. A possibility is to retract the leg just before touch down. Such a strategy is superfluous when compliant actuators are used. The control strategy should set an optimal compliance depending on hop-height/frequency or running speed, add energy to compensate for the friction and impact losses and one should add control to maintain stability.

Crucial for a robot to be ever allowed in a human environment will be a guaranteed safe human-robot interaction. In this field compliant actuation also can play a major role. Besides a soft skin the structure must behave inherently flexible to minimize the damage in case of an impact. Besides collision with other objects the robot must also be able to fall without getting damaged. Compliant actuators can reduce impact shocks and a strategy should detect if a fall occurs and position the robot in an optimal posture and compliance depending on the way the robot is falling.

The compliant actuator used during this project were the pleated pneumatic artificial muscles in an antagonistic setup. In this work some advantages of the muscles were stated. However, some disadvantages were also encountered. The muscles need valves, buffers, tubing and silencers which are quite complex, heavy and noisy. The bandwidth of the muscles is dependent on how fast air can be put in and out the muscles and this is yet not fast enough to reach walking speeds comparable with the speed of human walking. Especially depressurizing the muscle takes much time. The compliance range is limited and an antagonistic setup is not good for energy efficient walking. The PPAMs are also difficult to produce and the production time is long, although the newly developed 3rd generation copes with the problem. Compressed air is delivered by an external source in the lab. To make the robot autonomous the air compressor or tank should be taken onboard. Using batteries is far more easy. Although different designs of compliant actuators are currently under investigation, the ultimate design combining a stiffness range from completely stiff to zero stiffness, lightweight and compact and easy to control has not yet been invented. One can conclude that in the field of compliant actuation a lot of research is still possible, on the actuator itself, the applications, how to control the compliance, etc. Within the robotics community even a lot of misunderstanding between viewpoints exists concerning this field and the discussion has certainly not come to an end.

Despite the mentioned disadvantages, the complete study of Lucy has given important research results. Out of nothing a complete robot was built and programmed into a walking biped in a short period and with limited resources. This offered us many insights in different disciplines and gave interesting ideas on how a new biped should look like. In the meantime Lucy became world famous and hopefully the work inspired many other researchers.

Appendix A
Thermodynamic Model

In this section the first order differential equation describing the pressure changes inside the muscle valve system is formulated. The discussion is based on the works of Daerden (104) and Brun (66).

The first law of thermodynamics is applied to a muscle with its valve island of 6 on/off valves. The muscle itself and its tubing until the different input and exhaust valve orifices are taken as control volume V. Figure (A.1) gives a schematic representation where the two inlet valves and the four exhaust valves are respectively depicted as one inlet and one exhaust. The first law is given in its rate form and

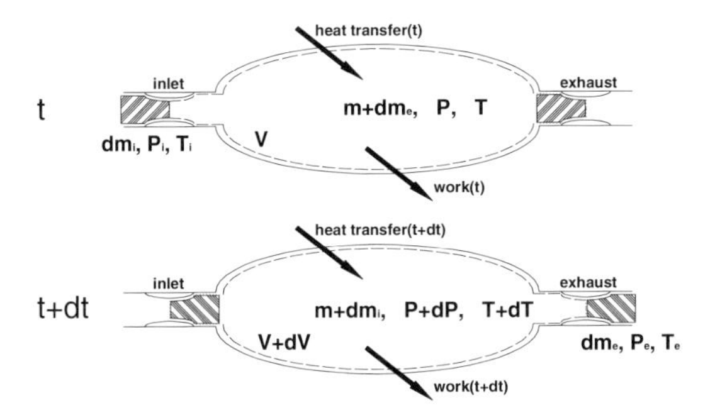

Fig. A.1 Muscle and valves on time step t and $t+dt$.

expresses that the variation of the total energy of an amount of fluid is equal to the sum of the work done by the exerted forces and the net heat transfer with the surrounding. Assuming a uniform thermodynamic state inside the control volume the first law of thermodynamics can be written as follows (variation referred to time):

$$dU + dE_k + dE_p = \delta W + \delta Q \tag{A.1}$$

with:

$$
\begin{aligned}
dU &= \text{variation of the fluid's total internal energy} \\
dE_k &= \text{variation of the fluid's total kinetic energy} \\
dE_p &= \text{variation of the fluid's total potential energy} \\
\delta W &= \text{work done by external forces} \\
\delta Q &= \text{the net transfer of heat across the boundary}
\end{aligned}
$$

The pressurized air can be regarded as an ideal gas for which the following relations hold:

$$PV = mrT \tag{A.2}$$

$$u = c_v(T - T_0) \tag{A.3}$$

$$h = c_p(T - T_0) \tag{A.4}$$

$$c_p = c_v + r \tag{A.5}$$

with:

$$P = \text{absolute pressure} \tag{A.6}$$

$$V = \text{air volume} \tag{A.7}$$

$$m = \text{air mass} \tag{A.8}$$

$$T = \text{temperature} \tag{A.9}$$

$$r = \text{dry air gas constant} = 287Jkg^{-1}K^{-1} \tag{A.10}$$

$$u = \text{specific internal energy} \tag{A.11}$$

$$h = \text{specific enthalpy} \tag{A.12}$$

$$c_v = \text{constant volume specific heat} = 718Jkg^{-1}K^{-1} \text{ for dry air at } 300K \tag{A.13}$$

$$c_p = \text{constant pressure specific heat} = 1005Jkg^{-1}K^{-1} \text{ for dry air at } 300K \tag{A.14}$$

$$T_0 = \text{reference temperature which is taken zero} \tag{A.15}$$

To calculate the different variations in equation A.1 for the open muscle-valve system, the constant mass $(m + dm_i + dm_e)$ is studied at two instant time steps t and $t + dt$ as depicted in figure (A.1). At time t, pressurized air with mass dm_i is about to enter the control volume V while mass $m + dm_e$ is inside this volume. At $t + dt$ mass dm_e is leaving while the mass inside the control volume is $m + dm_i$. Evaluating equation A.3 between the two time steps results in:

$$dU = [(m + dm_i)\, c_v\, (T + dT) + dm_e c_v T_e] - [(m + dm_e)\, c_v T + dm_i c_v T_i] \tag{A.16}$$

While neglecting second order terms, equation A.16 leads to:

$$dU = mc_v dT + dm_i c_v (T - T_i) + dm_e c_v (T_e - T) \tag{A.17}$$

Neglecting furthermore the kinetic energy of the air inside the muscle against the kinetic energy of the inlet and exhaust, the variation of kinetic and potential energy becomes:

$$dE_k = dm_e \frac{C_e^2}{2} - dm_i \frac{C_i^2}{2} \tag{A.18}$$

$$dE_p = dm_e g z_e - dm_i g z_i \tag{A.19}$$

The work exchanged with the environment, while assuming reversibility, is expressed as:

$$dW = -PdV + P_i dV_i - P_e dV_e \tag{A.20}$$

with the first term, the work done by the muscle and the other two terms associated with the work needed to transport dm_i and dm_e in and out the muscle volume.

Combining the first law of thermodynamics (A.1) with equations (A.17), (A.18), (A.19) and (A.20) gives:

$$\begin{aligned} mc_v dT + c_v T\,(dm_i - dm_e) &= -PdV \\ &\quad + dm_i \left(c_v T_i + P_i v_i + \frac{C_i^2}{2} + g z_i \right) \\ &\quad - dm_e \left(c_v T_e + P_e v_e + \frac{C_e^2}{2} + g z_e \right) + \delta Q \end{aligned} \tag{A.21}$$

with v_i and v_e the specific volume of inlet and exhaust. Taking into account conservation of mass and the definition of enthalpy:

$$dm = dm_i - dm_e \tag{A.22}$$

$$h = u + Pv \tag{A.23}$$

Differentiating the perfect gas law (A.2) gives:

$$d\,(PV) = PdV + VdP = mrdT + rTdm \tag{A.24}$$

Using (A.24), equation (A.21) can be transformed to:

$$\begin{aligned} \frac{c_v}{r} d\,(PV) &= -PdV \\ &\quad + dm_i \left(h_i + \frac{C_i^2}{2} + g z_i \right) \\ &\quad - dm_e \left(h_e + \frac{C_e^2}{2} + g z_e \right) + \delta Q \end{aligned} \tag{A.25}$$

Flows through small orifices, such as valves and tubes, are assumed to be adiabatic and since no mechanical work is exchanged with the surroundings, for these situations is stated:

$$h + \frac{C^2}{2} = constant \tag{A.26}$$

Thus for inlet and exhaust can be written:

$$h_i + \frac{C_i^2}{2} = h_s = c_p T_s \tag{A.27}$$

$$h_e + \frac{C_e^2}{2} = h = c_p T \tag{A.28}$$

with h_s and T_s the enthalpy and temperature of the pressurized air supply buffer, h and T are the enthalpy and temperature of the pressurized air inside the muscle volume. For equations (A.27) and (A.28) kinetic energy is neglected since the considered volumes are assumed large enough. Taking into account these two equations and the definition $\gamma = c_p/c_v$ and relation (A.5), the energy balance (A.25) can be rewritten in the following form, if potential energy of the air masses is neglected:

$$dP = -\frac{\gamma}{V}(PdV + rT_s dm_i - rT dm_e + (\gamma - 1)\delta Q) \tag{A.29}$$

If furthermore an adiabatic process is considered, $\delta Q = 0$, equation (A.29) becomes:

$$dP = \frac{\gamma}{V}(-PdV + rT_s dm_i - rT dm_e) \tag{A.30}$$

Expression (A.30) is valid for the so called isentropic process, where adiabatic and reversibility conditions are assumed. The non-ideal conditions can be represented in analogy with the polytropic process, by substituting γ with a polytropic coefficient n in equation (A.30) ($n = 1.2$):

$$dP = \frac{n}{V}(-PdV + rT_s dm_i - rT dm_e) \tag{A.31}$$

with dm_i and dm_e determined by air flows through the different inlet and exhaust valves and dependent on the number of valves that are opened.

Appendix B
Kinematics and Dynamics of the Biped Lucy during a Single Support Phase

B.1 Kinematics

The biped model during a single support phase is depicted in figure B.1. For the following derivations it is supposed that both legs are identical. Hereby assuming all inertial properties and the length of the upper and lower leg to be pairwise equal. l_i, m_i and I_i are respectively the length, mass and moment of inertia with respect to the local COG G_i of link i. The location of the COG's G_i are given by $J_1G_1 = J_6G_5 = \alpha l_1$, $J_2G_2 = J_5G_4 = \beta l_2$ and $J_3G_3 = \gamma l_3$ and for the foot $J_6G_6 = \sigma l_6$ where $0 < \alpha, \beta, \gamma, \sigma < 1$. The position of each link i is given by the angle θ_i, measured with respect to the horizontal axis.

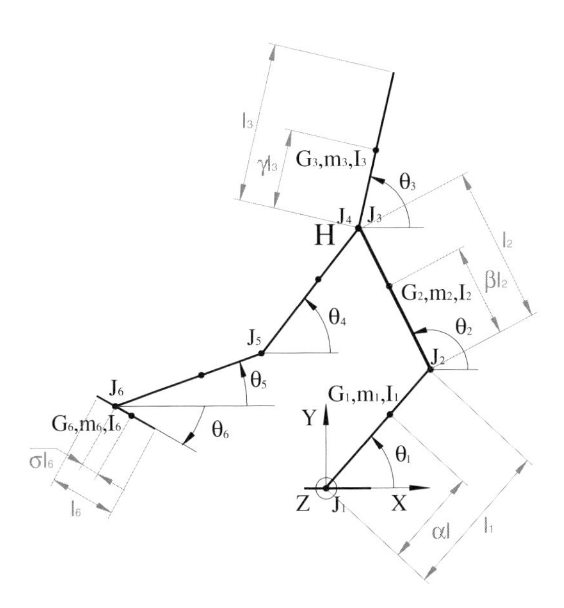

Fig. B.1 Model of the biped during a single support phase.

The hip takes a central position, so the location of the different COG's is calculated with reference to this point.

$$X_H = l_1 \cos\theta_1 + l_2 \cos\theta_2 \tag{B.1a}$$
$$Y_H = l_1 \sin\theta_1 + l_2 \sin\theta_2 \tag{B.1b}$$

The vectors defining the position of the local COG's of each of the five links are calculated as:

$$\overline{OG_1} = (X_H, Y_H)^T - (1-\alpha)\, l_1 \,(\cos\theta_1, \sin\theta_1)^T - l_2 \,(\cos\theta_2, \sin\theta_2)^T \tag{B.2a}$$
$$\overline{OG_2} = (X_H, Y_H)^T - (1-\beta)\, l_2 \,(\cos\theta_2, \sin\theta_2)^T \tag{B.2b}$$
$$\overline{OG_3} = (X_H, Y_H)^T + \gamma l_3 \,(\cos\theta_3, \sin\theta_3)^T \tag{B.2c}$$
$$\overline{OG_4} = (X_H, Y_H)^T - (1-\beta) l_2 \,(\cos\theta_4, \sin\theta_4)^T \tag{B.2d}$$
$$\overline{OG_5} = (X_H, Y_H)^T - (1-\alpha)\, l_1 \,(\cos\theta_5, \sin\theta_5)^T - l_2 \,(\cos\theta_4, \sin\theta_4)^T \tag{B.2e}$$
$$\begin{aligned}\overline{OG_6} = (X_H, Y_H)^T &+ \sigma l_6 \,(\cos\theta_6, \sin\theta_6)^T \\ &- l_1 \,(\cos\theta_5, \sin\theta_5)^T - l_2 \,(\cos\theta_4, \sin\theta_4)^T\end{aligned} \tag{B.2f}$$

The position of the global COG of the robot, stance foot not included, is given by:

$$\overline{OG} = (X_G, Y_G)^T \tag{B.3}$$

with:

$$\begin{aligned}X_G = X_H &+ a_1 \cos\theta_1 + a_2 \cos\theta_2 + a_3 \cos\theta_3 \\ &+ a_4 \cos\theta_4 + a_5 \cos\theta_5 + a_6 \cos\theta_6\end{aligned} \tag{B.3a}$$
$$\begin{aligned}Y_G = Y_H &+ a_1 \sin\theta_1 + a_2 \sin\theta_2 + a_3 \sin\theta_3 \\ &+ a_4 \sin\theta_4 + a_5 \sin\theta_5 + a_6 \sin\theta_6\end{aligned} \tag{B.3b}$$

and:

$$\begin{aligned}a_1 &= -(1-\alpha)\,\eta_1 l_1 \\ a_2 &= -\left[\eta_1 + (1-\beta)\,\eta_2\right] l_2 \\ a_3 &= \gamma \eta_3 l_3 \\ a_4 &= -\left[\eta_1 + \eta_6 + (1-\beta)\,\eta_2\right] l_2 \\ a_5 &= -\left[\eta_6 + (1-\alpha)\,\eta_1\right] l_1 \\ a_6 &= \sigma \eta_6 l_6\end{aligned}$$

and:

$$\eta_i = \frac{m_i}{2(m_1 + m_2) + m_3 + m_6}$$

The first and second derivative of (B.3a) and (B.3b), which are required for the derivation of the dynamic model and the ZMP, are straightforward and thus not explicitly listed here.

B.2 Dynamics

With the swing foot included, the robot has 6 DOF during the single support phase if the robot is assumed to move only in the sagittal plane. These degrees of freedom are represented by the 6-dimensional vector:

$$\mathbf{q} = \left[\theta_1\, \theta_2\, \theta_3\, \theta_4\, \theta_5\, \theta_6\right]^T \tag{B.4}$$

The dynamics are represented by 6 equations of motion of which the i th equation can be written with the Lagrange formulation as:

$$\frac{d}{dt}\left\{\frac{\partial K}{\partial \dot{q}_i}\right\} - \frac{\partial K}{\partial q_i} + \frac{\partial U}{\partial q_i} = Q_i \quad (i = 1 \ldots 6) \tag{B.5}$$

with K and U, respectively the total kinetic and gravitational energy of the robot, Q_i are the generalized forces associated with the generalized coordinates q_i.

The total kinetic energy can be found by the summation of the separate kinetic energy values of each link:

$$K = \sum_{i=1}^{6} K_i = \frac{1}{2}\sum_{i=1}^{6}\left(m_i v_{G_i}^2 + I_i \dot{\theta}_i^2\right) \tag{B.6}$$

with $\bar{v}_{G_i} = \left(\dot{X}_{Gi}, \dot{Y}_{Gi}\right)^T$ the velocity of the COG of link i and $\dot{\theta}_i$ the angular velocity. The expression of the total kinetic energy is quite long and is not explicitly listed here.

The gravitational (potential) energy is given by:

$$U = MgY_G \tag{B.7}$$

The generalized forces are the different net torques acting on each link of the robot (see figure B.2):

$$\mathbf{Q} = \tau = \begin{bmatrix} \tau_1 \\ \tau_2 \\ \tau_3 \\ \tau_4 \\ \tau_5 \\ \tau_6 \end{bmatrix} = \begin{bmatrix} \tau_{K_S} - \tau_{A_S} \\ \tau_{H_S} - \tau_{K_S} \\ -\tau_{H_S} - \tau_{H_a} \\ \tau_{H_a} - \tau_{K_a} \\ \tau_{K_a} - \tau_{A_a} \\ \tau_{A_a} \end{bmatrix} \tag{B.8}$$

The H, K and A stands for "Hip", "Knee" and "Ankle" respectively, a stands for "air", and s for "stance". Expression (B.8) gives the relations between net torques and applied joint torques.

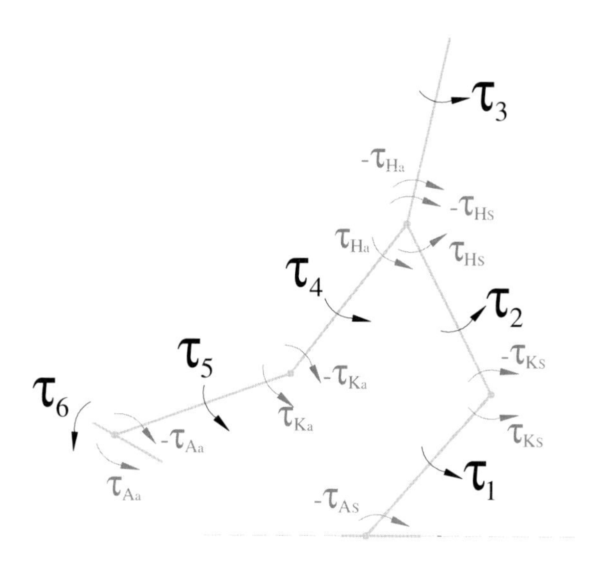

Fig. B.2 Definition of net torques and joint torques.

The 6 equations of motion (B.5) can be written in the following form (376):

$$D(\mathbf{q})\ddot{\mathbf{q}} + C(\mathbf{q},\dot{\mathbf{q}})\dot{\mathbf{q}} + G(\mathbf{q}) = \tau \tag{B.9}$$

with $D(\mathbf{q})$ the inertia matrix, $C(\mathbf{q},\dot{\mathbf{q}})$ the centrifugal/coriolis matrix, $G(\mathbf{q})$ the gravitational torque vector and τ the net torque vector.

The inertia matrix can be calculated with the following relation to the kinetic energy:

$$K = \frac{1}{2}\dot{\mathbf{q}}^T D(\mathbf{q})\dot{\mathbf{q}} \tag{B.10}$$

The elements of the centrifugal/coriolis matrix c_{kj} can be found with the following expression (376):

$$c_{kj} = \sum_{i=1}^{6} c_{ijk}\dot{\theta}_i = \sum_{i=1}^{6} \frac{1}{2}\left\{\frac{\partial d_{kj}}{\partial \theta_i} + \frac{\partial d_{ki}}{\partial \theta_j} - \frac{\partial d_{ij}}{\partial \theta_k}\right\}\dot{\theta}_i \tag{B.11}$$

with c_{ijk} the so called Christoffel symbols and d_{ij} the elements of the inertia matrix $D(\mathbf{q})$. The elements of the gravitational torque vector g_i are given by:

$$g_i = \frac{\partial U}{\partial q_i} \tag{B.12}$$

As a result all the parameters of the dynamic model are given below:

- Inertia matrix $D(\mathbf{q})$:

$$
\begin{aligned}
d_{11} &= I_1 + l_1^2\left[\left(1+\alpha^2\right)m_1 + 2m_2 + m_3 + m_6\right] \\
d_{12} &= l_1 l_2\left[m_1 + (1+\beta)m_2 + m_3 + m_6\right]\cos(\theta_1-\theta_2) = d_{21} \\
d_{13} &= l_1 l_3 \gamma m_3 \cos(\theta_1-\theta_3) = d_{31} \\
d_{14} &= l_1 l_2\left[(\beta-1)m_2 - m_1 - m_6\right]\cos(\theta_1-\theta_4) = d_{41} \\
d_{15} &= l_1^2\left[(\alpha-1)m_1 - m_6\right]\cos(\theta_1-\theta_5) = d_{51} \\
d_{16} &= l_1 l_6 m_6 \sigma \cos(\theta_1-\theta_6) = d_{61} \\
d_{22} &= I_2 + l_2^2\left[m_1 + \left(1+\beta^2\right)m_2 + m_3 + m_6\right] \\
d_{23} &= l_2 l_3 \gamma m_3 \cos(\theta_2-\theta_3) = d_{32} \\
d_{24} &= l_2^2\left[(\beta-1)m_2 - m_1 - m_6\right]\cos(\theta_2-\theta_4) = d_{42} \\
d_{25} &= l_1 l_2\left[(\alpha-1)m_1 - m_6\right]\cos(\theta_2-\theta_5) = d_{52} \\
d_{26} &= l_2 l_6 m_6 \sigma \cos(\theta_2-\theta_6) = d_{62} \\
d_{33} &= I_3 + \gamma^2 l_3^2 m_3 \\
d_{34} &= 0 = d_{43} \\
d_{35} &= 0 = d_{53} \\
d_{36} &= 0 = d_{63} \\
d_{44} &= I_2 + l_2^2\left[m_1 + (1-\beta)^2 m_2 + m_6\right] \\
d_{45} &= l_1 l_2\left[(1-\alpha)m_1 + m_6\right]\cos(\theta_4-\theta_5) = d_{54} \\
d_{46} &= -l_2 l_6 m_6 \sigma \cos(\theta_4-\theta_6) = d_{64} \\
d_{55} &= I_1 + l_1^2\left[m_1(1-\alpha)^2 + m_6\right] \\
d_{56} &= -l_1 l_6 m_6 \sigma \cos(\theta_5-\theta_6) = d_{65} \\
d_{66} &= I_6 + l_6^2 m_6 \sigma^2
\end{aligned}
$$

- Centrifugal/coriolis matrix $C(\mathbf{q},\dot{\mathbf{q}})$:

$$
\begin{aligned}
c_{11} &= 0 = c_{22} = c_{33} = c_{44} = c_{55} = c_{66} \\
c_{12} &= l_1 l_2\left[m_1 + (1+\beta)m_2 + m_3 + m_6\right]\sin(\theta_1-\theta_2)\,\dot{\theta}_2 \\
c_{13} &= l_1 l_3 \gamma m_3 \sin(\theta_1-\theta_3)\,\dot{\theta}_3 \\
c_{14} &= -l_1 l_2\left[m_1 + (1-\beta)m_2 + m_6\right]\sin(\theta_1-\theta_4)\,\dot{\theta}_4 \\
c_{15} &= -l_1^2\left[(1-\alpha)m_1 + m_6\right]\sin(\theta_1-\theta_5)\,\dot{\theta}_5 \\
c_{16} &= l_1 l_6 m_6 \sigma \sin(\theta_1-\theta_6)\,\dot{\theta}_6 \\
c_{21} &= -l_1 l_2\left[m_1 + (1+\beta)m_2 + m_3 + m_6\right]\sin(\theta_1-\theta_2)\,\dot{\theta}_1 \\
c_{23} &= l_2 l_3 \gamma m_3 \sin(\theta_2-\theta_3)\,\dot{\theta}_3 \\
c_{24} &= -l_2^2\left[m_1 + (1-\beta)m_2 + m_6\right]\sin(\theta_2-\theta_4)\,\dot{\theta}_4
\end{aligned}
$$

$$
\begin{aligned}
c_{25} &= -l_1 l_2 \left[(1-\alpha)m_1 + m_6\right] \sin(\theta_2 - \theta_5)\,\dot{\theta}_5 \\
c_{26} &= l_2 l_6 m_6 \sigma \sin(\theta_2 - \theta_6)\,\dot{\theta}_6 \\
c_{31} &= -l_1 l_3 \gamma m_3 \sin(\theta_1 - \theta_3)\,\dot{\theta}_1 \\
c_{32} &= -l_2 l_3 \gamma m_3 \sin(\theta_2 - \theta_3)\,\dot{\theta}_2 \\
c_{34} &= 0 = c_{35} = c_{43} = c_{53} = c_{63} = c_{36} \\
c_{41} &= l_1 l_2 \left[m_1 + (1-\beta)m_2 + m_6\right] \sin(\theta_1 - \theta_4)\,\dot{\theta}_1 \\
c_{42} &= l_2^2 \left[m_1 + (1-\beta)m_2 + m_6\right] \sin(\theta_2 - \theta_4)\,\dot{\theta}_2 \\
c_{45} &= l_1 l_2 \left[(1-\alpha)m_1 + m_6\right] \sin(\theta_4 - \theta_5)\,\dot{\theta}_5 \\
c_{46} &= -l_2 l_6 m_6 \sigma \sin(\theta_4 - \theta_6)\,\dot{\theta}_6 \\
c_{51} &= l_1^2 \left[(1-\alpha)m_1 + m_6\right] \sin(\theta_1 - \theta_5)\,\dot{\theta}_1 \\
c_{52} &= l_1 l_2 \left[(1-\alpha)m_1 + m_6\right] \sin(\theta_2 - \theta_5)\,\dot{\theta}_2 \\
c_{54} &= -l_1 l_2 \left[(1-\alpha)m_1 + m_6\right] \sin(\theta_4 - \theta_5)\,\dot{\theta}_4 \\
c_{56} &= -l_1 l_6 m_6 \sigma \sin(\theta_5 - \theta_6)\,\dot{\theta}_6 \\
c_{61} &= -l_1 l_6 m_6 \sigma \sin(\theta_1 - \theta_6)\,\dot{\theta}_1 \\
c_{62} &= -l_2 l_6 m_6 \sigma \sin(\theta_2 - \theta_6)\,\dot{\theta}_2 \\
c_{64} &= l_2 l_6 m_6 \sigma \sin(\theta_4 - \theta_6)\,\dot{\theta}_4 \\
c_{65} &= l_1 l_6 m_6 \sigma \sin(\theta_5 - \theta_6)\,\dot{\theta}_5
\end{aligned}
$$

- Gravitational torque vector $G(\mathbf{q})$:

$$
\begin{aligned}
g_1 &= \left[(\alpha+1)m_1 + 2m_2 + m_3 + m_6\right] g l_1 \cos\theta_1 \\
g_2 &= \left[m_1 + (\beta+1)m_2 + m_3 + m_6\right] g l_2 \cos\theta_2 \\
g_3 &= \gamma m_3 g l_3 \cos\theta_3 \\
g_4 &= \left[-m_1 + (\beta-1)m_2 - m_6\right] g l_2 \cos\theta_4 \\
g_5 &= \left[(\alpha-1)m_1 - m_6\right] g l_1 \cos\theta_5 \\
g_6 &= g l_6 m_6 \sigma \cos\theta_6
\end{aligned}
$$

Appendix C
Details of the Electronics

C.1 Joint Micro-controller Board

Figure C.1 gives a detailed overview of the micro-controller board which is provided for each modular unit. This micro-controller board regulates muscle pressure with the bang-bang control structure. Furthermore, it handles sensory inputs originating from two pressure sensors and an encoder, and provides a buffered interface between the central PC and the local micro-controller. The same board architecture is also used for an extra micro-controller, which handles additional sensory information such as absolute robot position, supply pressure conditions, ground reaction forces and control of the treadmill.

The core of the joint controller board is the MC68HC916Y3 micro-controller of Motorola. It has a 16 bit central processor unit, CPU, and a separate processor, TPU, which is designed to handle sensory input and control output without disturbing the CPU.

The micro-controller unit can be debugged and programmed via the serial SDI interface which is a commercially available device. A 10 pin connector is provided to link the essential pins to the SDI debugger module. This interface has only been used during the development of the micro-controller board. Currently, the micro-controllers are programmed via the 16-bit communication interface.

This interface is created with a dual ported RAM unit. This unit provides a buffered structure which communicates with the Cypress micro-controller communication interface board (see C.7). Two dual ported ram chips IDT7130SA (8 bit wide) are used to create the 16-bit parallel bus interface. Each chip has 1*Kbyte* of memory, the first chip is used to store the lowest byte of the 16-bit data, while the other stores the highest byte. The memory is physically divided into a read data block and a write data block by connecting the $R/\overline{W}$ signal to address line number 8 of the dual ported RAM memory. The highest address line is not used, which means that two memory storage places are provided for 256 16*bit* wide data. Due to the divided structure into a read and write block, it is never possible to access a memory place from both sides simultaneously, therefore the BUSY and INT pins of the dual ported RAM units are not used.

The connector to the USB interface board redirects the pins of port PF which can be used to generate interrupts on the CPU (MC68HC916Y3) and give acknowledge signals to the communication master. E.g. the Cypress USB micro-controller, which is the communication master controlling the communication sampling rate, generates an interrupt on the CPU of all the Motorola micro-controllers each communication sample. Furthermore these pins are used to reset all the Motorola CPUs and in the other direction, to acknowledge to the communication master that the specific Motorola CPU is ready for a read or write action.

One connector is provided for the interface to the sensors and the valves. These valves are controlled by several TPU signals. The micro-controller board provides 6 separate signals to control the 6 valves of a valve island, but currently only 4 of them are used since three exhaust valves are switched together. The 3 incremental encoder channels are also connected with the TPU, which presents a position signal to the CPU without demanding any processor time. Additionally, one of the two main channel of the encoder are linked with a secondary TPU pin in order to estimate angular joint rotation speed. This speed is determined by a time measurement between two neighboring encoder flanks. The 12-bit digital signals of the two pressure sensors are linked to the micro-controller via the serial SPI interface. Finally, port G is connected with 8 LEDs which are used to visualize the different operation modes of the robot.

Resetting the controller can be done by a local button on the micro-controller board or by the USB micro-controller via the dual ported RAM units. The local reset and micro-controller initialization scheme uses an AND-port (chip 4023) structure as clearly explained in the data sheets. Furthermore are provided an oscillation circuit to generate the clock for the CPU, two RS232 interfaces and a flash EEPROM programming circuit, all described in the data sheets.

The communication software is programmed into the flash EEPROM and works with two essential modes: program and run mode. These modes are selected by the first word of the communication data block, which come with $32 bytes$ each sample. Program mode is selected to load the micro-controller with the specific low-level controller program, such as e.g. the bang-bang controller, and in the run mode this downloaded program is executed while exchanging necessary control data with the central PC. So there is no fixed controller design programmed in the controller but it is downloaded each time the robot is initialized. This creates a fast and flexible experimental low-level control board for which different controller strategies can be implemented easily.

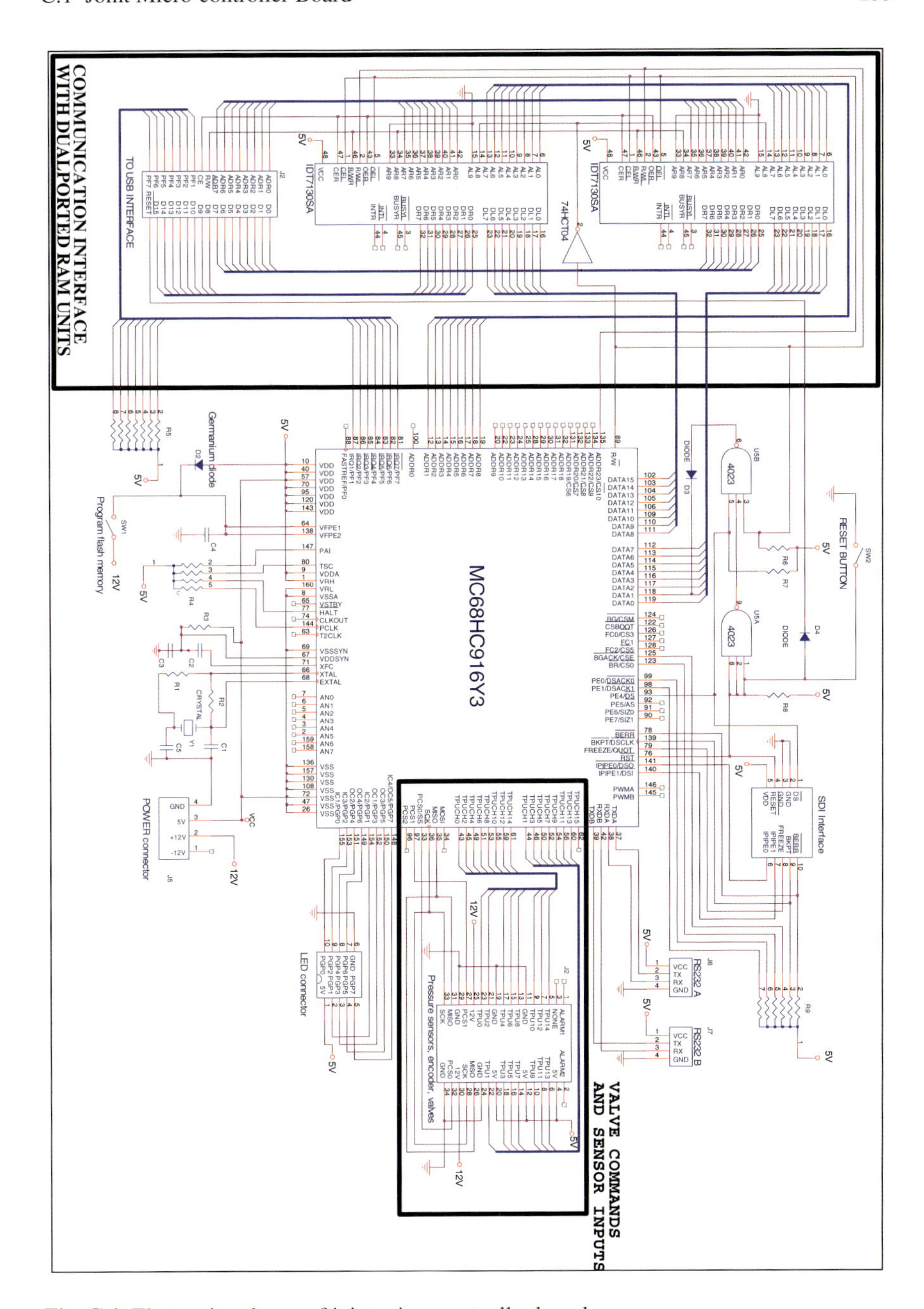

Fig. C.1 Electronic scheme of joint micro-controller board.

C.2 Speed-up Circuitry

In order to enhance the opening time of the Matrix valves, the manufacturer proposes a speed-up in tension circuitry. With a temporal $24V$ during a period of $2.5ms$ and a remaining $5V$ the opening time of the valves is said to be $1ms$. But during practical tests more than double values for the opening time were recorded. The opening tension is therefore increased to $36V$, but the time during which this voltage is applied is decreased to the actual opening time of $1ms$, such that the valves do not get overheated.

Figure C.2 gives the complete electronic scheme of the speed-up circuitry. Four identical schemes are provided, two for inlet and two for exhaust valves, of which one circuit commands three exhaust valves to open and close simultaneously. For each circuitry two LED's are provided in order to visualize valve action, one of them only lights up when the increased voltage is applied. These LEDs are important to check if the pressure control block is properly working. For each circuitry, the micro-controller commands a valve via discrete $5V$ on/off signals. These signals directly activate mosfet Q1 (IRF530) in order to apply $5V$ over the valve. The same signal passes parallel through a one-shot (74LS123) in order to increase the applied voltage over the valve during the first $1ms$ of valve activation. The output of the one-shot therefore temporally activates mosfet Q2 (IRF610) which on its turn commands mosfet Q3 (IRF9540) to branch the $36V$ supply to a valve. Whenever the micro-controller commands a valve to close, by disabling mosfet Q1, the discharge path is connected to the increased supply source via diode D2. This provides a fast discharge of the electromagnetic energy stored in the valve, which results in a faster closing time.

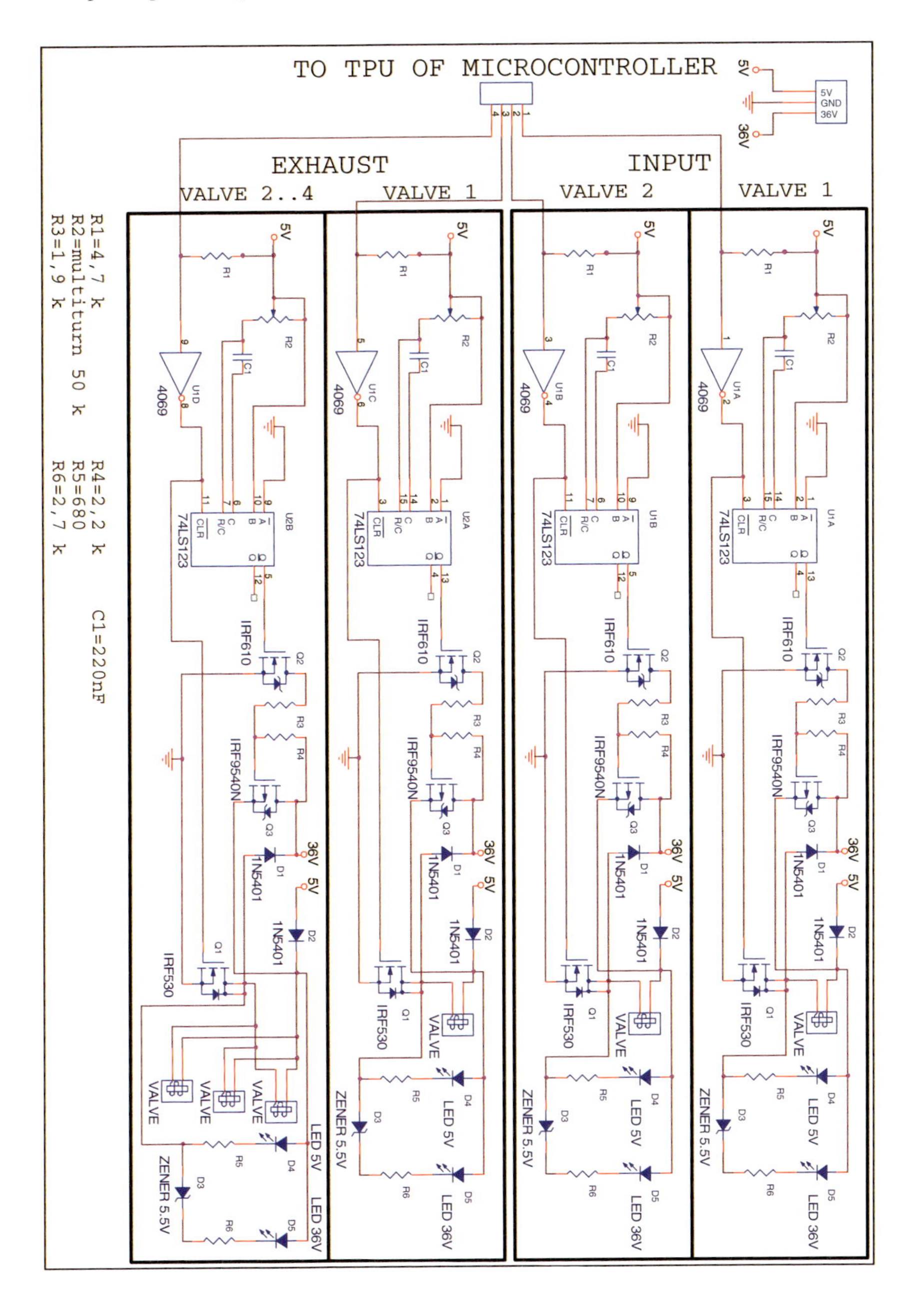

Fig. C.2 Electronic scheme of the speed-up circuitry.

C.3 Pressure Sensor

Figure C.3 depicts the electronic scheme which conditions the pressure sensor signal. The most important component is the absolute pressure sensor, CPC100AFC, from Honeywell. This sensor measures absolute pressure values up to $100psi$ ($6.9bar$) and has an accuracy of about $20mbar$. Approximately $100mV$ for each $100psi$ is generated, meaning $14.5mV$ for $1bar$. The output of the pressure sensor is amplified by a differential amplifier. The gain of this amplifier is approximately 63.2. In order to avoid as much as possible noise generation, the amplified pressure signal (V_0) is immediately digitized by a $12bit$ analog to digital converter. A stable reference voltage for this converter is locally generated by a cascade circuit of two zener diodes. The negative input (-IN) of the AD-converter is augmented with a fixed voltage to roughly compensate atmospheric pressure. The AD-converter chip communicates with the micro-controller unit by a serial SPI interface, which is typically used for communication between chips and micro-controllers. A comparator $LM324$ is provided to generate an alarm signal in order to protect the muscle against pressure overload. This signal is not treated by the micro-controller, but immediately acts on the central pressure supply valve (see 2.7.1.2). Whenever the muscle gauge pressure exceeds approximately $4.2bar$, the pressure supply is cut-off. The pressure sensor circuit is calibrated each time the robot is initialized. This calibration is performed via an other pneumatic calibration circuit with an additional pressure sensor. In order to pass through the entrance of a muscle, the diameter of the sensor and its electronics is made small ($12mm$).

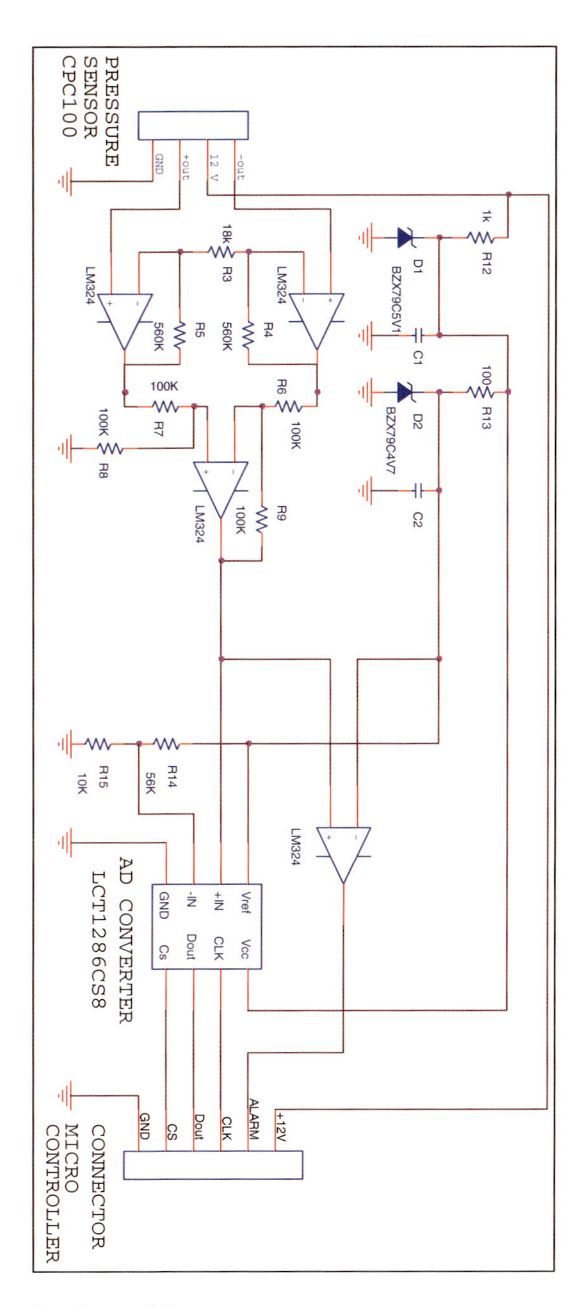

Fig. C.3 Electronic scheme of the pressure sensor.

C.4 Foot Measurement Board

An electronic scheme of the foot measurement board is shown in figure C.4. The ground reaction forces are measured by load cells of Transducer Techniques (THA-250-Q). A commercial strain gauge amplifier of RS-components (846-171) is used to amplify the signal of the full bridge circuit. The chip requires also a negative supply voltage of $-10V$. To avoid the necessity of a new power supply a DC/DC-converter of Traco Power (TEN 10-2422) is used which makes out of $24V$ $+/-12V$. The foot board has one such DC/DC-converter for the gauge amplifiers of the rear and front load cell. The amplified force signal is digitized by a 12 bit AD-converter (LTC 1286). A stable reference voltage for this converter is locally generated by zener diodes. The AD-converter chip communicates with the micro-controller unit by a serial SPI interface. Each foot has also a mechanical switch to detect if the foot

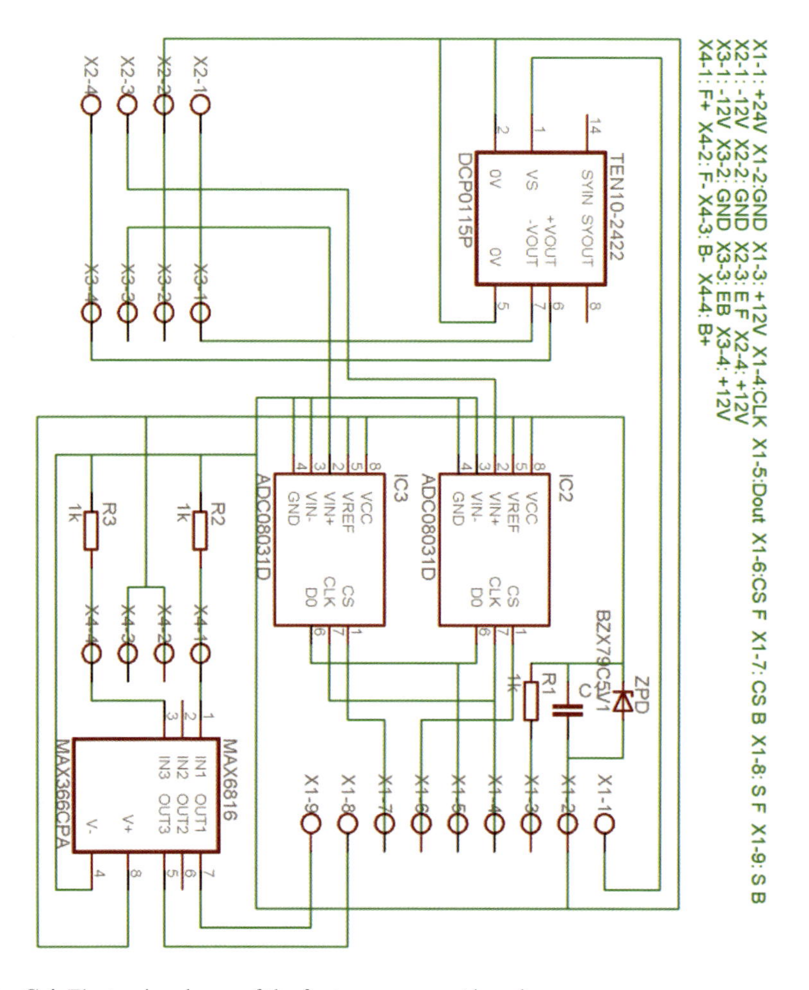

Fig. C.4 Electronic scheme of the foot measurement board.

is on the ground or not. To provide a clean interface with the digital system switch debouncers (MAX6816) are used.

C.5 Treadmill Control Board

The treadmill is powered by a 3 phase synchronous AC Motor controlled by the frequency inverter ACS 350 from ABB. This motor drive contains a vector control to provide enough torque at low rotation speeds. The steering signal and the measured rotation speed of the motor are treated by a separate electrical board which can be seen in figure C.5 This board contains opto-couplers so in case of a fault like an overvoltage on one side, the other side is not corrupted, especially to protect the low voltage electronics of the robot. This board is also connected to the emergency buttons: if an emergency button is pressed the treadmill stops automatically.

The treadmill control board (figure C.5) connects the 7th micro-controller with the frequency inverter ACS 350, which controls the motor of the treadmill. Essential are the opto-couplers 6N139 for galvanic separation of the robot electronics and the frequency inverter. A PWM signal coming from the TPU unit of the 7th micro-controller represents the speed signal. This signal is inverted (74HCT04) and feeds a LED and the opto-coupler 6N139. 4 emergency stops can be connected to this board. These signals are merged by OR-gates and if one of the emergency stops is pressed the power supply to the opto-coupler is turned off. After the opto-coupler a first order filter, formed by a capacity and resistor, makes an analogues signal between 0-10V for the frequency inverter. MC14538B is a monostable multivibrator. When there are no pulses the output is driven so the Darlington transistor is low and the frequency inverter is stopped. When pulses occur they are lengthened by the multivibrator so they form a continuous high signal and the frequency inverter is started. The drive has one programmable transistor output. In this application the frequency inverter is programmed to give the speed signal. Through an opto-coupler the signal is sent to the TPU unit of the 7th micro-controller. A manual switch makes it possible to reverse the rotation speed.

The part of the frequency inverter is fed by the frequency inverter itself. The board is designed so it consumes less than 200mA, the maximum current it can supply.

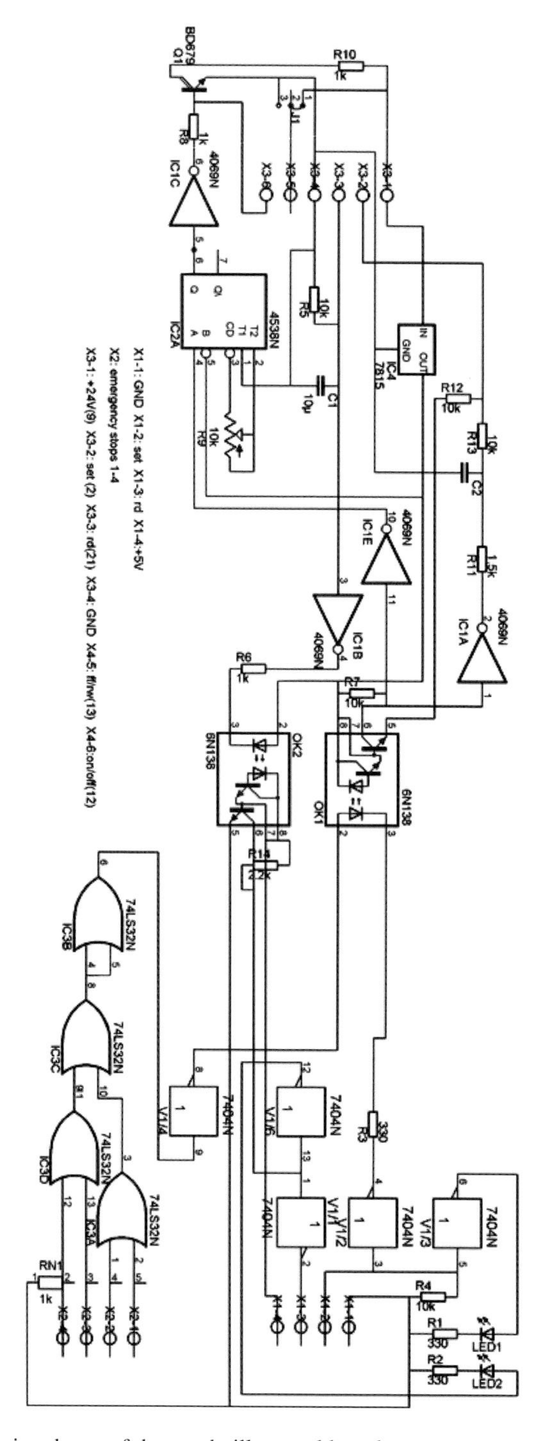

Fig. C.5 Electronic scheme of the treadmill control board.

C.6 Safety Board

The safety board is provided in order to control the supply pressure flow. It will cut-off the supply pressure in case an emergency situation is met. It can also select a lower calibration supply pressure required for the calibration of the 12 muscle pressure sensors. Figure C.6 shows the electronic scheme of the safety board. There are three valves which control the supply pressure. Opening valve 1 connects the robot to the high supply pressure and valve 2 introduces a lowered calibration pressure. Both valves are activated by a transistor circuit for which signals S1 or S2 have to be logic zero in order to open valve 1 or valve 2 respectively. If these signals are high, than valve 3 is opened in order to depressurize the robot. This happens when the robot is not working or when a pressure alarm or emergency stop is activated. A pressure alarm is induced by the pressure sensors in the muscles, whenever the pressure exceeds approximately $4.5bar$ gauge pressure. In this case a rising flank on the alarm signal switches the output of a D flip-flop to low logic state. The flip-flop is used to remember this emergency state and close the pressure valve until a manual reset is given on the safety board. All alarm signals have their own flip-flop structure with an additional LED such that is can be easily detected in which muscle the alarm signal was generated. An OR structure on all the flip-flop outputs in combination with 4 mechanical emergency stops depressurizes the robot whenever one of them alerts for a dangerous situation. Selection between the high or initialization pressure is done by two external signals, which are commanded by the extra micro-controller.

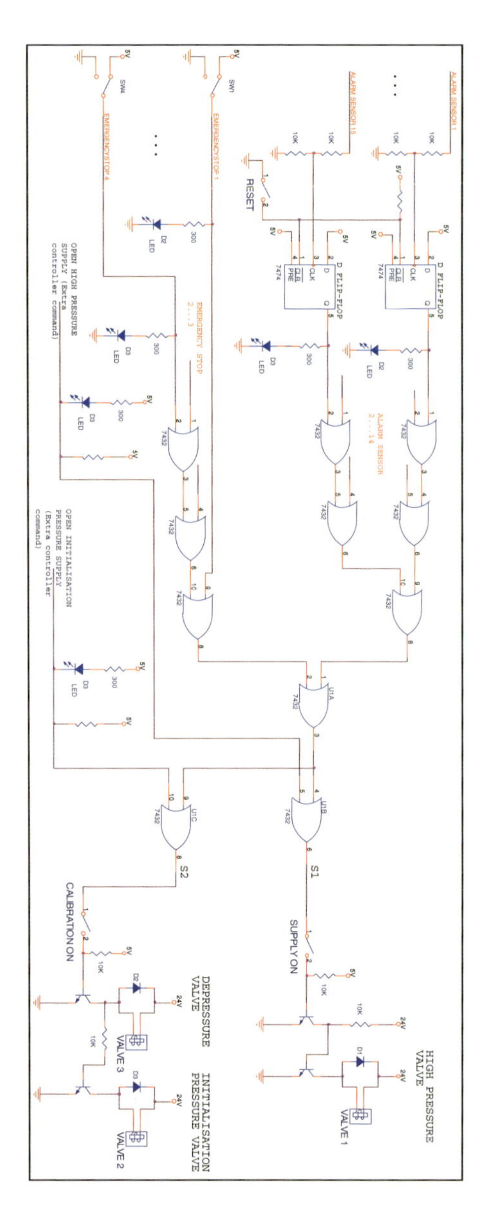

Fig. C.6 Electronic scheme of the safety board.

C.7 Cypress Communication Interface

C.7.1 Why USB 2.0?

Since a lot of extensive calculations are required due to the model based control algorithms, a central PC is used. Therefore a fast communication line between PC and robot hardware is provided. A fast communication line could be an extension of the PC bus by means of a parallel data communication, but this kind of communication is only suitable for short distance applications. For larger distances (several meters) serial communication protocols are preferable. This section will deal about the electronics needed to make the connection and the design choices made to handle the communication between the central PC and the 7 micro-controllers. This section will not deal about how the USB protocol works. For more information (25; 35; 28) is suggested.

Table C.1 shows an overview of some common interfaces.

RS-232 is not fast enough, FireWire is limited too much in allowable distance. For this application it was chosen to use a USB 2.0 communication interface, which has a data transfer rate of $480 Mbit/s$.

Other advantages for choosing an USB protocol are:

- A USB device can be plugged in anytime, even while the PC is turned on.
- When the PC detects that a USB device has been plugged in, it automatically interrogates the device to learn its capabilities and requirements. From this information, the PC automatically loads the devices driver into the operating system. When the device is unplugged, the operating system automatically logs it off and unloads its driver.
- USB devices do not use DIP switches, jumpers, or configuration programs. There is never an IRQ, DMA, memory, or I/O conflict with a USB device. USB expansion hubs make the bus simultaneously available to dozens of devices.
- Single connector type, the USB defines a single connector used to attach any USB device. Additional connectors can be added with USB hubs.

Table C.1 Comparison between different interfaces

Interface	Type	max # peripherals	transmission speed	max distance
USB	serial	127	USB 1.0: 1.5Mbits/s, low speed USB 1.1: 12Mbits/s, full speed USB 2.0: 480Mbits/s, high speed	5m (with hubs up to 30m)
RS-232	serial	2	20 - 115kbit/s	15-20m
IEEE-1394 (FireWire)	serial	64	400Mbit/s IEEE-1394b: 3.2Gbit/s	4m
Ethernet	serial	1024	10Mbit/s 1Gbit/s 100Mbit/s	4m

C.7.2 EZ-USB FX2

Since the local Motorola controllers (6 joint controllers+1 extra controller) have a $16bit$ parallel communication bus via the dual ported RAM units, the serial USB bulk communication data blocks have to be divided into 7 blocks of $16bit$ parallel data. Therefore an extra micro-controller, EZ-USB FX2 from Cypress Semiconductors, is provided to act only as data transfer agent. The main reasons to choose the EZ-USB FX2 were:

- An integrated, high-performance CPU based on the industry-standard 8051 processor.
- A soft (RAM-based) architecture that allows unlimited configuration and upgrades. Full USB throughput. USB devices that use EZ-USB chips are not limited by number of endpoints, buffer sizes, or transfer speeds.
- Automatic handling of most of the USB protocol.
- No external power supply, The $3.3V$ power supply can be delivered by the 5V power available at the USB connector (which the USB Specification allows to be as low as $4.4V$).

This controller runs at $48Mhz$ and is able to transfer the serial data block of $226bytes$ to the peripheral $16bit$ data bus in less than $50\mu s$. Additional to the Cypress development board, an electronic interface has been created to connect the peripheral bus of the Cypress micro-controller to the different dual ported RAM units. Figure C.7 gives the electronic scheme of the interface. Since the Cypress controller works at $3V$ supply voltage level and the dual ported RAM units at $5V$, all lines connecting both parts are buffered via octal supply translating transceiver chips, 74LVC4245. These have a tristate when not enabled, this is important especially for connecting the data lines FD[i] of the Cypress controller to data lines D[i] of the dual ported RAM units. Two chips, U1 and U2, are foreseen for the 16 bit data lines, which work in both directions. The address lines are buffered with U3 which only translates in one direction as is the same for chip U4. The latter connects port PE of the Cypress controller to the other micro-controllers in order to give communication commands. These are: selection of a specific dual ported RAM unit by means of the line decoder chip U6, directing the R/W signal, global reset by software via pin PE5 and two extra general purpose control pins connected to PF1 and PF2 of the Motorola controller. These PF port pins can be controlled interrupt driven. In the other direction, pins PF3 of all the Motorola micro-controllers are connected separately to port PA of the Cypress controller. The pins PF4 of all Motorola controllers are connected together via an AND gate to pin PA7. These signals are used as communication acknowledgement signals, knowing that the Cypress controller is the bus master. Furthermore, a dip switch is provided to act on pin PF5 in order to select between two working modes. Finally, a general purpose interrupt can be generated manually on pin PF7 of all controllers and a manual global reset button is provided also.

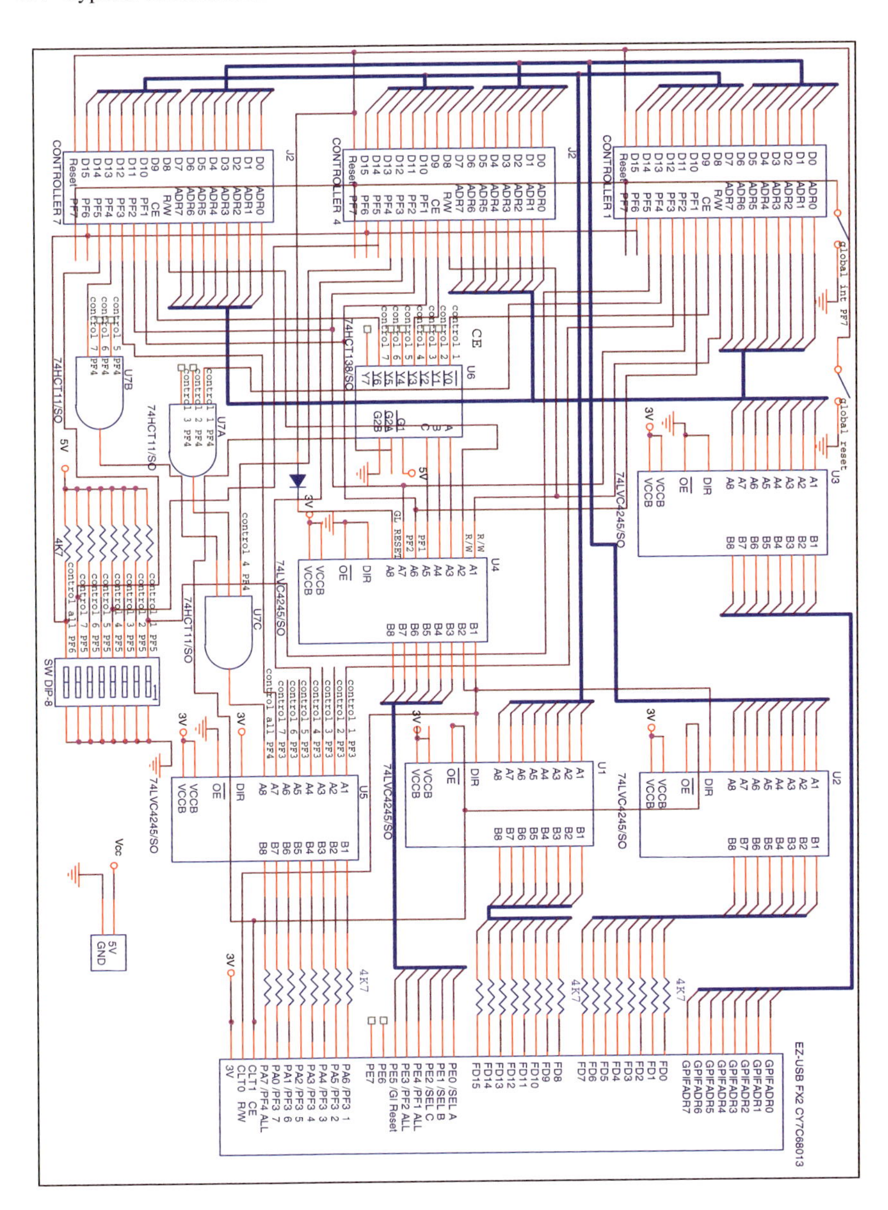

Fig. C.7 Electronic scheme of the cypress communication interface.

C.7.3 USB Transfer Types

USB defines four transfer types. These match the requirements of different data types delivered over the bus.

- Bulk Transfers
- Interrupt Transfers
- Isochronous Transfers
- Control Transfers

The EZ-USB FX2 is configured as Bulk Transfer for the robot application. Bulk data is bursty, traveling in packets of $512 bytes$ at high speed. The most important reason to choose for this type is that it has guaranteed accuracy, due to an automatic retry mechanism for erroneous data. The host schedules bulk packets when there is available bus time.

C.7.4 EZ-USB FX2 Architecture

The FX2 packs all the intelligence required by a USB peripheral interface into a compact integrated circuit. As figure C.8 illustrates, an integrated USB transceiver connects to the USB bus pins D+ and D-. A Serial Interface Engine (SIE) decodes and encodes the serial data and performs error correction, bit stuffing, and the other signaling-level tasks required by USB. Ultimately, the SIE transfers parallel data to and from the USB interface.

The high-level USB protocol is not bandwidth-critical, so the FX2s CPU is well-suited for handling host requests over the control endpoint. However, the data rates offered by USB 2.0 are too high for the CPU to process the USB data directly. For this reason, the CPU is not in the high bandwidth data path between endpoint FIFOs and the external interface. Instead, the CPU simply configures the interface, then "gets out of the way" while the unified FX2 FIFOs move the data directly between the USB and the external interface. To do this the General Programmable Interface (GPIF) is used. This internal FX2 timing generator serves as an internal master, interfacing directly to the FIFOs and generating user-programmed control signals for the interface to external logic.

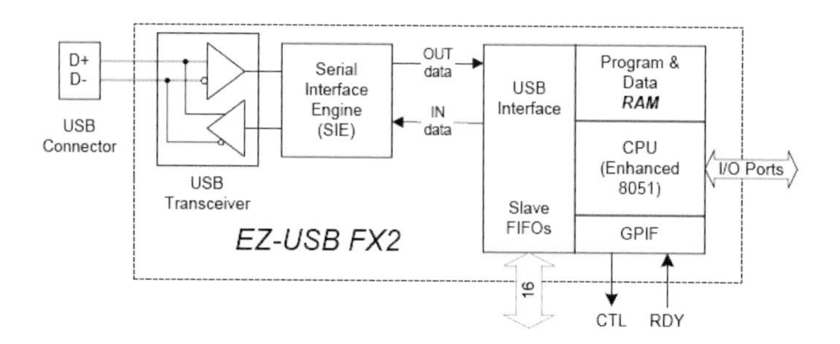

Fig. C.8 FX2 Simplified Block Diagram (28).

C.7.5 FX2 Endpoint Buffers

The USB Specification defines an endpoint as a source or sink of data. Since USB is a serial bus, a device endpoint is actually a FIFO which sequentially empties or fills with USB data bytes. The host selects a device endpoint by sending a 4-bit address and a direction bit. Therefore, USB can uniquely address 32 endpoints, IN0 through IN15 and OUT0 through OUT15.

From the FX2s point of view, an endpoint is a buffer full of bytes received or held for transmission over the bus. The FX2 reads host data from an OUT endpoint buffer, and writes data for transmission to the host to an IN endpoint buffer. The terms "IN" and "OUT" are from the viewpoint of the PC, because the PC is always the master in a USB system. USB devices respond to host requests. USB devices cannot send information among themselves, as they could if USB were a peer-to-peer topology.

EP0 is the default CONTROL endpoint, a bidirectional endpoint that uses a single 64-byte buffer for both IN and OUT data.

Endpoints 2, 4, 6 and 8 are the large, high bandwidth, data moving endpoints. In this application the endpoint 2 is used as OUT endpoint and 6 as IN endpoint.

C.7.6 Firmware

Because the FX2s configuration is soft, one chip can take on the identities of multiple distinct USB devices. The functionality of the controller, called the firmware, isn't stored in the memory of the chip itself, but on the central PC. This has as advantage that it is easily adapted. When first plugged into USB, the FX2 enumerates automatically and downloads firmware and USB descriptor tables over the USB cable. Next, the FX2 enumerates again, this time as a device defined by the downloaded information. This patented two-step process, called ReNumeration^{TM}, happens instantly when the device is plugged in. The new programmed FX2 is visible in the "Device manager".

C.7.7 Driver

To communicate between the FX2 and the PC the proper device driver has to be installed. A device driver is the software that Windows or a Windows based application uses to interact with a piece of hardware. As a result the application doesn't need to bother about protocols, physical connections, signals and can stay platform independent.

Because the FX2 is not a standard piece of hardware, a class driver can't be used. Examples of class drivers are the human interface device (HID) class, which supports devices like mice, joysticks, and keyboards. Another is the monitor class, which controls image position, size, and alignment on video displays. Custom drivers are an alternative to class drivers.

Because we didn't want to write our own driver with for example Driver Developer's Kit (DDK), the generic driver "WinRT for USB" of BSQUARE is used. Using this, developers can write application-level Win32 hardware control programs for USB hardware and eliminate the need for device driver toolkits or custom device driver development.

Communicating with the FX2 is then possible with easy to understand functions as for example "WinRTBulkTransfer" and "FindBulkEndpoints".

Appendix D
Publication List Related to Lucy

- Trajectory Planning for the Walking Biped 'Lucy'
  VERMEULEN Jimmy, VERRELST Bjorn, VANDERBORGHT Bram, LEFEBER Dirk
  International Journal of Robotics Research, Vol. 25, No. 9, 2006, pp. 867-887
- Controlling a Bipedal Walking Robot Actuated by Pleated Pneumatic Artificial Muscles
  VANDERBORGHT Bram, VERRELST Bjorn, VAN HAM Ronald, LEFEBER Dirk
  Robotica, Vol. 24, No. 4, July 2006, pp. 401-410
- Second Generation Pleated Pneumatic Artificial Muscle and Its Robotic Applications
  VERRELST Bjorn, VAN HAM Ronald, VANDERBORGHT Bram, LEFEBER Dirk, DAERDEN Frank, VAN DAMME Michael
  Advanced Robotics, Vol. 20 No. 7, 2006, pp. 783-805
- Exploiting Natural Dynamics to Reduce Energy Consumption by Controlling the Compliance of Soft Actuators
  VANDERBORGHT Bram, VERRELST Bjorn, VAN HAM Ronald, VAN DAMME Michael, LEFEBER Dirk, MEIRA Y DURAN Bruno, BEYL Pieter
  International Journal of Robotic Research, Vol. 25, No. 4, April 2006, pp. 343-358
- A real-time joint trajectory planner for dynamic walking bipeds in the sagittal plane
  VERMEULEN Jimmy, VERRELST Bjorn, LEFEBER Dirk, KOOL Patrick, VANDERBORGHT Bram
  Robotica, Vol. 23, No. 06, November 2005, pp. 669-680
- Control Architecture for the Pneumatically Actuated Dynamic Walking Biped 'Lucy'
  VERRELST Bjorn, VANDERBORGHT Bram, VERMEULEN Jimmy, VAN HAM Ronald, NAUDET Joris, LEFEBER Dirk
  Mechatronics, Vol. 15, No. 6, July 2005, pp. 703-729

- The Pneumatic Biped 'LUCY' Actuated with Pleated Pneumatic Artificial Muscles
  VERRELST Bjorn, VAN HAM Ronald, VANDERBORGHT Bram, DAERDEN Frank, LEFEBER Dirk, VERMEULEN Jimmy
  Autonomous robots, Vol. 18, No. 2, March 2005, pp. 201-213
- Motion Generation and Control for the Pneumatic Biped "Lucy"
  VERRELST Bjorn, VERMEULEN Jimmy, VANDERBORGHT Bram, VAN HAM Ronald, NAUDET Joris, LEFEBER Dirk, DAERDEN Frank, VAN DAMME Michael
  International Journal of Humanoid Robotics (IJHR), Vol. 3, No. 1, 2006, pp. 1-35
- Fast and Accurate Pressure Control Using On-Off Valves
  VAN HAM Ronald, DAERDEN Frank, VERRELST Bjorn, VANDERBORGHT Bram, LEFEBER Dirk
  International Journal of Fluid Power, No. 6, 2005, pp. 53-58
- Treadmill walking of the pneumatic biped Lucy: Walking at different speeds and step-lengths Authors: VANDERBORGHT Bram, VERRELST BJORN, VAN HAM Ronald, VAN DAMME Michael, VERSLUYS Rino, LEFEBER DIRK Reference: International Applied Mechanics, vol. 44, n. 7, 2008, pp. 134 - 142
- Objective locomotion parameters based inverted pendulum trajectory generator Authors: VANDERBORGHT Bram, VERRELST BJORN, VAN HAM Ronald, VAN DAMME Michael, LEFEBER DIRK Reference: Robotics and Autonomous Systems, vol. 56, n. 9, 2008, pp. 738 - 750
- Development of a compliance controller to reduce energy consumption for bipedal robots Authors: VANDERBORGHT Bram, VERRELST BJORN, VAN HAM Ronald, VAN DAMME Michael, BEYL Pieter, LEFEBER DIRK Reference: Autonomous Robots, vol. 15, n. 6, 2008, pp. 419 - 434
- Overview of the Lucy Project: Dynamic Stabilization of a Biped Powered by Pneumatic Artificial Muscles Authors: VANDERBORGHT Bram, VAN HAM Ronald, VERRELST BJORN, VAN DAMME Michael, LEFEBER DIRK Reference: Advanced Robotics, vol. 22, n. 10, 2008, pp. 1027 - 1051
- Dynamically Walking over Large Obstacles by a Humanoid Robot Authors: STASSE Olivier, VERRELST BJORN, VANDERBORGHT Bram, YOKOI Kazuhito Reference: IEEE Transactions on Robotics, vol. 25, n. 4, 2009, pp. 960 - 967
- Comparison of Mechanical Design and Energy Consumption of Adaptable, Passive-compliant Actuators Authors: VANDERBORGHT Bram, VAN HAM Ronald, LEFEBER DIRK, Sugar Thomas, Hollander Kevin Reference: International Journal of Robotics Research, vol. 28, n. 1, 2009, pp. 90 - 103

For a publication list including conference papers please visit
http://mech.vub.ac.be/bram.htm

References

[1] http://support.sony-europe.com/aibo/2_0_news.asp?news=13
[2] Aldebaran robotics, http://www.aldebaran-robotics.com/
[3] Anybots, http://anybots.com
[4] Biped robot entertainment robo-one official site, http://www.robo-one.com/
[5] Development of a humanoid robot based on technical transfer from AIST, http://www.aist.go.jp/aist_e/latest_research/2006/20060713/20060713.html
[6] Development of waseda robot -the study of biomechanisms at (the late) kato laboratory, http://www.humanoid.waseda.ac.jp/booklet/katobook.html
[7] The federation of international robot-soccer association (FIRA), http://www.fira.net
[8] Festo ag & co., http://www.festo.com/
[9] Hitachi medical corporation, http://www.hitachimedicalsystems.com/
[10] Hitec robotics, http://www.hitecrobotics.com/
[11] KineoWorks is a software development kit dedicated to motion planning and marketed by kineo CAM, http://www.kineocam.com/
[12] Kondo Kagaku Co., Ltd. KHR-1, http://www.kondo-robot.com
[13] Merlin robotics, http://www.merlinrobotics.co.uk/
[14] Modular controller architecture version2 (mca2), http://www.mca2.org/
[15] Network based humanoid robot - mahru, ahra, http://humanoid.kist.re.kr/eng/
[16] Nuvo - the first humanoid robot in the world "for home use", http://nuvo.jp/nuvo_home_e.html
[17] Official home page of honda's humanoid robot. http://world.honda.com/ASIMO/
[18] Overview of the partner robots of toyota, http://www.toyota.co.jp/en/special/robot/
[19] Pal technology, http://www.pal-robotics.com/reem.html
[20] Population division of the department of economic and social affairs of the united nations secretariat, world population prospects: The 2004 revision and world urbanization prospects: The 2003 revision, http://esa.un.org/unpp
[21] Press release: World's first running humanoid robot, http://www.sony.net/SonyInfo/News/Press/200312/03-060E
[22] Robocup, http://www.robocup.org
[23] Sarcos, http://www.sarcos.com

[24] Shadow robot company, `http://www.shadowrobot.com/`
[25] Universal serial bus specification version 2.0, `http://www.usb.org`
[26] ZMP Inc., `http://www.zmp.co.jp/`
[27] Summary report on technology strategy for creating a "robot society" in the 21st century. Tech. rep., Japan Robot Society (2001)
[28] EZ-USB FX2 Technical Reference Manuel. Cypress Semiconductor (2003)
[29] Adamczyk, P.G., Collins, S.H., Kuo, A.D.: The advantages of a rolling foot in human walking. Journal of Experimental Biology 209, 3953–3963 (2006)
[30] Ahn, C.K., Lee, M.C., Go, S.J.: Development of a biped robot with toes to improve gait pattern. In: IEEE/ASME International Conference on Advanced Intelligent Mechatronics (AIM 2003), vol. 2, pp. 729–734 (2003)
[31] Albert, A., Gerth, W.: New path planning algorithms for higher gait stability of a bipedal robot. In: 4th International Conference on Climbing and Walking Robots (CLAWAR 2001), pp. 521–528 (2001)
[32] Alexander, R.M.: Exploring Biomechanics: Animals in Motion. Scientific American Library (1992)
[33] Alexander, R.M.: Human elasticity. Physics Education 29, 358–362 (1994)
[34] Alexander, R.M.: Physiology: Enhanced: Walking made simple. Science 308(5718), 58–59 (2005)
[35] Anderson, D., Dzatko, D.: Universal Serial Bus System Architecture. Addison-Wesley, Reading (2001)
[36] Anderson, F.C., Pandy, M.G.: Dynamic optimization of human walking. Journal of Biomechanical Engineering 123(5), 81–390 (2001)
[37] Aoi, S., Tsuchiya, K.: Locomotion control of a biped robot using nonlinear oscillators. Autonomous Robots 19(3), 219–232 (2005)
[38] Aristotle: De Anima (On the soul), vol. 1 406b15-22 (350 BC)
[39] Aristotle: The Politics, vol. 1253b33-1254a1 (350 BC)
[40] Asimov, I.: I, Robot. Doubleday & Company, New York (1950)
[41] Asimov, I.: Robots and Empire. Grafton Books, London (1985)
[42] Azevedo, C., Andreff, N., Arias, S.: BIPedal walking: from gait design to experimental analysis. Mechatronics 14/6, 639–665 (2004)
[43] Azevedo, C., Poignet, P., Espiau, B.: Artificial locomotion control from human to robots. Robotics and Autonomous Systems 47/4, 203–223 (2004)
[44] Baillie, J.C.: URBI: towards a universal robotic low-level programming language. In: IEEE/RSJ International Conference on Intelligent Robots and Systems (IROS 2005), pp. 820–825 (2005)
[45] Bares, J.E.: Dante II: Technical description, results, and lessons learned. The International Journal of Robotics Research 18(7), 621–649 (1999)
[46] Basmajian, J., De Luca, C.: Muscles alive: their function revealed by electromyography, p. 1964 (1985)
[47] Basmajian, J.V., Tuttle, R.: EMG of locomotion in gorilla and man. In: Control of Posture and Locomotion, pp. 599–609. Plenum Press, New York (1973)
[48] Bauby, C.E., Kuo, A.D.: Active control of lateral balance in human walking. Journal of Biomechanics 33, 1433–1440 (2000)
[49] Behnke, S., Müller, J., Schreiber, M.: Toni: A Soccer Playing Humanoid Robot. LNAI, pp. 59–70. Springer, Heidelberg (2006)
[50] Behnke, S., Schreiber, M., Stückler, J., Renner, R., Strasdat, H.: See, walk, and kick: Humanoid robots start to play soccer. In: IEEE-RAS International Conference on Humanoid Robots, pp. 497–503 (2006)

[51] Beletskii, V., Berbyuk, V., Samsonov, V.: Parametric optimization of motions of a bipedal walking robot. Mechanics of Solids 17(1), 24–35 (1982)
[52] Berns, K., Braun, T.: Design concept of a human-like robot head. In: IEEE-RAS International Conference on Humanoid Robots, pp. 32–37 (2005)
[53] Beyl, P., Naudet, J., Vanderborgt, B., Van Damme, M., Van Ham, R., Verrelst, B., Lefeber, D.: Mechanical design of a step rehabilitation robot prototype: Dimensioning of compliant actuators. In: International Conference on Climbing and Walking Robots (CLAWAR 2006), pp. 454–460 (2006)
[54] Beyl, P., Van Damme, M., Van Ham, R., Vanderborght, B., Lefeber, D.: Design and control of a lower limb exoskeleton for robot-assisted gait training. Applied Bionics and Biomechanics 6(2), 229–243 (2009)
[55] Bicchi, A., Tonietti, G.: Fast and soft arm tactics: Dealing with the safety-performance trade-off in robot arms design and control. IEEE Robotics and Automation Magazine 11, 22–33 (2004)
[56] Bicchi, A., Tonietti, G., Bavaro, M., Piccigallo, M.: Variable stiffness actuators for fast and safe motion control. In: Int. Symposium Robotics Research, pp. 100–110 (2003)
[57] Blajer, W., Schiehlen, W.: Walking without impacts as a motion/force control problem. Journal of Dynamic Systems, Measurement and Control 114, 660–665 (1992)
[58] Bluethmann, W., Ambrose, R., Diftler, M., Askew, S., Huber, E., Goza, M., Rehnmark, F., Lovchik, C., Magruder, D.: Robonaut: A robot designed to work with humans in space. Autonomous Robots 14(2-3), 1573–7527 (2003)
[59] Bluethmann, W., Ambrose, R., Diftler, M., Huber, E., Fagg, A., Rosenstein, M., Platt, R., Grupen, R., Breazeal, C., Brooks, A., Lockerd, A., Peters, R., Jenkins, O., Mataric, M., Bugajska, M.: Building an autonomous humanoid tool user. In: IEEE/RAS International Conference on Humanoid Robots, pp. 402–421 (2004)
[60] Breazeal, C.: Emotion and sociable humanoid robots. International Journal of Human computer Studies 59(1-2), 119–155 (2003)
[61] Breazeal, C.: Designing Sociable Robots. The MIT Press, Cambridge (2004)
[62] Breazeal, C., Buchsbaum, D., Gray, J., Gatenby, D., Blumberg, B.: Learning from and about others: Towards using imitation to bootstrap the social understanding of others by robots. Artificial Life 11(1-2), 31–62 (2005)
[63] Brooks, A., Kaupp, T., Makarenko, A., Williams, S., Oreback, A.: Towards component-based robotics. In: IEEE/RSJ International Conference on Intelligent Robots and Systems (IROS 2005), pp. 163–168 (2005)
[64] Brooks, R.: Prospects for human level intelligence for humanoid robots. In: International Symposium on Humanoid Robots (HURO 1996) (1996)
[65] Brown, R.H., Schneider, S.C., Mulligan, M.G.: Analysis of algorithms for velocity estimation from discreteposition versus time data. IEEE Transactions on Industrial Electronics 39(1), 11–19 (1992)
[66] Brun, X.: Commendes linéaires et non linéaires en electropneumatique, méthodologies et applications. Ph.D. thesis, Institut National des Science Appliquees de Lyon (1999)
[67] Bruyninckx, H.: Open robot control software: the OROCOS project. In: IEEE International Conference on Robotics and Automation (ICRA 2001), vol. 3 (2001)
[68] Bruyninckx, H., Soetens, P., Koninckx, B.: The real-time motion control core of the orocos project. In: IEEE International Conference on Robotics and Automation (ICRA 2003), vol. 2, pp. 2766–2771 (2003)
[69] Buehler, M., Playter, R., Raibert, M.: Robots step outside. In: Int. Symp. Adaptive Motion of Animals and Machines (AMAM), Ilmenau, Germany, pp. 1–4 (2005)

[70] Burkhard, H., Duhaut, D., Fujita, M., Lima, P., Murphy, R., Rojas, R.: The road to RoboCup 2050. IEEE Robotics & Automation Magazine 9(2), 31–38 (2002)
[71] Buschmann, T., Lohmeier, S., Ulbrich, H., Pfeiffer, F.: Modeling and simulation of a biped robot. In: IEEE International Conference on Robotics and Automation (ICRA 2006), pp. 2673–2678 (2006)
[72] Caballero, R., Akinfiev, T., Montes, H., Manzano, C., Armada, M.: Design of the SMART actuated ROBICAM biped robot. In: International Conference on Climbing and Walking Robots (CLAWAR 2002), pp. 409–416 (2002)
[73] Caballero, R., Alarcon, P., Armada, M.: Development and experimental evaluation of sensorial system for SILO-2 biped robot. In: International Conference on Climbing and Walking Robots (CLAWAR 2006), pp. 386–395 (2006)
[74] Caballero, R., Armada, M.A., Akinfiev, T.: Robust cascade controller for nonlinearly actuated biped robots: Experimental evaluation. The International Journal of Robotics Research 23(10-11), 1075–1095 (2004)
[75] Cabas, L., Cabas, R., Staroverov, P., Arbulu, M., Kaynov, D., Pérez, C., Balaguer, C.: Challenges in the design of the humanoid robot R.H.1. In: International Conference on Climbing and Walking Robots (CLAWAR 2006), pp. 318–323 (2006)
[76] Cabodevilla, G., Chaillet, N., Abba, G.: Energy-minimized gait for a biped robot. In: Proceedings Fachgespräch Autonome Mobile Systemer, pp. 90–99 (1995)
[77] Caldwell, D., Medrano-Cerda, G., Bowler, C.: Investigation of bipedal robot locomotion using pneumatic muscle actuators. In: IEEE International Conference on Robotics and Automation (ICRA 1997), pp. 799–804 (1997)
[78] Caldwell, D., Medrano-Cerda, G., Goodwin, M.: Control of pneumatic muscle actuators. IEEE Control Systems Magazine 15(1), 40–48 (1995)
[79] Caldwell, D., Tsagarakis, N., Badihi, D., Medrano-Cerda, G.: Pneumatic muscle actuator technology: a light weight power system for a humanoid robot. In: IEEE International Conference on Robotics and Automation (ICRA 1998), vol. 4, pp. 3053–3058 (1998)
[80] Caldwell, G.: Natural and artificial muscle elements as robot actuators. Mechatronics 3(3), 269–283 (1993)
[81] Capi, G., Nasu, Y., Barolli, L.K.M.: Real time gait generation for autonomous humanoid robots: A case study for walking. Robotics and Autonomous Systems 42(2), 107–116 (2003)
[82] Cavagna, G., Thys, H., Zamboni, A.: The sources of external work in level walking and running. Journal of Physiology 262, 639–657 (1976)
[83] Cavagna, G.A., Saibene, F.P., Margaria, R.: Mechanical work in running. Journal of Applied Physiology 19, 249–256 (1964)
[84] Chakarov, D.: Some approaches for passive compliance realization for precize opperation manipulators. In: Bulgarian Conference on Bionics, Biomechanics and Mechatronics, pp. 88–91 (2004)
[85] Channon, P., Hopkins, S., Pham, D.: Derivation of optimal walking motion for a bipedal walking robot. Robotica 10, 165–172 (1992)
[86] Cheng, G., Hyon, S.H., Jun Morimoto, A.U., Colvin, G., Scroggin, W., Jacobsen, S.C.: CB: A humanoid research platform for exploring neuroscience. In: IEEE-RAS International Conference on Humanoid Robots, pp. 182–187 (2006)
[87] Cherfas, J.: The difficulties of darwinism. New Scientist 102, 29 (1984)
[88] Chestnutt, J., Kuffner, J., Nishiwaki, K., Kagami, S.: Planning biped navigation strategies in complex environments. In: IEEE International Conference on Humanoid Robots (2003)

[89] Chevallereau, C., Abba, G., Aoustin, Y., Plestan, F., Westervelt, E., Canudas-De-Wit, C., Grizzle, J.: RABBIT: a testbed for advanced control theory. IEEE Control Systems Magazine 23(5), 57–79 (2003)

[90] Chevallereau, C., Aoustin, Y.: Optimal running trajectories for a biped. In: International Conference on Climbing and Walking Robots (CLAWAR 1999), pp. 559–570 (1999)

[91] Chevallereau, C., Aoustin, Y.: Optimal reference trajectories for walking and running of a biped robot. Robotica 19, 557–569 (2001)

[92] Chew, C., Pratt, G.: Frontal plane algorithms for dynamic bipedal walking. Robotica 22(01), 29–39 (2004)

[93] Chew, C.M., Pratt, G.A.: Frontal plane algorithms for dynamic bipedal walking. In: IEEE International Conference on Robotics & Automation (ICRA 2003), pp. 45–50 (2003)

[94] Cho, D.: Wrestle a robot. New Scientist 191(2564), 39 (2006)

[95] Choi, C., Tsao, T.C.: H_∞ preview control for discrete-time systems. Journal of Dynamic Systems, Measurement, and Contro 123(1), 117–124 (2001)

[96] Clarke, R.: Asimov's laws of robotics: implications for information technology-part I. Computer 26(12), 53–61 (1993)

[97] Clarke, R.: Asimov's laws of robotics: implications for information technology-part II. Computer 27(1), 57–66 (1994)

[98] Colbrunn, R., Nelson, G., Quinn, R.: Design and control of a robotic leg with braided pneumaticactuators. In: IEEE/RSJ International Conference on Intelligent Robots and Systems (IROS 2001), vol. 2, pp. 992–998 (2001)

[99] Collins, S., Ruina, A.: A bipedal walking robot with efficient and human-like gait. In: IEEE International Conference on Robotics and Automation (ICRA 2005), pp. 1983–1988 (2005)

[100] Collins, S.H., Ruina, A., Tedrake, R., Wisse, M.: Efficient bipedal robots based on passive-dynamic walkers. Science 18(307), 1082–1085 (2005)

[101] Colon, E.: CoRoBa, a multi mobile robot control and simulation framework. Ph.D. thesis, Vrije Universiteit Brussel - Koninklijke Militaire School (2006)

[102] Cote, C., Letourneau, D., Michaud, F., Valin, J.M., Brosseau, Y., Raievsky, C., Lemay, M., Tran, V.: Code reusability tools for programming mobile robots. In: IEEE/RSJ International Conference on Intelligent Robots and Systems (IROS 2004), vol. 2, pp. 1820–1825 (2004)

[103] Cupec, R., Lorch, O., Schmidt, G.: Vision-guided humanoid walking - concepts and experiments. In: Workshop on Robotics in Alpe-Adria-Danube Region (RAAD 2003), Cassino, Italy (2003)

[104] Daerden, F.: Conception and realization of pleated pneumatic artificial muscles and their use as compliant actuation elements. Ph.D. thesis, Vrije Universiteit Brussel (1999)

[105] Daerden, F., Lefeber, D.: Pneumatic artificial muscles: actuators for robotics and automation. European Journal of Mechanical and Environmental Engineering 47(1), 11–21 (2002)

[106] Daerden, F., Lefeber, D., Verrelst, B., Van Ham, R.: Pleated pneumatic artificial muscles: compliant robotic actuators. In: IEEE/RSJ International Conference on Intelligent Robots and Systems (IROS 2001), vol. 4, pp. 1958–1963 (2001)

[107] Dasgupta, A., Nakamura, Y.: Making feasible walking motion of humanoid robots from human motion capture data. In: IEEE International Conference on Robotics and Automation (ICRA 1999), Detroit, USA, vol. 2, pp. 1044–1049 (1999)

[108] David, A., Bruneau, O.: Dynamic stabilization of an under-actuated robot using dynamic effects of the legs and the trunk. In: IEEE/RSJ International Conference on Intelligent Robots and Systems (IROS 2006), pp. 898–903 (2006)

[109] Davis, S., Canderle, J., Artrit, P., Tsagarakis, N., Caldwell, D.: Enhanced dynamic performance in pneumatic muscle actuators. In: IEEE International Conference on Robotics and Automation (ICRA 2002), vol. 3, pp. 2836–2841 (2002)

[110] Davis, S., Tsagarakis, N., Canderle, J., Caldwell, D.G.: Enhanced modelling and performance in braided pneumatic muscle actuators. The International Journal of Robotic Research 22(22), 213–227 (2003)

[111] De Schutter, J.: A study of active compliant motion control methods for rigid manipulators based on a generic scheme. In: IEEE International conference on Robotics and Automation (ICRA 1987), pp. 1060–1065 (1987)

[112] Denk, J., Schmidt, G.: Synthesis of walking primitive databases for biped robots in 3D-environments. In: IEEE International Conference on Robotics and Automation (ICRA 2003), vol. 1, pp. 1343–1349 (2003)

[113] Dertien, E.: Dynamic walking with dribbel. IEEE Robotics & Automation Magazine 13(3), 118–122 (2006)

[114] Dickinson, M.H., Farley, C.T., Full, R.J., Koehl, M.A.R., Kram, R., Lehman, S.: How animals move: An integrative view. Science 288, 100–106 (2000)

[115] Djoudi, D., Chevallereau, C.: Feet can improve the stability property of a control law for a walking robot. In: IEEE International Conference on Robotics and Automation (ICRA 2006), pp. 1206–1212 (2006)

[116] Donelan, J.M., Kram, R., Kuo, A.D.: Mechanical work for step-to-step transitions is a major determinant of the metabolic cost of human walking. The Journal of Experimental Biology 205, 3717–3727 (2002)

[117] Einstein, J., Pawlik, G.: Feasibility of bipedal locomotion in robotic systems: Pneumatics vs. electronics. In: Proceedings 5th International Symposium on Measurement and Control in Robotics, Smolenice, Slovakia, pp. 453–458 (1995)

[118] Elftman, H.: The function of arms in walking. Human Biology 11, 529–535 (1939)

[119] Endo, G., Morimoto, J., Nakanishi, J., Cheng, G.: An empirical exploration of a neural oscillator for biped locomotion control. In: IEEE International Conference on Robotics and Automation (ICRA 2004), vol. 3, pp. 3036–3042 (2004)

[120] English, C., Russell, D.: Mechanics and stiffness limitations of a variable stiffness actuator for use in prosthetic limbs. Mechanism and Machine Theory 34(1), 7–25 (1999)

[121] Fallis, G.T.: Walking toy (improvement in walking toys). U. S. Patent, No. 376, 588 (1888)

[122] Farley, C., Gonzalez, O.: Leg stiffness and stride frequency in human running. Journal of Biomechanics 29(2), 181–186 (1996)

[123] Farley, C.T., Houdijk, H.H.P., Van Strien, C., Louie, M.: Mechanism of leg stiffness adjustment for hopping on surfaces of different stiffnesses. Journal of Applied Physiology 85(3), 1044–1055 (1998)

[124] Fernández, R., Hespanha, J., Akinfiev, T., Armada, M.: Nonlinear control for the dual smart drive using backstepping and a time-optimal reference. Autonomous Robots 19(3), 233–255 (2005)

[125] Fernández-Madrigal, J.A., Galindo, C., González, J., Cruz-Martín, E., Cruz-Martí, A.: A software engineering approach for the development of heterogeneous robotic applications. Robotics and Computer-Integrated Manufacturing (2007) (in press)

[126] Ferris, D.P., Farley, C.T.: Interaction of leg stiffness and surface stiffness during human hopping. Journal of Applied Physiology 82(1), 15–22 (1997)

[127] Festo: Information brochure Fluidic Muscle DMSP with pressed-on connections
[128] Figliolini, G., Ceccarelli, M.: Walking programming for an electropneumatic biped robot. Mechatronics 9(8), 941–964 (1999)
[129] Figliolini, G., Ceccarelli, M.: Climbing stairs with EPWAR2 biped robot. In: IEEE International Conference on Robotics and Automation (ICRA 2001), pp. 4116–4121 (2001)
[130] Figliolini, G., Ceccarelli, M.: EP-WAR3 biped robot for climbing and descending stairs. Robotica 22, 405–417 (2004)
[131] Fujiwara, K., Kajita, S., Harada, K., Kaneko, K., Morisawa, M., Kanehiro, F., Nakaoka, S., Hirukawa, H.: Towards an optimal falling motion for a humanoid robot. In: IEEE-RAS International Conference on Humanoid Robots, pp. 524–529 (2006)
[132] Fujiwara, K., Kanehiro, F., Kajita, S., Hirukawa, H.: Safe knee landing of a human-size humanoid robot while falling forward. In: IEEE/RSJ International Conference on Intelligent Robots and Systems (IROS 2004), pp. 503–508 (2004)
[133] Fujiwara, K., Kanehiro, F., Saito, H., Kajita, S., Harada, K., Hirukawa, H.: Falling motion control of a humanoid robot trained by virtual supplementary tests. In: IEEE International Conference on Robotics and Automation (ICRA 2004), pp. 1077–1082 (2004)
[134] Gates, B.: A robot in every home. Scientific American 296(1), 58–65 (2007)
[135] Geng, T., Porr, B., Florentin, W.: Fast biped walking with a sensor-driven neuronal controller and real-time online learning. International Journal of Robotics Research 25(3), 243–259 (2006)
[136] Gerecke, M., Albert, A., Hofschulte, J., Strasser, R., Gerth, W.: Towards an autonomous, bipedal service robot. In: Tagungsband 10 Jahre Fraunhofer IFF, Magdeburg, pp. 163–168 (2002)
[137] Geyer, H., Seyfarth, A., Blickhan, R.: Compliant leg behaviour explains basic dynamics of walking and running. Proceedings of the royal society B 273(1603), 2861 (2006)
[138] Ghan, J., Kazerooni, H.: System identification for the berkeley lower extremity exoskeleton (BLEEX). In: IEEE International Conference on Robotics and Automation (ICRA 2006), pp. 3477–3484 (2006)
[139] Goddard, R., Hemami, H., Weimer, F.: Biped side step in the frontal plane. IEEE Transactions on Automatic Control 28(2), 179–187 (1983)
[140] Gorce, P., Guihard, M.: On dynamic control of pneumatic bipeds. Journal of Robotic Systems 15(7), 421–433 (1998)
[141] Goswami, A.: Postural stability of biped robots and the foot-rotation indicator (FRI) point. The International Journal of Robotics Research 18(6), 523–533 (1999)
[142] Gould, S.J.: The Panda's Thumb (1980)
[143] Spampinato, G., Muscato, G.: Control architecture and walking strategy for a pneumatic biped robot. In: International Conference on Climbing and Walking Robots, CLAWAR 2005 (2005)
[144] Guan, Y., Neo, E., Yokoi, K., Tanie, K.: Stepping over obstacles with humanoid robots. IEEE Transactions on Robotics 22(5), 958–973 (2006)
[145] Guccione, S., Muscato, G., Spampinato, G.: A human inspired robotic leg: Design, control and mechanical realization. In: IEEE International Conference on Humanoid Robots (2003)
[146] Guizzo, E., Goldstein, H.: The rise of the body bots [robotic exoskeletons]. IEEE Spectrum 42(10), 50–56 (2005)

[147] Gutmann, J.S., Fukuchi, M., Fujita, M.: A modular architecture for humanoid robot navigation. In: IEEE-RAS International Conference on Humanoid Robots, pp. 26–31 (2005)

[148] Habumuremyi, J.C., Kool, P., Doroftei, I., Baudoin, Y.: Method to find dynamic parameters of a system using neuro-fuzzy techniques and its application to a pantograph-based leg of a robot AMRU5. In: International Symposium on Measurement and Control in Robotics (2005)

[149] Hansen, A., Childress, D., Miff, S., Gard, S., Mesplay, K.: The human ankle during walking: implications for design of biomimetic ankle prostheses. Journal of Biomechanics 37(10), 1467–1474 (2004)

[150] Harada, K., Hirukawa, H., Kanehiro, F., Fujiwara, K., Kaneko, K., Kajita, S., Nakamura, M.: Dynamical balance of a humanoid robot grasping an environment. In: IEEE/RSJ International Conference on Intelligent Robots and Systems (IROS 2004), vol. 2, pp. 1167–1173 (2004)

[151] Harada, K., Kajita, S., Kaneko, K., Hirukawa, H.: ZMP analysis for arm/leg coordination, pushing manipulation by humanoid considering two-kinds of ZMPs. In: International Conference on Intelligent Robots and Systems (IROS 2003), pp. 75–81 (2003)

[152] Hardt, M., Kreutz-Delgado, K., Helton, J., Stryk, O.V.: Obtaining minimum energy biped walking gaits with symbolic models and numerical optimal control. In: Workshop - Biomechanics Meets Robotics, Modelling and Simulation of Motion (1999)

[153] He, J., Kram, R., McMahon, T.: Mechanics of running under simulated low gravity. Journal of Applied Physiology 71(3), 863–870 (1991)

[154] Hemami, H., Wyman, B.: Modelling and control of constrained dynamic systems with application to biped locomotion in the frontal plane. IEEE Transactions on Automatic Control 24(4), 526–535 (1979)

[155] Hemker, T., Sakamoto, H., Stelzer, M., von Strykpp, O.: Hardware-in-the-loop optimization of the walking speed of a humanoid robot. In: International Conference on Climbing and Walking Robots (CLAWAR 2006), pp. 614–623 (2006)

[156] Hildebrand, M.: Walking and running. In: Functional Vertebrate Morphology, pp. 38–57. Belknap, Harvard Univ. Press, Cambridge (1985)

[157] Hildebrandt, A., Sawodny, O., Neumann, R., Hartmann, A.: Cascaded control concept of a robot with two degrees of freedom driven by four artificial pneumatic muscle actuators. In: American Control Conference, vol. 1, pp. 680–685 (2005)

[158] Hirai, K.: The honda humanoid robot: development and future perspective. Industrial Robot 26(4), 260–266 (1999)

[159] Hirai, K., Hirose, M., Haikawa, Y., Takenaka, T.: The development of honda humanoid robot. In: IEEE International Conference on Robotics and Automation (ICRA 1998), vol. 2, pp. 1321–1326 (1998)

[160] Hirose, S., Yokota, S., Torii, A., Ogata, M., Suganuma, S., Takita, K., Kato, K.: Quadruped walking robot centered demining system - development of TITAN-IX and its operation. In: IEEE International Conference on Robotics and Automation (ICRA 2005), pp. 1284–1290 (2005)

[161] Hirukawa, H.: Humanoid robotics projects in japan. In: Workshop ICRA 2006, Understanding Humanoid Robots (2006)

[162] Hirukawa, H., Hattori, S., Harada, K., Kajita, S., Kaneko, K., Kanehiro, F., Fujiwara, K., Morisawa, M.: A universal stability criterion of the foot contact of legged robots - adios ZMP. In: IEEE International Conference on Robotics and Automation (ICRA 2006), pp. 1976–1983 (2006)

[163] Hirukawa, H., Kanehiro, F., Kajita, S., Fujiwara, K., Yokoi, K., Kaneko, K., Harada, K.: Experimental evaluation of the dynamic simulation of biped walking of humanoid robots. In: IEEE International Conference on Robotics and Automation (ICRA 2003), vol. 2, pp. 1640–1645 (2003)

[164] Hirukawaa, H., Kanehiroa, F., Kanekoa, K., Kajitaa, S., Fujiwaraa, K., Kawaia, Y., Tomitaa, F., Hiraia, S., Taniea, K., Isozumib, T., Akachib, K., Kawasakib, T., Otab, S., Yokoyamac, K., Handac, H., Fukased, Y., Ichiro Maedad, J., Nakamurae, Y., Tachie, S., Inoue, H.: Humanoid robotics platforms developed in HRP. Robotics and Autonomous Systems 48(4), 165–175 (2004)

[165] Hobbelen, D., de Boer, T., Wisse, M.: System overview of bipedal robots flame and tulip: Tailor-made for limit cycle walking. In: IEEE/RSJ International Conference on Intelligent Robots and Systems (IROS2008), pp. 2486–2491 (2008)

[166] Hobbelen, D., Wisse, M.: Active lateral foot placement for 3D stabilization of a Limit Cycle Walker prototype. International Journal of Humanoid Robotics (IJHR) 6(1), 93–116 (2009)

[167] Hodgins, J., Koechling, J., Raibert, M.: Running experiments with a planar biped. In: Giralt, G. (ed.) Proceedings 3rd International Symposium on Robotics Research, pp. 349–355. MIT Press, Cambridge (1986)

[168] Hodgins, J., Raibert, M.: Biped gymnastics. International Journal of Robotics Research (Special Issue on Legged Locomotion) 9(2), 115–132 (1990)

[169] Hofschulte, J., Seebode, M., Gerth, W.: Parallel manipulator hip joint for a bipedal robot. In: International Conference on Climbing and Walking Robots (CLAWAR 2004), pp. 601–609 (2004)

[170] Höhn, O., Gacnik, J., Gerth, W.: Detection and classification of posture instabilities of bipedal robots. In: International Conference on Climbing and Walking Robots (CLAWAR 2005), pp. 409–416 (2005)

[171] Höhn, O., Gerth, W.: Probabilistic balance monitoring for bipedal robots. In: International Conference on Climbing and Walking Robots (CLAWAR 2006), pp. 435–442 (2006)

[172] Hollander, K., Sugar, T.: Concepts for compliant actuation in wearable robotic systems. In: US-Korea Conference on Science, Technology and Entrepreneurship, UKC 2004 (2004)

[173] Hollander, K., Sugar, T., Herring, D.: Adjustable robotic tendon using a "jack spring"^{TM}. In: 9th International Conference on Rehabilitation Robotics (ICORR 2005), pp. 113–118 (2005)

[174] Hollerbach, J., Hunter, I., Ballantyne, J.: A comparative analysis of actuator technologies for robotics. The Robotics Review 2, 299–342 (1991)

[175] Honda Motor Co., Ltd., New asimo - running at 6km/h (2005), `http://world.honda.com/HDTV/ASIMO/New-ASIMO-run-6kmh/`

[176] Hood, M.: Running against the wind [sports prosthetics]. IEEE Spectrum 42(6), 13–14 (2005)

[177] Hornyak, T.: Android science, pp. 32–34. Scientific American (2006)

[178] Hosoda, K., Takuma, T., Ishikawa, M.: Design and control of a 3D biped robot actuated by antagonistic pairs of pneumatic muscles. In: International Symposium on Adaptive Motion in Animals and Machines (2005)

[179] Hosoda, K., Takuma, T., Nakamoto, A.: Design and control of 2D biped that can walk and run with pneumatic artificial muscles. In: IEEE-RAS International Conference on Humanoid Robots, pp. 284–289 (2006)

[180] Hosoda, K., Takuma, T., Nakamoto, A., Hayashi, S.: Biped robot design powered by antagonistic pneumatic actuators for multi-modal locomotion. Robotics and Autonomous Systems 56(1), 46–53 (2007)

[181] Huang, Q., Kajita, S., Koyachi, N., Kaneko, K., Yokoi, K., Arai, H., Komoriya, K., Tanie, K.: High stability, smooth walking pattern for a biped robot. In: IEEE International Conference on Robotics and Automation (ICRA 1999), vol. 1, pp. 65–71 (1999)

[182] Huang, Q., Nakamura, Y.: Sensory reflex control for humanoid walking. IEEE Transactions on Robotics 21(5), 977–984 (2005)

[183] Huang, Q., Yokoi, K., Kajita, S., Kaneko, K., Arai, H., Koyachi, N., Tanie, K.: Planning walking patterns for a biped robot. IEEE Transactions on Robotics and Automation 17(3), 280–289 (2001)

[184] Hunter, I., Lafontaine, S.: A comparison of muscle with artificial actuators. In: IEEE Solid-State Sensor and Actuator Workshop, pp. 178–185 (1992)

[185] Hurst, J.W., Chestnutt, J., Rizzi, A.: An actuator with mechanically adjustable series compliance, CMU-RI-TR-04-24 (2004)

[186] Hurst, J.W., Chestnutt, J., Rizzi, A.A.: An actuator with physically variable stiffness for highly dynamic legged locomotion, new orleans, usa. In: IEEE International Conference on Robotics and Automation (ICRA 2004), pp. 4662–4667 (2004)

[187] Hurst, J.W., Rizzi, A.: Physically variable compliance in running. In: International Conference on Climbing and Walking Robots (CLAWAR 2004), pp. 123–134 (2004)

[188] Hurst, J.W., Rizzi, A.A.: Series compliance for robot actuation: Application on the electric cable differential leg. IEEE Robotics & Automation Magazine 15(3), 2008 (2008)

[189] Hyon, S., Abe, S., Emura, T.: Development of a biologically-inspired biped robot KenkenII. In: Japan-France Congress on Mechatronics & 4th Asia-Europe Congress on Mechatronics, pp. 404–409 (2003)

[190] Hyon, S., Emura, T., Mita, T.: Dynamics-based control of one-legged hopping robot. Journal of Systems and Control Engineering 217(2), 83–98 (2003)

[191] Hyon, S.H., Cheng, G.: Gravity compensation and full-body balancing for humanoid robots. In: IEEE-RAS International Conference on Humanoid Robots, pp. 214–221 (2006)

[192] Ikemata, Y., Sano, A., Fujimoto, H.: A physical principle of gait generation and its stabilization from mechanism of fixed point. In: IEEE International Conference on Robotics and Automation (ICRA 2006), pp. 836–841 (2006)

[193] Inaba, Kagami, Nishiwaki: Robot Anatomy, ISBN4-00-011247-3

[194] Inoue, H., Hirukawa, H.: Explorations of humanoid robot applications. In: IEEE-RAS International Conference on Humanoid Robots (2001)

[195] International Standard ISO6358: Pneumatic fluid power - method of test - determination of flow rate characteristics of components using compressible fluids (1989)

[196] Ishida, T., Kuroki, Y., Yamaguchi, J.: Mechanical system of a small biped entertainment robot. In: IEEE/RSJ International Conference on Intelligent Robots and Systems (IROS 2003), vol. 2, pp. 1129–1134 (2003)

[197] Ishida, T., Kuroki, Y., Yamaguchi, J., Fujita, M., Doi, T.: Motion entertainment by a small humanoid robot based on OPEN-R. In: IEEE/RSJ International Conference on Intelligent Robots and Systems (IROS 2001), vol. 2, pp. 1079–1086 (2001)

[198] Ishiguro, H., Asada, M., Shapiro, S.C., Thielscher, M., Breazeal, C., Mataric, M.J., Ishida, H.: Trends and controversies: Human-inspired robots. IEEE Intelligent Systems 21(4), 74–85 (2006)

[199] Itoh, K., Miwa, H., Matsumoto, M., Zecca, M., Takanobu, H., Roccella, S., Carrozza, M., Dario, P., Takanishi, A.: Various emotional expressions with emotion expression humanoid robot WE-4RII. In: IEEE Technical Exhibition Based Conference on Robotics and Automation (TExCRA 2004), pp. 35–36 (2004)

[200] Iwata, H., Hoshino, H., Morita, T., Sugano, S.: A physical interference adapting hardware system using MIA arm and humanoid surface covers. In: IEEE/RSJ International Conference on Intelligent Robots and Systems (IROS 1999), vol. 2, pp. 1216–1221 (1999)

[201] Jalón, J.G., Bayo, E.: Kinematic and Dynamic Simulation of Multibody Systems The Real-Time Challenge. Springer, Heidelberg (1994)

[202] Japan Robot Society: Summary report on technology strategy for creating a robot society in the 21st century

[203] Jezernik, S., Colombo, G., Keller, T., Frueh, H., Morari, M.: Robotic orthosis lokomat: A rehabilitation and research tool. Neuromodulation 6(2), 108–115 (2003)

[204] Jha, R.K., Singh, B., Pratihar, D.K.: Online stable gait generation of a two-legged robot using a genetic-fuzzy system. Robotics and Autonomous Systems 53(1), 15–35 (2005)

[205] Jung, H., Seo, Y., Ryoo, M., Yang, H.: Affective communication multimodality for the humanoid robot AMI. In: IEEE-RAS/RSJ International Conference on Humanoid Robots (2004)

[206] Kagami, S., Kitagawa, T., Nishiwaki, K., Sugihara, T.M.I., Inoue, H.: A fast dynamically equilibrated walking trajectory generation method of humanoid robot. Autonomous Robots 12(1), 71–82 (2002)

[207] Kagami, S., Mochimaru, M., Ehara, Y., Miyata, N., Nishiwaki, K., Kanade, T., Inoue, H.: Measurement and comparison of human and humanoid walking. In: IEEE International Symposium on Computational Intelligence in Robotics and Automation (ICRA 2003), vol. 2, pp. 918–922 (2003)

[208] Kagami, S., Nishiwaki, K., Kuffner, J., Kuniyoshi, Y., Inaba, M., Inoue, H.: Online 3D vision, motion planning and bipedal locomotion control coupling system of humanoid robot: H7. In: IEEE/RSJ International Conference on Intelligent Robots and System (IROS 2002), vol. 3, pp. 2557–2562 (2002)

[209] Kagami, S., Nishiwaki, K., Kuffner, J.J., Kuniyoshi, Y., Inaba, M., Inoue, H.: Online 3d vision, motion planning and bipedal locomotion control coupling system of humanoid robot: H7. In: IEEE/RSJ International Conference on Intelligent Robots and System (IROS 2002), vol. 3, pp. 2557–2562 (2002)

[210] Kagami, S., Nishiwaki, K., Sugihara, T., Kuffner, J., Inaba, M., Inoue, H.: Design and implementation of software research platform for humanoid robotics: H6. In: IEEE International Conference on Robotics and Automation (ICRA 2001), vol. 3, pp. 2431–2436 (2001)

[211] Kajita, S.: and Kanehiro, F., Kaneko, K., Fujiwara, K., Yokoi, K., Hirukawa, H.: A realtime pattern generator for biped walking. In: IEEE International Conference on Robotics and Automation (ICRA 2002), vol. 1, pp. 31–37 (2002)

[212] Kajita, S., Kanehiro, F., Kaneko, K., Fujiwara, K., Harada, K., Yokoi, K., Hirukawa, H.: Biped walking pattern generation by using preview control of zero-moment point. In: IEEE International Conference on Robotics and Automation (ICRA 2003), vol. 2, pp. 1620–1626 (2003)

[213] Kajita, S., Kaneko, K., Harada, K., Kanehiro, F., Fujiwara, K., Hirukawa, H.: Biped walking on a low friction floor. In: IEEE/RSJ International Conference on Intelligent Robots and Systems (IROS 2004), vol. 4, pp. 3546–3552 (2004)

[214] Kajita, S., Matsumoto, O., Saigo, M.: Real-time 3D walking pattern generation for a biped robot with telescopic legs. In: IEEE International Conference on Robotics and Automation (ICRA 2001), pp. 2299–2306 (2001)

[215] Kajita, S., Nagasaki, T., Kaneko, K., Yokoi, K., Tanie, K.: A hop towards running humanoid biped. In: IEEE International Conference on Robotics and Automation (ICRA 2004), pp. 629–635 (2004)

[216] Kajita, S., Nagasaki, T., Kaneko, K., Yokoi, K., Tanie, K.: A Running Controller of Humanoid Biped HRP-2LR. In: IEEE International Conference on Robotics and Automation (ICRA 2005), pp. 616–622 (2005)

[217] Kajita, S., Nagasaki, T., Yokoi, K., Kaneko, K., Tanie, K.: Running pattern generation for a humanoid robot. In: IEEE International Conference on Robotics and Automation (ICRA 2002), vol. 3, pp. 2755–2761 (2002)

[218] Kajita, S., Tani, K.: Study of dynamic biped locomotion on rugged terrain - derivation and application of the linear inverted pendulum mode. In: Proceedings 1991 IEEE International Conference on Robotics and Automation, Sacramento, California, USA, pp. 1405–1411 (1991)

[219] Kanda, T., Miyashita, T., Osada, T., Haikawa, Y., Ishiguro, H.: Analysis of humanoid appearances in human-robot interaction. In: IEEE/RSJ International Conference on Intelligent Robots and Systems (IROS 2005), pp. 899–906 (2005)

[220] Kanehira, N., Kawasaki, T., Ohta, S., Ismumi, T., Kawada, T., Kanehiro, F., Kajita, S., Kaneko, K.: Design and experiments of advanced leg module (HRP-2L) for humanoid robot (HRP-2) development. In: IEEE/RSJ International Conference on Intelligent Robots and System (IROS 2002), vol. 3, pp. 2455–2460 (2002)

[221] Kanehiro, F., Hirukawa, H., Kajita, S.: OpenHRP: Open architecture humanoid robotics platform. The International Journal of Robotics Research 23(2), 155–165 (2004)

[222] Kanehiro, F., Hirukawa, H., Kaneko, K., Kajita, S., Fujiwara, K., Harada, K., Yokoi, K.: Locomotion planning of humanoid robots to pass through narrow spaces. In: IEEE International Conference on Robotics and Automation (ICRA 2004), vol. 1, pp. 604–609 (2004)

[223] Kaneko, K., Harada, K., Kanehiro, F., Miyamori, G., Akachi, K.: Humanoid robot hrp-3. In: IEEE/RSJ International Conference on Intelligent Robots and Systems (IROS 2008), pp. 2471–2478 (2008)

[224] Kaneko, K., Kanehiro, F., Kajita, S., Hirukawa, H., Kawasaki, T., Hirata, M., Akachi, K., Isozumi, T.: Humanoid robot hrp-2. In: IEEE International Conference on Robotics and Automation (ICRA 2004), vol. 2, pp. 1083–1090 (2004)

[225] Kaneko, K., Kanehiro, F., Kajita, S., Morisawa, M., Fujiwara, K., Harada, K., Hirukawa, H.: Slip observer for walking on a low friction floor. In: IEEE/RSJ International Conference on Intelligent Robots and Systems (IROS 2005), pp. 634–640 (2005)

[226] Kaneko, K., Kanehiro, F., Kajita, S., Morisawa, M., Fujiwara, K., Harada, K., Hirukawa, H.: Motion suspension system for humanoids in case of emergency real-time motion generation and judgment to suspend humanoid. In: IEEE/RSJ International Conference on Intelligent Robots and Systems (IROS 2006), pp. 5496–5501 (2006)

[227] Kaneko, K., Kanehiro, F., Morisawa, M., Miura, K., Nakaoka, S., Kajita, S.: Cybernetic human hrp-4c. In: IEEE-RAS International Conference on Humanoid Robots (Humanoids 2009), pp. 7–14 (2009)

[228] Kanzaki, S., Okada, K., Inaba, M.: Bracing behavior in humanoid through preview control of impact disturbance. In: IEEE-RAS International Conference on Humanoid Robots, pp. 301–306 (2005)
[229] Katayama, T., Itoh, T., Ogawa, M., Yamamoto, H.: Optimal tracking control of a heat exchanger with change in load condition. In: Proceedings of the 29th IEEE Conference on Decision and Control, vol. 3, pp. 1584–1589 (1990)
[230] Katayama, T., Ohki, T., Inoue, T., Kato, T.: Design of an optimal controller for a discrete time system subject to previewable demand. Int. J. Control 41, 677–699 (1985)
[231] Kato, I.: Development of WABOT 1. Biomechanism 2, 173–214 (1973)
[232] Kato, I., Mori, Y., Masuda, T.: Pneumatically powered artificial legs walking automatically under various circumstances. In: Proceedings of the 4th International Conference on External Control of Human Extremities, pp. 458–470 (1972)
[233] Kawamoto, H., Sankai, Y.: Power assist method based on phase sequence and muscle force condition for hal. Journal Advanced Robotics 19(7), 717–734 (2005)
[234] Kawamura, S., Yamamoto, T., Ishida, D., Ogata, T., Nakayama, Y., Tabata, O., Sugiyama, S.: Development of passive elements with variable mechanical impedance for wearable robots. In: IEEE International Conference on Robotics and Automation (ICRA 2002), vol. 1, pp. 248–253 (2002)
[235] Kazerooni, H., Steger, R., Huang, L.: Hybrid control of the berkeley lower extremity exoskeleton (bleex). The International Journal of Robotics Research 25, 561–573 (2006)
[236] Kearney, R., Hunter, I.: System identification of human joint dynamics. Critical reviews in biomedical engineering 18(1), 55–87 (1990)
[237] Kerscher, T., Albiez, J., Zoellner, J.M., Dillmann, R.: Biomechanical inspired control for elastic legs. In: International conference on Climbing and Walking Robots (CLAWAR 2006), pp. 592–597 (2006)
[238] Kho, J.W., Lim, D.C., Kuc, T.Y.: Implementation of an intelligent controller for biped walking robot using genetic algorithm. In: IEEE International Symposium on Industrial Electronics, pp. 49–54 (2006)
[239] Khraief, N., M'Sirdi, N., Spong, M.: Nearly passive dynamic walking of a biped robot. In: European Control Conference (2003)
[240] Kikuuwe, R., Fujimoto, H.: Proxy-based sliding mode control for accurate and safe position control. In: IEEE International Conference on Robotics and Automation (ICRA 2006), pp. 25–30 (2006)
[241] Kim, J.H., Oh, J.H.: Walking control of the humanoid platform khr-1 based on torque feedback control. In: IEEE International Conference on Robotics and Automation (ICRA 2004), vol. 1, pp. 623–628 (2004)
[242] Kim, J.Y., Park, I.W., Lee, J., Kim, M.S., Cho, B.K., Oh, J.H.: System design and dynamic walking of humanoid robot KHR-2. In: IEEE International Conference on Robotics & Automation (ICRA 2005), pp. 1443–1448 (2005)
[243] Kingsley, D.A., Quinn, R.D., Ritzmann, R.E.: A cockroach inspired robot with artificial muscles. In: Proceedings of the International Symposium on Adaptive Motion of Animal and Machines, Kyoto, Japan (2003)
[244] Komatsu, T., Usui, M.: Dynamic walking and running of a bipedal robot using hybrid central pattern generator method. In: IEEE International Conference Mechatronics and Automation (ICMA 2005), vol. 2, pp. 987–992 (2005)
[245] Komi, P., Fukashiro, S., Jarvinen, M.: Biomechanical loading of achilles tendon during normal locomotion. Clinics in sports medicine 11(3), 521–531 (1992)

[246] Komura, T., Leung, H., Kudoh, S., Kuffner, J.: A feedback controller for biped humanoids that can counteract large perturbations during gait. In: IEEE International Conference on Robotics and Automation (ICRA 2005), pp. 1989–1995 (2005)
[247] Konczak, J.: On the notion of motor primitives in humans and robots. In: International Workshop on Epigenetic Robotics: Modeling Cognitive Development in Robotic Systems, pp. 47–53 (2005)
[248] Kram, R., Domingo, A., Ferris, D.: Effect of reduced gravity on the preferred walk-run transition speed. Journal of Experimental Biology 200, 821–826 (1997)
[249] Krishna, M., Bares, J., Mutschler, E.: Tethering system design for dante II. In: IEEE International Conference on Robotics and Automation (ICRA 1997), vol. 2, pp. 1100–1105 (1997)
[250] Kudoh, S., Komura, T.: C2 continuous gait-pattern generation for biped robots. In: IEEE/RSJ International Conference on Intelligent Robots and Systems (IROS 2003), vol. 2, pp. 1135–1140 (2003)
[251] Kuitunen, S., Komi, P., Kyrolainen, H.: Knee and ankle joint stiffness in sprint running. Medicine & Science in Sports & Exercise 34(1), 166–173 (2002)
[252] Kuo, A.D., Donelan, J.M., Ruina, A.: Energetic consequences of walking like an inverted pendulum: step-to-step transitions. Exercise and sport sciences reviews 33(2), 88–97 (2005)
[253] Kurazume, R., Tanaka, S., Yamashita, M., Hasegawa, T.: Straight legged walking of a biped robot. In: IEEE/RSJ International Conference on Intelligent Robots and Systems (IROS 2005), pp. 3095–3101 (2005)
[254] Kuroki, Y., Blank, B., Mikami, T., Mayeux, P., Miyamoto, A., Playter, R., Nagasaka, K., Raibert, M., Nagano, M., Yamaguchi, J.: Motion creating system for a small biped entertainment robot. In: IEEE/RSJ International Conference on Intelligent Robots and Systems (IROS 2003), vol. 2, pp. 1394–1399 (2003)
[255] Kuroki, Y., Fujita, M., Ishida, T., Nagasaka, K., Yamaguchi, J.: A small biped entertainment robot exploring attractive applications. In: IEEE International Conference on Robotics and Automation (ICRA 2003), vol. 1, pp. 471–476 (2003)
[256] Kurzweil, R.: The Age of Intelligent Machines. The MIT Press, Cambridge (1990)
[257] Kusuda, Y.: The humanoid robot scene in japan. Industrial Robot 29(5), 412–419 (2002)
[258] Kusuda, Y.: How japan sees the robotics for the future: observation at the world expo 2005. Industrial Robot 33(1), 11–18 (2006)
[259] Lee, J.: Developement of a human-riding humanoid robot HUBO FX-1. In: SICE - ICASE International Joint Conference 2006 (2006)
[260] Lee, S.H., Song, J.B.: Acceleration estimator for low-velocity and low-accelerationregions based on encoder position data. IEEE/ASME Transactions on Mechatronics 6(1), 58–64 (2001)
[261] Liu, Q., Huang, Q., Zhang, W., Wang, X., Wu, C., Li, D., Li, K.: Manipulation of a humanoid robot by teleoperation. In: Fifth World Congress on Intelligent Control and Automation (WCICA 2004), vol. 6, pp. 4894–4898 (2004)
[262] Cabas, L.M., Torre, S., Cabas, R., Kaynov, D., Arbulú, M., Staroverov, P., Balaguer, C.: Mechanical design and dynamic analysis of the humanoid robot Rh-0. In: International Conference on Climbing and Walking Robots, CLAWAR 2005 (2005)
[263] Löffler, K., Geinger, M., Pfeiffer, F.: Sensors and control concept of walking "johnnie". The International Journal of Robotics Research 22(3-4), 229–239 (2003)

[264] Loffler, K., Gienger, M., Pfeiffer, F.: Sensor and control design of a dynamically stable biped robot. In: IEEE International Conference on Robotics and Automation (ICRA 2003), vol. 1, pp. 484–490 (2003)
[265] Loffler, K., Gienger, M., Pfeiffer, F., Ulbrich, H.: Sensors and control concept of a biped robot. IEEE Transactions on Industrial Electronics 51(5), 972–980 (2004)
[266] Lohmeier, S., Buschmann, T., Ulbrich, H., Pfeiffer, F.: Modular joint design for performance enhanced humanoid robot LOLA. In: IEEE International Conference on Robotics and Automation (ICRA 2006), pp. 88–93 (2004)
[267] Lohmeier, S., Buschmann, T., Ulbrich, H., Pfeiffer, F.: Modular joint design for performance enhanced humanoid robot LOLA. In: IEEE International Conference on Robotics and Automation (ICRA 2006), pp. 88–93 (2006)
[268] Lorch, O., Albert, A., Denk, J., Gerecke, M., Cupec, R., Seara, J., Gerth, W., Schmidt, G.: Experiments in vision-guided biped walking. In: IEEE/RSJ International Conference on Intelligent Robots and System (IROS 2002), vol. 3, pp. 2484–2490 (2002)
[269] Manoonpong, P., Geng, T., Wörgötter, F.: Exploring the dynamic walking range of the biped robot RunBot with an active upper-body component. In: IEEE-RAS International Conference on Humanoid Robots, pp. 418–424 (2006)
[270] Mao, Y., Wang, J., Jia, P., Li, S., Qiu, Z., Zhang, L., Han, Z.: A reinforcement learning based dynamic walking control. In: IEEE International Conference on Robotics and Automation, ICRA 2007 (2007)
[271] Mao, Y., Wang, J., Li, S., Han, Z.: Energy-efficient control of pneumatic muscle actuated biped robot joints. In: 6th World Congress on Intelligent Control and Automation, pp. 8881–8885 (2006)
[272] Matsubara, T., Morimoto, J., Nakanishi, J., Aki Sato, M., Doya, K.: Learning CPG-based biped locomotion with a policy gradient method. Robotics and Autonomous Systems 54(11), 911–920 (2006)
[273] Matsui, T., Hirukawa, H., Ishikawa, Y., Yamasaki, N., Kagami, S., Kanehiro, F., Saito, H., Inamura, T.: Distributed real-time processing for humanoid robots. In: IEEE International Conference on Embedded and Real-Time Computing Systems and Applications (RTCSA 2005), pp. 205–210 (2005)
[274] McGeer, T.: Powered flight, child's play, silly wheels, and walking machines. In: IEEE International Conference on Robotics and Automation (ICRA 1989), pp. 1592–1597 (1989)
[275] McGeer, T.: Passive bipedal running. Proceedings Royal Society of London: Biological Sciences, 107–134 (1990)
[276] McGhee, R.B., Frank, A.A.: On the stability properties of quadruped creeping gaits. Journal of Mathematic Biosciences 3, 331–351 (1968)
[277] McMahon, T.A., Valiant, G., Frederick, E.C.: Groucho running. Journal of Applied Physiology 62(6), 2326–2337 (1987)
[278] Merlin Systems Corp. Ltd.: Flow-Controlled Air Muscle Datasheet v1.0
[279] Metta, G., Fitzpatrick, P., Natale, L.: YARP: yet another robot platform. International Journal on Advanced Robotics Systems 3(1), 43–48 (2006)
[280] Mianzo, L., Peng, H.: Lq and h_∞ preview control for a durability simulator. In: Proceedings of the American Control Conference, vol. 1, pp. 699–703 (1997)
[281] Mianzo, L., Peng, H.: A unified framework for LQ and H_∞ preview control algorithms. In: Proceedings of the 37th IEEE Conference on Decision and Control, vol. 3, pp. 2816–2821 (1998)

[282] Michel, P., Chestnutt, J., Kuffner, J., Kanade, T.: Vision-guided humanoid footstep planning for dynamic environments. In: IEEE-RAS International Conference on Humanoid Robots, pp. 13–18 (2005)
[283] Migliore, S.A., Brown, E.A., DeWeerth, S.P.: Biologically inspired joint stiffness control. In: IEEE International Conference on Robotics and Automation (ICRA 2005), pp. 4519–4524 (2005)
[284] Minato, T., Shimada, M., Itakura, S., Lee, K., Ishiguro, H.: Does gaze reveal the human likeness of an android? In: The 4nd International Conference on Development and Learning, pp. 106–111 (2005)
[285] Mita, T., Yamaguchi, T., Kashiwase, T., Kawase, T.: Realization of a high speed biped using modern control theory. International Journal of Control 40(1), 107–119 (1984)
[286] Mitobe, K., Capi, G., Nasu, Y.: A new control method for walking robots based on angular momentum. Mechatronics 14, 163–174 (2004)
[287] Miwa, H., Itoh, K., Matsumoto, M., Zecca, M., Takanobu, H., Roccella, S., Carrozza, M., Dario, P., Takanishi, A.: Effective emotional expressions with emotion expression humanoid robot WE-4RII. In: IEEE/RSJ International Conference on Intelligent RObots and Systems (IROS 2004), pp. 2203–2208 (2004)
[288] Miyakoshi, S., Taga, G., Kuniyoshi, Y., Nagakubo, A.: Three dimensional bipedal stepping motion using neural oscillators-towards humanoid motion in the real world. In: IEEE/RSJ International Conference on Intelligent Robots and Systems (IROS 1998), vol. 1, pp. 84–89 (1998)
[289] Morimoto, J., Endo, G., Nakanishi, J., Hyon, S., Cheng, G., Bentivegna, D., Atkeson, C.: Modulation of simple sinusoidal patterns by a coupled oscillator model for biped walking. In: IEEE International Conference on Robotics and Automation (ICRA 2006), pp. 1579–1584 (2006)
[290] Morisawa, M., Kajita, S., Harada, K., Fujiwara, K., Kanehiro, F., Kaneko, K., Hirukawa, H.: Emergency stop algorithm for walking humanoid robots. In: IEEE/RSJ International Conference on Intelligent Robots and Systems (IROS 2005), pp. 2109–2115 (2005)
[291] Morisawa, M., Kajita, S., Kaneko, K., Harada, K., Kanehiro, F., Fujiwara, K., Hirukawa, H.: Pattern generation of biped walking constrained on parametric surface. In: IEEE International Conference on Robotics and Automation (ICRA 2005), pp. 2405–2410 (2005)
[292] Morita, T., Sugano, S.: Development of a new robot joint using a mechanical impedance adjuster. In: IEEE International Conference on Robotics and Automation (ICRA 1995), vol. 3, pp. 2469–2475 (1995)
[293] Morris, B., Westervelt, E., Chevallereau, C., Buche, G., Grizzle, J.W.: Achieving bipedal running with rabbit: six steps towards infinity. In: Fast Motions in Biomechanics and Robotics, Heidelberg, Allemagne (2005)
[294] Nadjar-Gauthier, N., Cherrid, H., Cadiou, J.C.: A new second order sliding mode control for the experimental walking of an electro-pneumatic biped robot. In: Proceedings of the 5th International Conference on Climbing and Walking Robots and the Support Technologies for Mobile Machines, Paris, France, pp. 93–100 (2002)
[295] Nagasaka, K., Inaba, M., Inoue, H.: Stabilization of dynamic walk on a humanoid using torso position compliance control. In: Proceedings of 17th Annual Conference on Robotics Society of Japan, pp. 1193–1194 (1999) (in Japanese)
[296] Nakanishi, J., Morimoto, J., Endo, G., Cheng, G., Schaal, S., Kawato, M.: learning from demonstration and adaptation of biped locomotion. Robotics and Autonomous Systems 47(2-3), 79–91 (2004)

[297] Napoleon, N.S., Sampei, M.: Balance control analysis of humanoid robot based on zmp feedback control. In: IEEE/RSJ International Conference on Intelligent Robots and System (IROS 2002), vol. 3, pp. 2437–2442 (2002)
[298] Nishiwaki, K., Kagami, S.: High frequency walking pattern generation based on preview control of ZMP. In: IEEE International Conference on Robotics and Automation (ICRA 2006), pp. 2667–2672 (2006)
[299] Nishiwaki, K., Kagami, S., Kuniyoshi, Y., Inaba, M., Inoue, H.: Online generation of humanoid walking motion based on a fast generation method of motion pattern that follows desired ZMP. In: IEEE/RSJ International Conference on Intelligent Robots and System (IROS 2002), vol. 3, pp. 2684–2689 (2002)
[300] Nishiwaki, K., Kagami, S., Kuniyoski, Y., Inaba, M., Inoue, H.: Toe joints that enhance bipedal and fullbody motion of humanoid robots. In: IEEE International Conference on Robotics and Automation (ICRA 2002), pp. 3105–3110 (2002)
[301] Nishiwaki, K., Sugihara, T., Kagami, S., Kanehiro, F., Inaba, M., Inoue, H.: Design and development of research platform for perception-actionintegration in humanoid robot: H6. In: IEEE/RSJ International Conference on Intelligent Robots and Systems (IROS 2000), pp. 1559–1564 (2000)
[302] Nonami, K.: Development of mine detection robot COMET-II and COMET-III. In: Proceedings of the 41st SICE Annual Conference (SICE 2002), vol. 1, pp. 346–351 (2002)
[303] Nonami, K., Ikedo, Y.: Walking control of COMET-III using discrete time preview sliding mode controller. In: IEEE/RSJ International Conference on Intelligent Robots and Systems (IROS 2004), vol. 4, pp. 3219–3225 (2004)
[304] Nunez, V., Nadjar-Gauthier, N.: Humanoid vertical jump with compliant contact. In: International Conference on Climbing and Walking Robots (CLAWAR 2005), pp. 457–464 (2005)
[305] Ogura, Y., Aikawa, H., Shimomura, K., Kondo, H., Morishima, A., Lim, H.-O., Takanishi, A.: Development of a new humanoid robot WABIAN-2. In: IEEE International Conference on Robotics and Automation (ICRA 2006), pp. 76–81 (2006)
[306] Ogura, Y., Lim, H.-o., Takanishi, A.: Stretch walking pattern generation for a biped humanoid robot. In: IEEE/RSJ International Conference on Intelligent Robots and Systems (IROS 2003), vol. 1, pp. 352–357 (2003)
[307] Ogura, Y., Kataoka, T., Aikawa, H., Shimomura, K., Lim, H.-o., Takanishi, A.: Evaluation of various walking patterns of biped humanoid robot. In: IEEE International Conference on Robotics and Automation (ICRA 2005), pp. 603–608 (2005)
[308] Ogura, Y., Kataoka, T., Shimomura, K., Lim, H.-O., Takanishi, A.: A novel method of biped walking pattern generation with predetermined knee joint motion. In: IEEE/RSJ International Conference on Intelligent Robots and Systems (IROS 2004), vol. 3, pp. 2831–2836 (2004)
[309] Ogura, Y., Shimomura, K., Kondo, H., Morishima, A., Okubo, T., Momoki, S., Lim, H.-O., Takanishi, A.: Human-like walking with knee stretched, heel-contact and toe-off motion by a humanoid robot. In: IEEE/RSJ International Conference on Intelligent Robots and Systems (IROS 2006), pp. 3976–3981 (2006)
[310] Oh, J.H., Hanson, D., Kim, W.S., Han, I.Y., Kim, J.Y., Park, I.W.: Design of android type humanoid robot Albert HUBO. In: IEEE/RSJ International Conference on Intelligent Robots and Systems (IROS 2006), pp. 1428–1433 (2006)
[311] Oh, Y., Choi, Y., You, B.J., Oh, S.R.: Development of a biped humanoid robot: BabyBot. In: International Conference on Mechatronics and Information Technology (ICMIT 2003), pp. 690–695 (2003)

[312] Okada, K., Ogura, T., Haneda, A., Inaba, M.: Autonomous 3D walking system for a humanoid robot based on visual step recognition and 3D foot step planner. In: IEEE International Conference on Robotics and Automation (ICRA 2005), pp. 623–628 (2005)

[313] Okumura, Y., Tawara, T., Endo, K., Furuta, T., Shimizu, M.: Realtime ZMP compensation for biped walking robot using adaptive inertia force control. In: IEEE/RSJ International Conference on Intelligent Robots and Systems (IROS 2003), vol. 1, pp. 335–339 (2003)

[314] Olaru, I., Krut, S., Pierrot, F.: Novel Mechanical Design of Biped Robot SHERPA Using 2 DOF Cable Differential Modular Joints. In: IEEE/RSJ International Conference on Intelligent RObots and Systems (IROS 200), pp. 4463–4468 (2009)

[315] Ono, K., Liu, R.: Optimal biped walking locomotion solved by trajectory planning method. Transactions of the ASME Journal of Dynamic Systems, Measurement and Control 124, 554–565 (2002)

[316] Park, I.W., Kim, J.Y., Lee, J., Oh, J.H.: Mechanical design of humanoid robot platform KHR-3 (KAIST humanoid robot - 3: HUBO). In: IEEE-RAS International Conference on Humanoid Robots, pp. 321–326 (2005)

[317] Park, I.W., Kim, J.Y., Lee, J., Oh, J.H.: Online free walking trajectory generation for biped humanoid robot KHR-3(HUBO). In: IEEE International Conference on Robotics and Automation (ICRA 2006), pp. 1231–1236 (2006)

[318] Park, I.W., Kim, J.Y., Park, S.W., Oh, J.H.: Development of humanoid robot platform KHR-2 (KAIST humanoid robot-2). In: IEEE-RAS International Conference on Humanoid Robots, vol. 1, pp. 292–310 (2004)

[319] Park, J.: Fuzzy-logic zero-moment-point trajectory generation for reduced trunk motions of biped robots. Fuzzy Sets and Systems 134(1), 189–203 (2003)

[320] Park, J., Kim, K.: Biped robot walking using gravity-compensated inverted pendulum mode and computed torque control. In: IEEE International Conference on Robotics and Automation (ICRA 1998), vol. 4, pp. 3528–3533 (1998)

[321] Park, J., Rhee, Y.: ZMP trajectory generation for reduced trunk motions of biped robots. In: IEEE/RSJ International Conference on Intelligent Robots and Systems (IROS 1998), vol. 1, pp. 90–95 (1998)

[322] Park, J.H., Cho, H.C.: An online trajectory modifier for the base link of biped robots to enhance locomotion stability. In: IEEE International Conference on Robotics and Automation (ICRA 2000), vol. 4, pp. 3353–3358 (2000)

[323] Park, J.H., Chung, H.: ZMP compensation by online trajectory generation for biped robots. In: IEEE International Conference on Systems, Man, and Cybernetics (SMC 1999), vol. 4, pp. 960–965 (1999)

[324] Park, N.C., Yang, H.S., Park, H.W., Park, Y.P.: Position/vibration control of two-degree-of-freedom arms having one flexible link with artificial pneumatic muscle actuators. Robotics and Autonomous Systems 40(15), 239–253 (2002)

[325] Paynter, H.M.: Hyperboloid of revolution fluid-driven tension actuators and methods of making. US Patent No. 4 721 030 (1988)

[326] Peng, Z., Huang, Q., Zhao, X., Xiao, T., Li, K.: Online trajectory generation based on off-line trajectory for biped humanoid. In: IEEE International Conference on Robotics and Biomimetics (ROBIO 2004), pp. 752–756 (2004)

[327] Pfeiffer, F., Lffler, K., Gienger, M.: Humanoid robots. In: International Conference on Climbing and Walking Robots (CLAWAR 2003), pp. 505–516 (2003)

[328] Pfeiffer, F., Loffler, K., Gienger, M.: The concept of jogging johnnie. In: IEEE International Conference on Robotics and Automation (ICRA 2002), vol. 3, pp. 3129–3135 (2002)

[329] Playter, R., Raibert, M.: Control of a biped somersault in 3D. In: IEEE/RSJ International Conference on Intelligent Robots and Systems (IROS 1992), Raleigh, NC, USA, pp. 582–589 (1992)
[330] Plestan, F., Grizzle, J., Westervelt, E., Abba, G.: Stable walking of a 7-dof biped robot. IEEE Transactions on Robotics and Automation 19(4), 653–668 (2003)
[331] Pratt, G.: Legged robots at mit: what's new since raibert? IEEE Robotics & Automation Magazine 7(3), 15–19 (2000)
[332] Pratt, G., Williamson, M.M., Dilworth, P., Pratt, J., Wright, A.: Stiffness isn't everything. In: International Symposium on Experimental Robotics (ISER 1995), Stanford, California, pp. 253–262 (1995)
[333] Pratt, G.A., Williamson, M.M.: Series elastic actuators. In: IEEE International Workshop on Intelligent Robots and Systems (IROS 1995), Pittsburg, USA, pp. 399–406 (1995)
[334] Pratt, J.: Virtual model control of a biped walking robot. Master's thesis, Massachusetts Institute of Technology (1995)
[335] Pratt, J.: Exploiting inherent robustness and natural dynamics in the control of bipedal walking robots. Ph.D. thesis, Massachusetts Institute of Technology (2000)
[336] Pratt, J., Carff, J., Drakunov, S., Goswami, A.: Capture point: A step toward humanoid push recovery. In: IEEE-RAS International Conference on Humanoid Robots, pp. 200–207 (2006)
[337] Pratt, J., Chew, C.M., Torres, A., Dilworth, P., Pratt, G.: Virtual model control: An intuitive approach for bipedal locomotion. The International Journal of Robotics Research 20, 129–143 (2001)
[338] Pratt, J., Krupp, B.: Design of a bipedal walking robot. In: Society of Photo-Optical Instrumentation Engineers (SPIE) Conference Series, vol. 6962, p. 44 (2008)
[339] Quinn, R.D., Nelson, G., Bachmann, R., Kingsley, D., Offi, J., Ritzmann, R.E.: Insect designs for improved robot mobility. In: International Conference on Climbing and Walking Robots (CLAWAR 2001), pp. 69–76 (2001)
[340] Raibert, M.: Legged robots. Communications of the ACM 29, 499–514 (1986)
[341] Raibert, M.: Legged Robots That Balance. MIT Press, Cambridge (1986)
[342] Raibert, M.: Running with symmetry. International Journal of Robotics Research 5(4), 45–61 (1986)
[343] Raibert, M., Brown, H.J.: Experiments in balance with a 2D one-legged hopping machine. Journal of Dynamic Systems, Measurement and Control 106, 75–81 (1984)
[344] Raibert, M., Brown, H.J., Chepponis, M.: Experiments in balance with a 3D one-legged hopping machine. International Journal of Robotics Research 3(2), 75–92 (1984)
[345] Raibert, M., Chepponis Jr., M.H.B.: Running on four legs as though they were one. IEEE Transactions on Robotics and Automation 2(2), 70–82 (1986)
[346] Rao, C., Mitra, S.: Generalized Inverse of Matrices and its Applications. John Wiley & Sons, Inc., New York (1971)
[347] Rebula, J., Canas, F., Pratt, J., Goswami, A.: Learning capture points for humanoid push recovery. In: IEEE/RAS International Conference on Humanoid Robots (Humanoids 2007), pp. 65–72. Citeseer (2007)
[348] Ridderström, C.: Legged locomotion control - a literature study. Tech. rep., Mechatronics Lab, Department of Machine Design, Royal Institute of technology, Stockholm, Sweden (1999)

[349] Ridderström, C.: Legged locomotion control - a literature study. Tech. Rep. 27, Mechatronics Lab, Departement of Machine Design, Royal Institute of Technology, Stockholm, Sweden (1999)
[350] Righetti, L., Ijspeert, A.J.: programmable central pattern generators: an application to biped locomotion control. In: IEEE International Conference on Robotics and Automation (ICRA 2006), pp. 1585–1590 (2006)
[351] Ritter, H., Haschke, R., Koiva, R., Rothling, F., Steil, J.: A layered control architecture for imitation grasping with a 20-DOF pneumatic anthropomorphic hand. Internal report University of Bielefeld (2005)
[352] Rose, J., Gamble, J.: Human walking. Lippincott Williams & Wilkins (2006)
[353] Rosheim, M.E.: Leonardo's Lost Robots. Springer, Heidelberg (2006)
[354] Sabourin, C., Bruneau, O., Buche, G.: Control strategy for the robust dynamic walk of a biped robot. The International Journal of Robotics Research 25(9), 843–860 (2006)
[355] Sakagami, Y., Watanabe, R., Aoyama, C., Matsunaga, S., Higaki, N., Fujimura, K.: The intelligent asimo: System overview and integration. In: IEEE/RSJ International Conference on Intelligent Robots and System (IROS 2002), vol. 3, pp. 2478–2483 (2002)
[356] Sakka, S., Yokoi, K.: Humanoid vertical jumping based on force feedback and inertial forces optimization. In: IEEE International Conference on Robotics and Automation (ICRA 2005), pp. 3752–3757 (2005)
[357] Saldien, J., Goris, K., Yilmazyildiz, S., Verhelst, W., Lefeber, D.: On the design of the huggable robot probo. Journal of Physical Agents 2(2) (2008)
[358] Sandini, G., Metta, G., Vernon, D.: RobotCub: an open framework for research in embodied cognition. In: IEEE-RAS International Conference on Humanoid Robots, vol. 1, pp. 13–32 (2005)
[359] Schaal, S., Peters, J., Nakanishi, J., Ijspeert, A.: Learning movement primitives. In: International Symposium on Robotics Research (ISRR 2003). Springer, Heidelberg (2004)
[360] Schroder, J., Erol, D., Kawamura, K., Dillman, R.: Dynamic pneumatic actuator model for a model-based torque controller. In: IEEE International Symposium on Computational Intelligence in Robotics and Automation (CIRA 2003), vol. 1, pp. 342–347 (2003)
[361] Schuitema, E., Hobbelen, D., Jonker, P., Wisse, M., Karssen, J.: Using a controller based on reinforcement learning for a passive dynamic walking robot. In: IEEE-RAS International Conference on Humanoid Robots, pp. 232–237 (2005)
[362] Schuitema, E., Hobbelen, D., Jonker, P., Wisse, M., Karssen, J.: Using a controller based on reinforcement learning for a passive dynamic walking robot. In: IEEE-RAS International Conference on Humanoid Robots, pp. 232–237 (2005)
[363] Seara, J.F., Strobl, K.H., Schmidt, G.: Path-Dependent Gaze Control for Obstacle Avoidance in Vision Guided Humanoid Walking. In: IEEE International Conference on Robotics and Automation (ICRA 2003), Taipei, Taiwan, vol. 1, pp. 887–892 (2003)
[364] Seilacher, A.: Arbeitskonzept zur konstruktionsmorphologie. Lethaia 3, 393–396 (1970)
[365] Seilacher, A.: Self-organizing mechanisms in morphogenesis and evolution. In: Constructional Morphology and Evolution, pp. 251–271. Springer, Heidelberg (1991)
[366] Sekiguchi, A., Atobe, Y., Kameta, K., Tsumaki, Y., Nenchev, D.N.: A walking pattern generator around singularity. In: IEEE-RAS International Conference on Humanoid Robots, pp. 270–275 (2006)

[367] Sellaouti, R., Stasse, O., Kajita, S., Yokoi, K., Kheddar, A.: Faster and smoother walking of humanoid HRP-2 with passive toe joints. In: IEEE/RSJ International Conference on Intelligent Robots and Systems (IROS 2006), pp. 4909–4914 (2006)
[368] Shan, J., Nagashima, F.: Neural locomotion controller design and implementation for humanoid robot HOAP-1. In: Annual Conference of the Robotics Society of Japan (2002)
[369] Sheridan, T.: Three models of preview control. IEEE Transaction on Human Factors in Electronics HFE 7(2), 91–102 (1966)
[370] Shih, C.L.: Gait synthesis for the SD-2 biped robot to climb stairs. Robotica 15, 599–607 (1997)
[371] Shih, C.L., Gruver, W.: Control of a biped robot in the double-support phase. IEEE Transactions on Systems, Man and Cybernetics 22(4), 729–735 (1992)
[372] Slatter, R., Koenen, H.: Lightweight harmonic drive gears for service robots. Harmonic Drive AG
[373] Slotine, J.J.E., Li, W.: Applied Nonlinear Control. Prentice-Hall, Cambridge (1991)
[374] Song, S.M., Waldron, K.: Machines That Walk: The Adaptive Suspension Vehicle. MIT Press, Cambridge (1989)
[375] Spampinato, G., Muscato, G.: DIEES biped robot: A bio-inspired pneumatic platform for human locomotion analysis and stiffness control. In: IEEE-RAS International Conference on Humanoid Robots, pp. 478–483 (2006)
[376] Spong, M., Vidyasagar, M.: Robot Dynamics and Control. John Wiley and Sons, Inc., Chichester (1989)
[377] Stankevich, L.A.: Intellectual robots in russia: experience of development and robocup participation. In: SPECOM 2004, pp. 615–620 (2004)
[378] Stasse, O., Verrelst, B., Davison, A., Mansard, N., Vanderborght, B., Esteves, C., Saidi, F., Yokoi, K.: Integrating vision and walking to increase humanoid autonomy. In: IEEE International Conference on Robotics and Automation (ICRA 2007), Roma, Italia, pp. 2272–2273 (2007)
[379] Stasse, O., Verrelst, B., Vanderborght, B., Yokoi, K.: Dynamically walking over large obstacles by a humanoid robot. IEEE Transactions on Robotics 25(4), 960–967 (2009)
[380] Stilman, M., Kuffner, J.: Navigation among movable obstacles: Real-time reasoning in complex environments. International Journal of Humanoid Robotics 2(4), 479–504 (2005)
[381] Microsoft Robotics Studio, `http://msdn.microsoft.com/robotics`
[382] Sugahara, Y., Ohta, A., Hashimoto, K., Sunazuka, H., Kawase, M., Tanaka, C., Lim, H.-O., Takanishi, A.: Walking up and down stairs carrying a human by a biped locomotor with parallel mechanism. In: IEEE/RSJ International Conference on Intelligent Robots and Systems (IROS 2005), pp. 1489–1494 (2005)
[383] Sugihara, T., Nakamura, Y., Inoue, H.: Realtime humanoid motion generation through ZMP manipulation based on inverted pendulum control. In: IEEE International Conference on Robotics and Automation, ICRA 2002 (2002)
[384] Sulzer, J., Peshkin, M., Patton, J.: MARIONET: An exotendon-driven, rotary series elastic actuator for exerting joint torque. In: International Conference on Robotics for Rehabilitation (ICORR 2005), pp. 103–108 (2005)
[385] Swevers, J., Ganseman, C., Tukel, D., de Schutter, J., Van Brussel, H.: Optimal robot excitation and identification. IEEE Transactions on Robotics and Automation 13(5), 730–740 (1997)

[386] Tabata, O., Konishi, S., Cusin, P., Ito, Y., Kawai, F., Hirai, S., Kawamura, S.: Microfabricated tunable bending stiffness device. In: The Thirteenth Annual International Conference on Micro Electro Mechanical Systems (MEMS 2000), pp. 23–27 (2000)
[387] Taga, G.: A model of the neuro-musculo-skeletal system for human locomotion i: Emergence of basic gait. Biological Cybernetics 73(2), 97–111 (1995)
[388] Taga, G.: A model of the neuro-musculo-skeletal system for human locomotion II: Real-time adaptabilty under various constraints. Biological Cybernetics 73(2), 113–121 (1995)
[389] Taga, G., Yamaguchi, Y., Shimizu, H.: Self-organized control of bipedal locomotion by neural oscillators in unpredictable environment. Biological Cybernetics 65, 147–159 (1991)
[390] Tajima, D., Honda, D., Suga, K.: Fast Running Experiments Involving a Humanoid Robot. In: IEEE International Conference on Robotics and Automation (ICRA 2009), pp. 1571–1576 (2009)
[391] Tajima, R., Suga, K.: Motion having a flight phase: Experiments involving a one-legged robot. In: IEEE/RSJ International Conference on Intelligent Robots and Systems (IROS 2006), pp. 1726–1731 (2006)
[392] Takaba, K.: A tutorial on preview control systems. In: SICE 2003 Annual Conference, vol. 2, pp. 1388–1393 (2003)
[393] Takahashi, T., Kawamura, A.: Posture control using foot toe and sole for biped walking robot "ken". In: International Workshop on Advanced Motion Control, pp. 437–442 (2002)
[394] Takanishi, A., Ogura, Y., Itoh, K.: Some issues in humanoid robot design. In: International Symposium of Robotics Research, ISRR 2005 (2005)
[395] Takuma, T., Hosoda, K., Ogino, M., Asada, M.: Controlling walking period of a pneumatic muscle walker. In: International Conference on Climbing and Walking Robots (CLAWAR 2004), pp. 757–764 (2004)
[396] Takuma, T., Hosoda, K., Ogino, M., Asada, M.: Stabilization of quasi-passive pneumatic muscle walker. In: IEEE-RAS International Conference on Humanoid Robots, pp. 627–639 (2004)
[397] Tanaka, T., Takubo, T., Inoue, K., Arai, T.: Emergent stop for humanoid robots. In: IEEE/RSJ International Conference on Intelligent Robots and Systems (IROS 2006), pp. 3970–3975 (2006)
[398] Tang, Z., Zhou, C., Sun, Z.: Balance of penalty kicking for a biped robot. In: IEEE Conference on Robotics, Automation and Mechatronics, vol. 1, pp. 336–340 (2004)
[399] Tawara, T., Okumura, Y., Furuta, T., Shimizu, M., Shimomura, M., Endo, K., Kitano, H.: Morph: A desktop-class humanoid capable of acrobatic behavior. The International Journal of Robotics Research 23, 1097–1103 (2004)
[400] Tedrake, R.L.: Applied optimal control for dynamically stable legged locomotion. Ph.D. thesis, MIT (2004)
[401] Tellez, R., Ferro, F., Garcia, S., Gomez, E., Jorge, E., Mora, D., Pinyol, D., Oliver, J., Torres, O., Velazquez, J.: et al.: Reem-B: An autonomous lightweight human-size humanoid robot. In: IEEE-RAS International Conference on Humanoid Robots (Humanoids 2008), pp. 462–468 (2008)
[402] Tomizuka, M., Rosenthal, D.: On the optimal digital state feedback controller with integral and preview actions. Transactions of ASME, Journal of Dynamic Systems, Measurement and Control 101, 172–178 (1979)

[403] Tsagarakis, N., Sinclair, M., Becchi, F., Metta, G., Sandini, G., Caldwell, D.G.: Lower body design of the iCub a human-baby like crawling robot. In: IEEE-RAS International Conference on Humanoid Robots, pp. 450–455 (2006)

[404] Tsusaka, Y., Ota, Y.: Wire-driven bipedal robot. In: IEEE/RSJ International Conference on Intelligent Robots and Systems (IROS 2006), pp. 3958–3963 (2006)

[405] Tzafestas, S., Raibert, M., Tzafestas, C.: Robust sliding-mode control applied to a 5-link biped robot. Journal of Intelligent and Robotic Systems 15, 67–133 (1996)

[406] Ulaby, F.: The Legacy of Moore's Law. Proceedings of the IEEE 94(7), 1251–1252 (2006)

[407] University Archives and Records Center, University of Pennsylvania: Eadweard Muybridge Collection

[408] Utz, H., Sablatnog, S., Enderle, S., Kraetzschmar, G.: Miro - middleware for mobile robot applications. IEEE Transactions on Robotics and Automation 18(4), 493–497 (2002)

[409] Van Damme, M., Beyl, P., Vanderborght, B., Versluys, R., Van Ham, R., Vanderniepen, I., Daerden, F., Lefeber, D.: The safety of a robot actuated by pneumatic muscles - a case study. International Journal of Social Robotics (2010) (accepted for publication)

[410] Van Damme, M., Vanderborght, B., Verrelst, B., Van Ham, R., Daerden, F., Lefeber, D.: Proxy-based sliding mode control of a planar pneumatic manipulator. International Journal of Robotics Research 28(2), 266–284 (2009)

[411] Van Damme, M., Beyl, P., Vanderborght, B., Van Ham, R., Innes, V., Dirk, L.: Modeling hysteresis in pleated pneumatic artificial muscles. In: IEEE International Conferences on Robotics, Automation & Mechatronics (RAM), pp. 471–476 (2008)

[412] Van Ham, R.: Compliant actuation for biologically inspired bipedal walking robots. Ph.D. thesis, Vrije Universiteit Brussel (2006)

[413] Van Ham, R., Thomas, S., Vanderborght, B., Hollander, K., Lefeber, D.: Compliant actuator designs: Review of actuators with passive adjustable compliance/controllable stiffness for robotic applications. IEEE Robotics and Automation Magazine 16(3), 81–94 (2009)

[414] Van Ham, R., Vanderborght, B., Van Damme, M., Verrelst, B., Lefeber, D.: MACCEPA, the Mechanically Adjustable Compliance and Controllable Equilibrium Position Actuator: Design and Implementation in a Biped Robot. Robotics and Autonomous Systems 55(10), 761–768 (2007)

[415] Van Ham, R., Verrelst, B., Daerden, F., Vanderborght, B., Lefeber, D.: Fast and accurate pressure control using on-of valves. International Journal of Fluid Power 6(1), 53–58 (2005)

[416] Vandenhoudt, J.: PWM-sturing van een antagonistisch paar pneumatische artificile spieren. Master's thesis, Vrije Universiteit Brussel (2002)

[417] Vanderborght, B., Tsagarakis, N., Semini, C., Van Ham, R., Caldwell, D.: MACCEPA 2.0: Adjustable Compliant Actuator with Stiffening Characteristic for Energy Efficient Hopping. In: IEEE International Conference on Robotics and Automation (ICRA 2009), pp. 544–549 (2009)

[418] Vanderborght, B., Verrelst, B., Van Ham, R., Van Damme, M., Lefeber, D., Y Duran, M.B., Beyl, P.: Exploiting natural dynamics to reduce energy consumption by controlling the compliance of soft actuators. The International Journal of Robotics Research 25(4), 343–358 (2006)

[419] Vaughan, R., Gerkey, B., Howard, A.: On device abstractions for portable, reusable robot code. In: IEEE/RSJ International Conference on Intelligent Robots and Systems (IROS 2003), vol. 3, pp. 2421–2427 (2003)

[420] Verlinde, P., Acheroy, M., Baudoin, Y.: The belgian humanitarian demining project (hudem) and the european research context. European journal of mechanical and environmental engineering 46(2), 96–98 (2001)

[421] Vermeulen, J.: Trajectory generation for planar hopping and walking robots: An objective parameter and angular momentum approach. Ph.D. thesis, Vrije Universiteit Brussel (2004)

[422] Vermeulen, J., Verrelst, B., Lefeber, D., Kool, P., Vanderborght, B.: A real-time joint trajectory planner for dynamic walking bipeds in the sagittal plane. Robotica 23(6), 669–680 (2005)

[423] Verrelst, B., Stasse, O., Yokoi, K., Vanderborght, B.: Dynamically stepping over obstacles by the humanoid robot HRP-2. In: IEEE-RAS International Conference on Humanoid Robots, pp. 117–123 (2006)

[424] Verrelst, B., Van Ham, R., Vanderborght, B., Daerden, F., Lefeber, D.: The pneumatic biped LUCY actuated with pleated pneumatic artificial muscles. Autonomous Robots 18, 201–213 (2005)

[425] Verrelst, B., Van Ham, R., Vanderborght, B., Lefeber, D., Daerden, F., Van Damme, M.: Second generation pleated pneumatic artificial muscle and its robotic applications. Advanced Robotics 20(7), 783–805 (2006)

[426] Verrelst, B., Vanderborght, B., Stasse, O., Yokoi, K.: Stepping over large obstacles by the humanoid robot hrp-2. In: International Conference on Climbing and Walking Robots (CLAWAR 2006), pp. 747–754 (2006)

[427] Verrelst, B., Vermeulen, J., Vanderborght, B., Van Ham, R., Naudet, J., Lefeber, D., Daerden, F., Van Damme, M.: Motion generation and control for the pneumatic biped lucy. International Journal of Humanoid Robotics 25(4), 343–358 (2006)

[428] Verrelst, B., Yokoi, K., Stasse, O., Arisumi, H., Vanderborght, B.: Mobility of humanoid robots: Stepping over large obstacles dynamically. In: International Conference on Mechatronics and Automation (ICMA 2006), pp. 1072–1079 (2006)

[429] Versluy, R., Lenaerts, G., Van Damme, M., Jonkers, I., Desomer, A., Vanderborght, B.: Successful preliminary walking experiments on a transtibial amputee fitted with a powered prosthesis. Prosthetics and Orthotics International 33(4), 368–377 (2009)

[430] Versluys, R.: Study and development of articulated transtibial prostheses with adaptable compliance and push-off properties. Ph.D. thesis, Vrije Universiteit Brussel (2009)

[431] Versluys, R., Desomer, A., Lenaerts, G., Pareit, O., Vanderborght, B., Perre, G., Peeraer, L., Lefeber, D.: A biomechatronical transtibial prosthesis powered by pleated pneumatic artificial muscles. International Journal of Modelling, Identification and Control 4(4), 394–405 (2008)

[432] Veruggio, G.: The EURON roboethics roadmap. In: IEEE-RAS International Conference on Humanoid Robots, pp. 612–617 (2006)

[433] Visser, L., Carloni, R., Stramigioli, S.: A port-based comparison of variable stiffness actuators. In: IEEE International Conference on Robotics and Automation (ICRA 2010), accepted for publication (2010)

[434] Vukobratovic, M.: Legged Locomotion Robots and Anthropomorphic Mechanisms. Mihailo Pupin Institute, Beograd (1975)

[435] Vukobratovic, M., Borovac, B.: Zero-moment point - thirthy five years of its life. International Journal of Humanoid Robotics 1, 157–173 (2004)

[436] Vukobratovic, M., Borovac, B., Potkonjak, V.: Towards a unified understanding of basic notions and terms in humanoid robotics. Robotica 25(1), 87–101 (2005)

[437] Vukobratovic, M., Stepanenko, J.: On the stability of anthropomorphic systems. Mathematical Biosciences 15, 1–37 (1972)

[438] Wakimoto, S., Suzumori, K., Kanda, T.: Development of intelligent mckibben actuator. In: IEEE/RSJ International Conference on Intelligent Robots and Systems (IROS 2005), pp. 487–492 (2005)

[439] Walker, R.: Using air muscles for compliant bipedal and many-legged robotics. In: IEE Colloquium on Information Technology for Climbing and Walking Robots, pp. 3/1–3/3 (1996)

[440] Wang, L., Yu, Z., Meng, Q., Zhang, Z.: Influence analysis of toe-joint on biped gaits. In: International Conference on Mechatronics and Automation (ICMA 2007), pp. 1631–1635 (2006)

[441] Whittington, B., Silder, A., Heiderscheit, B., Thelen, D.: Passive elastic joint moments during humand walking. In: Dynamic Walking (2006)

[442] Wieber, P.B.: Trajectory free linear model predictive control for stable walking in the presence of strong perturbations. In: IEEE-RAS International Conference on Humanoid Robots, pp. 137–142 (2006)

[443] Wisse, M.: Essentials of dynamic walking: Analysis and design of two-legged robots. Ph.D. thesis, Technische Universiteit Delft (2004)

[444] Wisse, M., Hobbelen, D.G., Rotteveel, R.J., Anderson, S., Zeglin, G.J.: Ankle springs instead of arc-shaped feet for passive dynamic walkers. In: IEEE-RAS International Conference on Humanoid Robots, pp. 110–116 (2006)

[445] Wisse, M., Schwab, A., van der Linde, R., van der Helm, F.: How to keep from falling forward: elementary swing leg action for passive dynamic walkers. IEEE Transactions on Robotics and Automation 21(3), 393–401 (2005)

[446] Wisse, M., Schwab, A.L.: Skateboards, bicycles, and three-dimensional biped walking machines: Velocity-dependent stability by means of lean-to-yaw coupling. International Journal on Robotic Research 24(6), 417–429 (2005)

[447] Wisse, M., Schwab, A.L., Van Der Helm, F.C.T.: Passive dynamic walking model with upper body. Robotica 22(6), 681–688 (2004)

[448] Wisse, M., van Frankenhuyzen, J.: Design and construction of mike; a 2D autonomous biped based on passive dynamic walking. In: 2nd International Symposium on Adaptive Motion of Animals and Machines (2003)

[449] Xiao, T., Huang, Q., Li, J., Zhang, W., Li, K.: Trajectory calculation and gait change on-line for humanoid teleoperation. In: IEEE International Conference on Mechatronics and Automation (ICMA 2006), pp. 1614–1619 (2006)

[450] Yamaguchi, J., Nishino, D., Takanishi, A.: Realization of dynamic biped walking varying joint stiffness using antagonistic driven joints. In: IEEE International Conference on Robotics and Automation (ICRA 1998), vol. 3, pp. 2022–2029 (1998)

[451] Yamaguchi, J., Takanishi, A.: Design of biped walking robot having antagonistic driven joint using nonlinear spring mechanism. In: IEEE/RSJ International Conference on Intelligent Robots and Systems (IROS 1997), pp. 251–259 (1997)

[452] Yamaguchi, J., Takanishi, A.: Development of a biped walking robot having antagonistic driven joints using nonlinear spring mechanism. In: IEEE International Conference on Robotics and Automation (ICRA 1997), pp. 185–192 (1997)

[453] Yamakita, M., Kamamichi, N., Kozuki, T., Asaka, K., Luo, Z.W.: Control of biped walking robot with ipmc linear actuator. In: IEEE/ASME International Conference on Advanced Intelligent Mechatronics (ICMA 2005), pp. 48–53 (2005)

[454] Yamasaki, F., Miyashita, T., Matsui, T., Kitano, H.: PINO the humanoid that walk. In: IEEE-RAS International conference on Humanoid Robots, CD ROM (October 2000)

[455] Yang, H.S., Seo, Y.H., Chae, Y.N., Jeong, I.W., Kang, W.H., Lee, J.H.: Design and development of biped humanoid robot, AMI2, for social interaction with humans. In: IEEE-RAS International Conference on Humanoid Robots, pp. 352–357 (2006)
[456] Yang, J.S.: Adaptive control for a biped locomotion system. In: Proceedings of the 36th Midwest Symposium on Circuits and Systems, vol. 1, pp. 657–660 (1993)
[457] Yi, K.Y.: Walking of a biped robot with compliant ankle joints:implementation with kubca. In: IEEE Conference on Decision and Control, vol. 5, pp. 4809–4814 (2000)
[458] Yokoi, K., Kanehiro, F., Kaneko, K., Fujiwara, K., Kajita, S., Hirukawa, H.: A honda humanoid robot controlled by AIST software. In: IEEE/RAS International Conference on Humanoid Robots, Tokyo, Japan, pp. 259–264 (2001)
[459] Yokoi, K., Kanehiro, F., Kaneko, K., Kajita, S., Fujiwara, K., Hirukawa, H.: Experimental study of humanoid robot HRP-1S. The International Journal of Robotics Research 23(4-5), 351–362 (2004)
[460] Yoshida, E., Belousov, I., Esteves, C., Laumond, J.P.: Humanoid motion planning for dynamic tasks. In: IEEE-RAS International Conference on Humanoid Robots, pp. 1–6 (2005)
[461] Yoshida, E., Esteves, C., Sakaguchi, T., Laumond, J.P., Yokoi, K.: Smooth collision avoidance: Practical issues in dynamic humanoid motion. In: IEEE/RSJ International Conference on Intelligent Robots and Systems (IROS 2006), pp. 827–832 (2006)
[462] Yoshimi, T., Kawai, Y., Fukase, Y., Araki, H., Tomita, F.: Measurement of ground surface displacement using stereo vision and mechanical sensors on humanoid robots. In: IEEE International Conference on Multisensor Fusion and Integration for Intelligent Systems (MFI 2003), pp. 125–130 (2003)
[463] Zarrugh, M., Radcliffe, C.: Computer generation of human gait kinematics. Journal of biomechanics 12(2), 99–111 (1979)
[464] Zecca, M., Endo, N., Momoki, S., Itoh, K., Takanishi, A.: Design of the humanoid robot KOBIAN-preliminary analysis of facial and whole body emotion expression capabilities. In: IEEE-RAS International Conference on Humanoid Robots (Humanoids 2008), pp. 487–492 (2008)
[465] Zheng, Y., Hemami, H.: Mathematical modeling of a robot collision with its environment. Journal of Robotic Systems 2(3), 289–307 (1985)
[466] Zinn, M., Roth, B., Khatib, O., Salisbury, J.: A new actuation approach for human friendly robot design. The International Journal of Robotics Research 23(4-5), 379–398 (2004)
[467] Zinn, M., Khatib, O., Roth, B., Salisbury, J.: Playing it safe [human-friendly robots]. IEEE Robotics & Automation Magazine 11(2), 12–21 (2004)

Index

Springer Tracts in Advanced Robotics

Edited by B. Siciliano, O. Khatib and F. Groen

Further volumes of this series can be found on our homepage: springer.com

Vol. 62: Howard, A.; Iagnemma, K.; Kelly, A. (Eds.)
Field and Service Robotics
511 p. 2010 [978-3-642-13407-4]

Vol. 61: Mozos, Ó.M.
Semantic Labeling of Places with Mobile Robots
134 p. 2010 [978-3-642-11209-6]

Vol. 60: Zhu, W.-H.
Virtual Decomposition Control – Toward Hyper Degrees of Freedom Robots
443 p. 2010 [978-3-642-10723-8]

Vol. 59: Otake, M.
Electroactive Polymer Gel Robots – Modelling and Control of Artificial Muscles
238 p. 2010 [978-3-540-23955-0]

Vol. 58: Kröger, T.
On-Line Trajectory Generation in Robotic Systems – Basic Concepts for Instantaneous Reactions to Unforeseen (Sensor) Events
230 p. 2010 [978-3-642-05174-6]

Vol. 57: Chirikjian, G.S.; Choset, H.; Morales, M., Murphey, T. (Eds.)
Algorithmic Foundations of Robotics VIII – Selected Contributions of the Eighth International Workshop on the Algorithmic Foundations of Robotics
680 p. 2010 [978-3-642-00311-0]

Vol. 56: Buehler, M.; Iagnemma, K.; Singh S. (Eds.)
The DARPA Urban Challenge – Autonomous Vehicles in City Traffic
625 p. 2009 [978-3-642-03990-4]

Vol. 55: Stachniss, C.
Robotic Mapping and Exploration
196 p. 2009 [978-3-642-01096-5]

Vol. 54: Khatib, O.; Kumar, V.; Pappas, G.J. (Eds.)
Experimental Robotics: The Eleventh International Symposium
579 p. 2009 [978-3-642-00195-6]

Vol. 53: Duindam, V.; Stramigioli, S.
Modeling and Control for Efficient Bipedal Walking Robots
211 p. 2009 [978-3-540-89917-4]

Vol. 52: Nüchter, A.
3D Robotic Mapping
201 p. 2009 [978-3-540-89883-2]

Vol. 51: Song, D.
Sharing a Vision
186 p. 2009 [978-3-540-88064-6]

Vol. 50: Alterovitz, R.; Goldberg, K.
Motion Planning in Medicine: Optimization and Simulation Algorithms for Image-Guided Procedures
153 p. 2008 [978-3-540-69257-7]

Vol. 49: Ott, C.
Cartesian Impedance Control of Redundant and Flexible-Joint Robots
190 p. 2008 [978-3-540-69253-9]

Vol. 48: Wolter, D.
Spatial Representation and Reasoning for Robot Mapping
185 p. 2008 [978-3-540-69011-5]

Vol. 47: Akella, S.; Amato, N.; Huang, W.; Mishra, B.; (Eds.)
Algorithmic Foundation of Robotics VII
524 p. 2008 [978-3-540-68404-6]

Vol. 46: Bessière, P.; Laugier, C.; Siegwart R. (Eds.)
Probabilistic Reasoning and Decision Making in Sensory-Motor Systems
375 p. 2008 [978-3-540-79006-8]

Vol. 45: Bicchi, A.; Buss, M.; Ernst, M.O.; Peer A. (Eds.)
The Sense of Touch and Its Rendering
281 p. 2008 [978-3-540-79034-1]

Vol. 44: Bruyninckx, H.; Přeučil, L.; Kulich, M. (Eds.)
European Robotics Symposium 2008
356 p. 2008 [978-3-540-78315-2]

Vol. 43: Lamon, P.
3D-Position Tracking and Control
for All-Terrain Robots
105 p. 2008 [978-3-540-78286-5]

Vol. 42: Laugier, C.; Siegwart, R. (Eds.)
Field and Service Robotics
597 p. 2008 [978-3-540-75403-9]

Vol. 41: Milford, M.J.
Robot Navigation from Nature
194 p. 2008 [978-3-540-77519-5]

Vol. 40: Birglen, L.; Laliberté, T.; Gosselin, C.
Underactuated Robotic Hands
241 p. 2008 [978-3-540-77458-7]

Vol. 39: Khatib, O.; Kumar, V.; Rus, D. (Eds.)
Experimental Robotics
563 p. 2008 [978-3-540-77456-3]

Vol. 38: Jefferies, M.E.; Yeap, W.-K. (Eds.)
Robotics and Cognitive Approaches to
Spatial Mapping
328 p. 2008 [978-3-540-75386-5]

Vol. 37: Ollero, A.; Maza, I. (Eds.)
Multiple Heterogeneous Unmanned Aerial
Vehicles
233 p. 2007 [978-3-540-73957-9]

Vol. 36: Buehler, M.; Iagnemma, K.;
Singh, S. (Eds.)
The 2005 DARPA Grand Challenge – The Great
Robot Race
520 p. 2007 [978-3-540-73428-4]

Vol. 35: Laugier, C.; Chatila, R. (Eds.)
Autonomous Navigation in Dynamic
Environments
169 p. 2007 [978-3-540-73421-5]

Vol. 34: Wisse, M.; van der Linde, R.Q.
Delft Pneumatic Bipeds
136 p. 2007 [978-3-540-72807-8]

Vol. 33: Kong, X.; Gosselin, C.
Type Synthesis of Parallel
Mechanisms
272 p. 2007 [978-3-540-71989-2]

Vol. 30: Brugali, D. (Ed.)
Software Engineering for Experimental Robotics
490 p. 2007 [978-3-540-68949-2]

Vol. 29: Secchi, C.; Stramigioli, S.; Fantuzzi, C.
Control of Interactive Robotic Interfaces – A
Port-Hamiltonian Approach
225 p. 2007 [978-3-540-49712-7]

Vol. 28: Thrun, S.; Brooks, R.;
Durrant-Whyte, H. (Eds.)
Robotics Research – Results of the 12th
International Symposium ISRR
602 p. 2007 [978-3-540-48110-2]

Vol. 27: Montemerlo, M.; Thrun, S.
FastSLAM – A Scalable Method for the
Simultaneous Localization and Mapping
Problem in Robotics
120 p. 2007 [978-3-540-46399-3]

Vol. 26: Taylor, G.; Kleeman, L.
Visual Perception and Robotic Manipulation – 3D
Object Recognition, Tracking and Hand-Eye
Coordination
218 p. 2007 [978-3-540-33454-5]

Vol. 25: Corke, P.; Sukkarieh, S. (Eds.)
Field and Service Robotics – Results of the 5th
International Conference
580 p. 2006 [978-3-540-33452-1]

Vol. 24: Yuta, S.; Asama, H.; Thrun, S.;
Prassler, E.; Tsubouchi, T. (Eds.)
Field and Service Robotics – Recent Advances in
Research and Applications
550 p. 2006 [978-3-540-32801-8]

Vol. 23: Andrade-Cetto, J,; Sanfeliu, A.
Environment Learning for Indoor Mobile Robots
– A Stochastic State Estimation Approach
to Simultaneous Localization and Map Building
130 p. 2006 [978-3-540-32795-0]

Vol. 22: Christensen, H.I. (Ed.)
European Robotics Symposium 2006
209 p. 2006 [978-3-540-32688-5]

Vol. 21: Ang Jr., H.; Khatib, O. (Eds.)
Experimental Robotics IX – The 9th International
Symposium on Experimental Robotics
618 p. 2006 [978-3-540-28816-9]

Vol. 20: Xu, Y.; Ou, Y.
Control of Single Wheel Robots
188 p. 2005 [978-3-540-28184-9]

Vol. 19: Lefebvre, T.; Bruyninckx, H.;
De Schutter, J. Nonlinear Kalman Filtering
for Force-Controlled Robot Tasks
280 p. 2005 [978-3-540-28023-1]